Pesticidal Plants

Pesticidal Plants

From Smallholder Use to Commercialisation

Special Issue Editors

Philip C. Stevenson
Steven R. Belmain
Murray B. Isman

MDPI • Basel • Beijing • Wuhan • Barcelona • Belgrade

Special Issue Editors

Philip C. Stevenson
University of Greenwich
UK

Steven R. Belmain
University of Greenwich
UK

Murray B. Isman
University of British Columbia
Canada

Editorial Office
MDPI
St. Alban-Anlage 66
4052 Basel, Switzerland

This is a reprint of articles from the Special Issue published online in the open access journal *Plants* (ISSN 2223-7747) from 2019 to 2020 (available at: https://www.mdpi.com/journal/plants/special_issues/Pesticidal).

For citation purposes, cite each article independently as indicated on the article page online and as indicated below:

LastName, A.A.; LastName, B.B.; LastName, C.C. Article Title. *Journal Name* **Year**, *Article Number*, Page Range.

ISBN 978-3-03928-788-8 (Pbk)
ISBN 978-3-03928-789-5 (PDF)

Cover image courtesy of Philip C. Stevenson.

© 2020 by the authors. Articles in this book are Open Access and distributed under the Creative Commons Attribution (CC BY) license, which allows users to download, copy and build upon published articles, as long as the author and publisher are properly credited, which ensures maximum dissemination and a wider impact of our publications.

The book as a whole is distributed by MDPI under the terms and conditions of the Creative Commons license CC BY-NC-ND.

Contents

About the Special Issue Editors vii

Preface to "Pesticidal Plants" ix

Sara Vitalini, Francesca Orlando, Valentina Vaglia, Stefano Bocchi and Marcello Iriti
Potential Role of *Lolium multiflorum* Lam. in the Management of Rice Weeds
Reprinted from: *Plants* **2020**, *9*, 324, doi:10.3390/plants9030324 1

G. Mandela Fernández-Grandon, Steven J. Harte, Jaspher Ewany, Daniel Bray and
Philip C. Stevenson
Additive Effect of Botanical Insecticide and Entomopathogenic Fungi on Pest Mortality and the Behavioral Response of Its Natural Enemy
Reprinted from: *Plants* **2020**, *9*, 173, doi:10.3390/plants9020173 14

Angela G. Mkindi, Yolice L. B. Tembo, Ernest R. Mbega, Amy K. Smith, Iain W. Farrell,
Patrick A. Ndakidemi, Philip C. Stevenson and Steven R. Belmain
Extracts of Common Pesticidal Plants Increase Plant Growth and Yield in Common Bean Plants
Reprinted from: *Plants* **2020**, *9*, 149, doi:10.3390/plants9020149 28

María J. Pascual-Villalobos, Manuel Cantó-Tejero, Pedro Guirao and María D. López
Fumigant Toxicity in *Myzus persicae* Sulzer (Hemiptera: Aphididae): Controlled Release of (*E*)-anethole from Microspheres
Reprinted from: *Plants* **2020**, *9*, 124, doi:10.3390/plants9010124 39

Kelita Phambala, Yolice Tembo, Trust Kasambala, Vernon H. Kabambe, Philip C. Stevenson
and Steven R. Belmain
Bioactivity of Common Pesticidal Plants on Fall Armyworm Larvae (*Spodoptera frugiperda*)
Reprinted from: *Plants* **2020**, *9*, 112, doi:10.3390/plants9010112 50

Kateřina Kovaříková and Roman Pavela
United Forces of Botanical Oils: Efficacy of Neem and Karanja Oil against Colorado Potato Beetle under Laboratory Conditions
Reprinted from: *Plants* **2019**, *8*, 608, doi:10.3390/plants8120608 60

Angela G. Mkindi, Yolice Tembo, Ernest R. Mbega, Beth Medvecky, Amy Kendal-Smith,
Iain W. Farrell, Patrick A. Ndakidemi, Steven R. Belmain and Philip C. Stevenson
Phytochemical Analysis of *Tephrosia vogelii* across East Africa Reveals Three Chemotypes that Influence Its Use as a Pesticidal Plant
Reprinted from: *Plants* **2019**, *8*, 597, doi:10.3390/plants8120597 71

Liliana Ruiz-Vásquez, Matías Reina, Víctor Fajardo, Matías López and
Azucena González-Coloma
Insect Antifeedant Components of *Senecio fistulosus* var. *fistulosus*—Hualtata
Reprinted from: *Plants* **2019**, *8*, 176, doi:10.3390/plants8060176 82

Naomi B. Rioba and Philip C. Stevenson
Opportunities and Scope for Botanical Extracts and Products for the Management of Fall Armyworm (*Spodoptera frugiperda*) for Smallholders in Africa
Reprinted from: *Plants* **2020**, *9*, 207, doi:10.3390/plants9020207 94

Victor Jaoko, Clauvis Nji Tizi Taning, Simon Backx, Jackson Mulatya, Jan Van den Abeele, Titus Magomere, Florence Olubayo, Sven Mangelinckx, Stefaan P.O. Werbrouck and Guy Smagghe
The Phytochemical Composition of *Melia volkensii* and Its Potential for Insect Pest Management
Reprinted from: *Plants* **2020**, *9*, 143, doi:10.3390/plants9020143 . . . . **111**

Ricardo A. Rincón, Daniel Rodríguez and Ericsson Coy-Barrera
Botanicals Against *Tetranychus urticae* Koch Under Laboratory Conditions: A Survey of Alternatives for Controlling Pest Mites
Reprinted from: *Plants* **2019**, *8*, 272, doi:10.3390/plants8080272 . . . . **122**

About the Special Issue Editors

Philip C. Stevenson is a Professor of Plant Chemistry at the Natural Resources Institute in the University of Greenwich (UK) where he leads Chemical Ecology research. He holds a dual position as an NERC Merit researcher and Head of Biological Chemistry and In Vitro research at the Royal Botanic Gardens, Kew (UK).

Phil's research has focussed on the biological and ecological functions of plant chemicals and understanding how these compounds can be used in support of sustainable agriculture. His work includes research on pollen and nectar chemistry to determine their role in pollinator behaviour and health and behavioural ecology, natural pest resistance in crops to identify breeding traits and the optimisation of pesticidal plants (botanical insecticides) as benign and affordable alternatives to synthetic insecticides. Phil's research is or has been funded by UK Research and Innovation (BBSRC), the European Union, Darwin Initiative, USDA and NSF (USA) and the Pewter Sowerby Foundation. His research is published in over 100 international journal articles, including recent papers in *Science, Current Biology, Journal of Ecology, Ecological Monographs* and *Frontiers in Ecology and the Environment*. His work has been cited more than 5000 times. His international scientific role is represented through positions on the Editorial Boards of journals including Subject Editor at the *Bulletin of Entomological Research*, Regional Editor of *Biopesticides International* and the Editorial boards of *Crop Protection, Industrial Crops and Products, Plants* (MDPI) and *Plants, People, Planet*. Phil also advises the UK's Department for Environment Food and Rural Affairs. He is a Fellow of the Royal Entomological Society and Member of the British Ecological Society and International Society of Chemical Ecology.

Steven R. Belmain is a Professor of Ecology at the Natural Resources Institute in the University of Greenwich (UK) where he leads the institute's research on small mammals and wildlife management. Steve's research in the field of applied ecology has specialized in the sustainable management of vertebrates and invertebrates as crop pests and disease vectors affecting people living in rural and urban communities, improving the management of natural resources and advancing the livelihoods of the poor. His work on the optimizing the use of pesticidal plants in sub-Saharan Africa stretches back more than 20 years investigating their use in post-harvest insect pest management, their application to a variety of field crops targeting a range of arthropod pests and their integration in to agro-ecologically sustainable production systems. Further research interests include ecosystem services, ecological pest management, agroecology and ecological engineering, entomology, chemical ecology, eco-epidemiology of zoonoses, animal behaviour and small mammal ecology. As principal investigator Steve has led more than 15 multidisciplinary research projects collaborating with institutions in more than 50 countries across Europe, Africa and Asia with funding from the EU and European governments, UN agencies, the World Bank and charitable foundations. He is Associate Editor for the journal *Wildlife Research*, Fellow of the Royal Entomology Society, Fellow of the Higher Education Academy, Coordinating member of the International Society of Zoological Sciences, Committee member of the World Health Organization's Global Leptospirosis Environmental Action Network, Member of the British Ecology Society, and author of over 100 publications in peer-reviewed journals, books and book chapters.

Murray B. Isman was Professor of Entomology and Toxicology at the University of British Columbia, Vancouver, and former Dean of the Faculty of Land and Food Systems. He performed extensive research for over 35 years in the areas of insect toxicology and behavior, with particular emphasis on the discovery and development of botanical insecticides. Others areas of research included insect-plant chemical interactions, metabolism of plant toxins by insects, habituation to antifeedants and insect memory, and non-target effects on pollinators and fish. He has authored over 200 refereed publications (cited >20,000 times according to Google Scholar). Collaborative research with industry has culminated in the commercialization of botanical insecticides partly developed in his laboratory that are currently used in over a dozen countries, including the USA.

He is a former President of the International Society of Chemical Ecology, Phytochemical Society of North America, and the Entomological Society of British Columbia. Murray has served on the editorial boards of the *Journal of Economic Entomology, Journal of Pest Science, Bulletin of Entomological Research, Bulletin of Environmental Toxicology and Contamination*, and the *Journal of Chemical Ecology*, and is the founding Chief Specialty Editor (Pest Management) for Frontiers in Agronomy. He received the Entomological Society of Canada's Gold Medal for Excellence in Entomology, and is an elected Fellow of the Entomological Society of America and the Royal Entomological Society (London).

Preface to "Pesticidal Plants"

Global perceptions about pesticides are changing as a consequence of their environmental impacts, persistence, broad spectrum activities and non-target effects. As a result of this, pesticide regulations are changing in some regions. For example, Europe has limited the number of synthetic chemical products permitted for use in pest control. The increasing pressure on synthetic products has reinvigorated efforts to seek alternative pest management options, including new opportunities for plant-based solutions that are environmentally benign and tailored to different farmers' needs, from commercial to smallholder and subsistence farming.

This Special Issue captures some of the latest developments in research on pesticidal plants and botanical pesticides from fundamental aspects, including the identification of bioactive plant chemicals and the evaluation of their bioactivities against pests, mechanisms of activity, validation of use in small-scale systems and commercial-scale pest management, including in horticulture and disease vector control. Other work reports developments in the use of botanicals in rice weeds, combination biopesticides and the growth-promoting properties of some plant extracts when used to control pests. We also provide insight into how chemistry can vary dramatically and influence the value and effectiveness of botanical insecticides in different locations. Three reviews assess wider questions around the potential of plant-based pest management to address the global challenges of new, invasive and established pests and previously underexploited pesticidal species of plant.

Philip C. Stevenson, Steven R. Belmain, Murray B. Isman
Special Issue Editors

Article

Potential Role of *Lolium multiflorum* Lam. in the Management of Rice Weeds

Sara Vitalini [1,*,†], Francesca Orlando [2,†], Valentina Vaglia [3], Stefano Bocchi [3,‡] and Marcello Iriti [1,*,‡]

1. Department of Agricultural and Environmental Sciences, Università degli Studi di Milano, 20133 Milan, Italy
2. Department of Molecular and Translational Medicine (DMMT), Università degli Studi di Brescia, 25123 Brescia, Italy; francesca.orlando@unibs.it
3. Department of Environmental Science and Policy, Università degli Studi di Milano, 20133 Milan, Italy; valentina.vaglia@unimi.it (V.V.); stefano.bocchi@unimi.it (S.B.)
* Correspondence: sara.vitalini@unimi.it (S.V.); marcello.iriti@unimi.it (M.I.)
† Those authors contributed equally to this work.
‡ Those authors contributed equally to this work.

Received: 21 February 2020; Accepted: 29 February 2020; Published: 4 March 2020

Abstract: The phytotoxic relationships between crops and weeds can cover a role in weed management, reducing the use of chemical herbicides. Starting from the organic farmers' experience, the study aimed to define the inhibitory action of *Lolium multiflorum* Lam., used as a cover crop before rice sowing, against *Echinochloa oryzoides* (Ard.) Fritsch, one of the main rice weeds. In vitro 7-day assays were carried out in Petri dishes to compare the effect of different *L. multiflorum* Lam. parts, in the form of aqueous extract or powder, on the seed germination and seedling growth of *Oryza sativa* L. and *E. oryzoides* and to verify the hypothesis of a higher susceptibility of the weed. The total polyphenolic content, as the potential source of allelochemicals, in the *L. multiflorum* parts was measured. The results showed that both species suffer the phytotoxic action of *L. multiflorum*, but a more marked effect against *E. oryzoides* was recorded. In according with the polyphenol quantities, stem and inflorescence extracts showed the more significant species-specific inhibition. In all assays, the weed showed a stronger reduction in the root length and seedling vigor index, and, in some cases, also in the germination percentage and shoot length compared to rice.

Keywords: Italian ryegrass; barnyard grass; rice; cover crops; organic farming; weed control; phytotoxic activity

---

1. Introduction

Weeds cause severe crop losses in rice production worldwide. The yield reductions in flooded paddy fields are due to the presence of invasive aquatic and semi-aquatic species. Among them, the species of the *Echinochloa* genus are the most common weeds in wetlands and water-saturated conditions and are included in the list of the ten worst weeds in the world [1]. In particular, *E. oryzoides* (Ard.) Fritsch, known as early watergrass, is a rice mimic with very similar emergence and flowering times that allow it to achieve greater competitiveness than other *Echinochloa* species by influencing rice in its early growth stages [2,3]. Therefore, any intervention able to control the incidence of this weed is useful to give rice a competitive advantage.

Rice growers in temperate regions (Europe, US, Australia) are particularly attentive to weed control. Within an established model of industrial agriculture, based on monoculture and high-input systems, they face no option other than the application of synthetic herbicides because of the low knowledge of alternative agronomic practices and plant-based solutions, as well as the unfeasibility of hand-weeding due to the high labor costs [4]. However, the herbicide-based weed management has proved to be unable to solve the issues, leading to well documented resistance phenomena as in the case of *E. oryzoides* [2,5,6].

In addition, the special attention of the European Union to the risks and hazards for humans, animals and the environment associated with the use of chemical substances has led to the banning in its member states of many herbicides, such as oxadiazon-based plant protection products, a compound largely used in rice fields [7–9].

Given the above, there is the need to move toward new and more sustainable weed management strategies [10,11].

In this context, the long-term experience of the organic farmers could be used to recognize and set innovative and good practices, transferable somehow also to the conventional or integrated systems [12,13]. They paid particular attention to solve the weed issue identified as the main cause of yield variability and loss in the Mediterranean regions, which is the main challenge for organic rice production [14,15].

Weed control in organic farming is carried out through crop rotation and the use of cover crops, smother crops and green mulching, which are important for regulating the weed seed population in the soil and the plant population in the field [16,17]. In this regard, several factors influence the weed growth such as competition for space, water and nutrients, changes in temperature and shade as well as toxic microbial products, soil pH and release of allelochemicals [18]. Especially allelopathy, defined as the release of compounds from the living or dead tissues of a plant species with strong phytotoxic effect towards another one, is a phenomenon which deserves attention in the study of alternative pest management options, thanks to its potential action in weed suppression [19,20]. It is known that some species or varieties, used as cover crop or crop in rotation, produce relevant amounts of allelochemicals significantly affecting the weed germination and growth [21].

Allelopathic compounds could be used to control weeds in both organic and conventional agriculture: in the first case, through the direct cultivation of allelopathic species, respecting the ban on the use of any products for herbicide purposes and with the principle of low-external-input farming [22]; while, in the second case, through the marketing and use of plant-based herbicides able to support and integrate weed management, reducing the need for synthetic herbicides.

Particularly for rice, a participatory research carried out by Orlando and co-workers [23] with a small group of organic farmers in North Italy identified a strategy based on the cultivation of *Lolium multiflorum* Lam. as the most promising practice for weed control. In the study area (Po Valley, between Piedmont and Lombardy regions), characterized by a typical Mediterranean climate, 94% of national rice production is concentrated. The farming systems are mainly based on continuous flooding and a wide use of pesticides and herbicides that caused the highest groundwater and surface water pollution in the country [24]. *L. multiflorum*, known as Italian ryegrass, is used by rice growers during the winter season as a cover crop. Then, in May, the rice is sown directly among the standing plants of *L. multiflorum* and, subsequently, its biomass is mowed, chopped or rolled, producing green mulch. The farmer's empirical knowledge suggested to the researchers the existence of an allelopathic suppressive action of *L. multiflorum* versus *Echinochloa* spp., with a chemical inhibition of weed germination and growth, beyond the well-known competitive effect of green mulching.

Accordingly, the present study was aimed to verify the inhibitory activity of different organs of *L. multiflorum* against *E. oryzoides*. Two in vitro bioassays were carried out in order to evaluate the possible release of phytotoxic chemicals from the cover crop separately from other factors occurring simultaneously in the field able to influence the weed growth and from the complex dynamics of the soil seed bank. The impact of the *L. multiflorum* biomass aqueous extracts and its powder was assessed versus the germination and seedling growth of both *E. oryzoides* and *O. sativa* to highlight a potential species-specific action of *L. multiflorum*.

2. Results

2.1. Stem Effects

The obtained data showed a significant impact of the *L. multiflorum* stem aqueous extract on all the considered indices, except for the mean germination time (MGT), in both target species, but with a more evident effect against *E. oryzoides* than *O. sativa*, (p-values ≤ 0.05 for the interaction "species × *L. multiflorum* stem extract treatment") (Table 1).

Table 1. Germination indices measured for *E. oryzoides* and *O. sativa* under the effect of different concentrations of *L. multiflorum* stem extract.

Species	Stem Extract Concentration (%)	Germination (%)	MGT	Root Length (mm)	Shoot Length (mm)	SVI
E. oryzoides	0	100.0 ± 0.0 a	5.0 ± 0.0	56.2 ± 14.5 a	34.0 ± 3.9 a	9008 ± 1764 a
	1	83.2 ± 13.6 ab	5.8 ± 0.4	64.0 ± 4.0 a	28.0 ± 4.0 ab	6613 ± 657 b
	10	76.8 ± 17.4 ab	5.6 ± 0.5	62.0 ± 4.0 a	26.7 ± 0.6 b	5323 ± 227 b
	20	50.0 ± 35.6 bc	5.8 ± 0.5	58.0 ± 0.0 a	26.0 ± 0.0 b	2787 ± 0 c
	50	22.0 ± 22.8 c	5.0 ± 0.0	9.7 ± 6.7 b	22.0 ± 2.6 b	1198 ± 647 c
	100	20.0 ± 12.2 c	5.5 ± 0.6	4.0 ± 2.2 b	14.5 ± 1.0 c	482 ± 21 c
	F	13.756	2.892	36.382	22.701	43.232
	p-value	0.000 *	0.052	0.000 *	0.000 *	0.000 *
O. sativa	0	96.8 ± 4.6 a	5.0 ± 0.0	43.0 ± 13.5 a	19.8 ± 2.6 a	6072 ± 1525 a
	1	96.6 ± 3.5 a	5.0 ± 0.0	49.0 ± 1.0 a	19.0 ± 2.0 a	6364 ± 276 a
	10	96.6 ± 3.5 a	5.0 ± 0.0	52.0 ± 3.0 a	18.3 ± 5.5 a	6805 ± 1009 a
	20	96.6 ± 3.5 a	5.0 ± 0.7	47.0 ± 2.0 a	15.7 ± 1.5 ab	6042 ± 149 a
	50	79.0 ± 12.1 a	5.0 ± 0.0	20.2 ± 10.8 b	11.0 ± 1.6 bc	2564 ± 1235 b
	100	61.6 ± 21.1 b	5.2 ± 0.4	8.6 ± 2.1 b	8.6 ± 1.1 c	1003 ± 403 b
	F	9.922	0.286	19.413	15.045	24.718
	p-value	0.000 *	0.916	0.000 *	0.000 *	0.000 *
Interaction species × treatment						
	F	6.709	1.396	2.999	2.513	8.130
	p-value	0.000 *	0.252	0.025 *	0.05	0.000 *

Values are mean ± standard deviation, asterisk and different letters indicate statistically significant differences at p-value ≤ 0.05 among treatments in each species. F-value and p-value of the ANOVA test. MGT, mean germination time; SVI, seedling vigor index.

In particular, the extract, from 20% to 100% concentration, significantly reduced the *E. oryzoides* germination percentage (by 50%–80%), while *O. sativa* germination was affected only by 100% extract concentration with a 36.4% decrease compared to the control. Moreover, stem extract was able to inhibit *E. oryzoides* root and shoot elongation (up to 93% and 57%, respectively) by significantly lowering the seedling vigour index (SVI) values for all used concentrations (p-value = 0.000). Otherwise, only the treatments with 50% and 100% extract concentrations were effective on *O. sativa* whose SVI, root and shoot length were reduced by 58%–83%, 53%–80% and 44%–57%, respectively.

Bioassay carried out with stem powder provided less evident effects (Table 2).

Table 2. Germination indices measured for E. oryzoides and O. sativa under the effect of different quantity of L. multiflorum powdered stems.

Species	Powdered Stem Guantity (g/dm^2)	Germination (%)	MGT	Root Length (mm)	Shoot Length (mm)	SVI
E. oryzoides	0.00	100.0 ± 0.0	5.0 ± 0.0 a	92.7 ± 3.6 a	36.5 ± 1.5	12925 ± 371 a
	0.4	92.5 ± 9.6	5.3 ± 0.1 ab	43.1 ± 24.5 b	37.6 ± 3.7	7444 ± 2530 b
	0.8	95.0 ± 5.0	5.4± 0.3 b	38.4 ± 11.4 b	35.3 ± 2.9	6973 ± 1180 b
	F	1.964	7.226	20.427	0.780	23.126
	p-value	0.186	0.011 *	0.000 *	0.482	0.000 *
O. sativa	0.00	90.0 ± 10.0	5.0 ± 0.1	64.7 ± 3.9	27.6 ± 1.9	8297 ± 804
	0.4	90.0 ± 7.1	5.2 ± 0.1	55.8 ± 19.6	23.8 ± 4.1	7214 ± 2381
	0.8	88.8 ± 4.5	5.1 ± 0.1	54.3 ± 11.9	21.4 ± 5.4	6632 ± 1338
	F	0.118	2.000	0.878	2.923	1.321
	p-value	0.890	0.178	0.441	0.092	0.303
Interaction species × treatment						
	F	0.762	3.384	6.715	1.588	5.573
	p-value	0.478	0.052	0.005 *	0.227	0.011 *

Values are mean ± standard deviation, asterisk and different letters indicate statistically significant differences at p-value ≤ 0.05 among treatments in each species. F-value and p-value of the ANOVA test. MGT, mean germination time; SVI, seedling vigor index.

No significant results were detected for O. sativa in relation to the measured indices. Similarly, stems were not able to affect germination percentage and shoot growth of E. oryzoides. On the other hand, at 0.4 and 0.8 g/dm^2, the treatment increased its MGT by 8% and decreased the root length up to 59% by significantly influencing SVI, reduced by 42% and 46%, respectively. In this case, the interaction "species × L. multiflorum stem powder treatment" was significant (p-values ≤ 0.05) only for root length and SVI.

In general, the results obtained for the L. mutiflorum stems showed a higher susceptibility of the weed than the crop in their responses to the increasing concentrations (Tables 1 and 2).

2.2. Inflorescence Effects

Similarly to the stems, L. multiflorum inflorescence extract affected the seed development of both studied species showing a greater inhibitory action on E. oryzoides compared to O. sativa, particularly on the three seedling growth parameters, namely SVI, root and shoot length (p-values < 0.05 for the interaction "species × L. multiflorum inflorescence extract treatment") (Table 3). Otherwise, there is no preferential effect by the extract in reducing the germination of the one of the two species.

Their germination percentage was remarkably lowered by 50% and 100% extract concentrations (p-values = 0.000). In the first case, the germinated seeds of O. sativa and E. oryzoides were 38% and 18%, respectively, while 100% extract concentration was able to completely inhibit them (0% germination). In addition, the 50% extract concentration decreased root length of both species by 83% and 92% than controls, as well as their shoot length (−33% and −70%) by significantly reducing the corresponding SVI values (−86% and −95%). Notably, the E. oryzoides shoot elongation was also affected by inflorescence 20% extract concentration (−26%).

Like the extract, also L. multiflorum powdered inflorescences placed in direct contact with O. sativa and E. oryzoides seeds significantly affected all measured indices, except for MGT (Table 4).

Table 3. Germination indices measured for E. oryzoides and O. sativa under the effect of different concentrations of L. multiflorum inflorescence extract.

Species	Inflorescence Extract Concentration (%)	Germination (%)	Mean Germination Time	Root Length (mm)	Shoot Length (mm)	Seedling Vigor Index
E. oryzoides	0	100.0 ± 0.0 a	5.0 ± 0.0	66.2 ± 14.5 a	34.0 ± 3.9 a	9008 ± 1764 a
	1	83.2 ± 7.5 a	5.6 ± 0.5	76.0 ± 1.0 a	33.3 ± 2.0 a	7733 ± 264 a
	10	90.2 ± 5.6 a	5.8 ± 0.4	68.0 ± 9.0 a	29.3 ± 2.5 ab	8405 ± 978 a
	20	94.8 ± 3.0 a	5.6 ± 0.5	62.0 ± 4.0 a	25.0 ± 0.6 b	9255 ± 772 a
	50	18.0 ± 21.7 b	5.3 ± 0.6	4.3 ± 0.6 b	10.3 ± 2.1 c	407 ± 276 b
	100	0.0 ± 0.0 c	n.d.	n.d.	n.d.	n.d.
	F	100.627	2.179	27.971	39.963	32.699
	p-value	0.000 *	0.113	0.000 *	0.000 *	0.000 *
O. sativa	0	96.8 ± 4.6 a	5.0 ± 0.0	43.0 ± 13.5 a	19.8 ± 2.6 a	6072 ± 1525 a
	1	96.8 ± 5.6 a	4.6 ± 0.5	44.3 ± 1.5 a	17.7 ± 0.6 a	5794 ± 508 a
	10	95.0 ± 5.4 a	4.6 ± 0.5	40.3 ± 3.5 a	18.3 ± 1.5 a	4825 ± 222 a
	20	88.4 ± 2.6 a	5.0 ± 0.7	48.7 ± 1.5 a	15.2 ± 4.1 ab	5806 ± 327 a
	50	37.0 ± 20.1 b	5.2 ± 0.4	7.2 ± 3.4 b	13.3 ± 2.5 b	825 ± 554 b
	100	0.0 ± 0.0 c	n.d.	n.d.	n.d.	n.d.
	F	101.674	1.385	21.825	3.155	27.790
	p-value	0.000 *	0.275	0.000 *	0.048*	0.000 *
Interaction species × treatment						
	F	1.39	2.764	4.45	17.641	4.656
	p-value	0.265	0.051	0.007 *	0.000 *	0.006 *

Values are mean ± standard deviation, asterisk and different letters indicate statistically significant differences at p-value ≤ 0.05 among treatments in each species. F-value and p-value of the ANOVA test. MGT, mean germination time; SVI, seedling vigor index.

Table 4. Germination indices measured for E. oryzoides and O. sativa under the effect of different quantity of L. multiflorum powdered inflorescences.

Species	Powdered Inflorescence Quantity (g/dm^2)	Germination (%)	MGT	Root Length (mm)	Shoot Length (mm)	SVI
E. oryzoides	0.00	100.0 ± 0.0 a	5.0 ± 0.1	90.6 ± 5.9 a	36.2 ± 1.7 a	12688 ± 647 a
	0.4	56.4 ± 35.1 b	5.2 ± 0.2	9.8 ± 9.9 b	28.8 ± 3.7 b	2649 ± 627 b
	0.8	22.0 ± 31.9 b	5.6 ± 0.8	2.9 ± 0.8 b	14.9 ± 5.0 c	920 ± 57 c
	F	10.165	2.121	175.721	33.891	436.013
	p-value	0.003 *	0.182	0.000 *	0.000 *	0.000 *
O. sativa	0.00	92.0 ± 8.4 a	5.0 ± 0.0	63.2 ± 5.1 a	25.4 ± 1.3 a	8122 ± 355 a
	0.4	72.5 ± 10.9 b	5.1 ± 0.1	12.6 ± 8.3 b	18.8 ± 6.4 b	2401 ± 1489 b
	0.8	30.0 ± 12.2 c	5.0 ± 0.0	3.5 ± 1.1 c	11.3 ± 1.4 c	466 ± 223 c
	F	44.506	2.889	160.545	16.963	99.309
	p-value	0.000 *	0.095	0.000 *	0.000 *	0.000 *
Interaction species × treatment						
	F	2.292	3.357	15.724	2.014	21.465
	p-value	0.127	0.055	0.000 *	0.160	0.000 *

Values are mean ± standard deviation, asterisk and different letters indicate statistically significant differences at p-value ≤ 0.05 among treatments in each species. F-value and p-value of the ANOVA test. MGT, mean germination time; SVI, seedling vigor index.

The germination percentage decreased by 21%–67% and 44%–78%, respectively; the root length by 80%–94% and 89%–97%, shoot length by 26%–56% and 21%–59%, SVI by 70%–94% and 79%–93%, due to both used quantities (0.4 and 0.8 g/dm^2). The species showed a similar response to the treatments both as regards the germination percentage and the shoot length (p-values > 0.05). Accordingly, the interaction "species × L. multiflorum inflorescence powder treatment" was significant only for root length and SVI (*p*-values = 0.000) confirming the tendency towards greater susceptibility of *E. oryzoides* shown by the previous results.

2.3. Root Effects

Unlike stems and inflorescences, *L. multiflorum* root extract was not able to affect, at any used concentration, both MGT and germination percentage in the studied species. Furthermore, only 50% and 100% extract concentrations showed cases of significant impact on other considered indices (Table 5).

Table 5. Germination indices measured for *E. oryzoides* and *O. sativa* under the effect of different concentrations of *L. multiflorum* root extract.

Species	Root Extract Concentration (%)	Germination (%)	MGT	Root Length (mm)	Shoot Length (mm)	SVI
E. oryzoides	0	100.0 ± 0.0	5.0 ± 0.0	66.2 ± 14.5 a	34.0 ± 3.9 a	9008 ± 1764 a
	1	93.2 ± 7.5	5.6 ± 0.5	68.3 ± 3.5 a	32.0 ± 1.0 a	8740 ± 120 a
	10	91.6 ± 2.6	5.6 ± 0.5	63.7 ± 1.5 a	32.3 ± 0.6 a	8787 ± 123 a
	20	93.2 ± 4.6	5.4 ± 0.5	67.0 ± 7.0 a	32.0 ± 2.0 a	9557 ± 537 a
	50	98.0 ± 4.5	5.2 ± 0.4	46.8 ± 4.0 b	32.2 ± 4.1 a	7734 ± 1040 a
	100	92.0 ± 8.4	5.4 ± 0.5	21.8 ± 2.9 c	24.8 ± 2.7 b	4321 ± 820 b
	F	1.870	1.171	22.461	5.315	14.805
	p-value	0.140	0.352	0.000 *	0.004 *	0.000 *
O. sativa	0	96.8 ± 4.6	5.0 ± 0.0	43.0 ± 13.5 a	19.8 ± 2.6 a	6072 ± 1525 a
	1	100.0 ± 0.0	5.0 ± 0.0	43.0 ± 4.0 a	20.7 ± 3.5 ab	6323 ± 752 a
	10	95.0 ± 5.4	4.6 ± 0.5	48.7 ± 1.5 a	25.3 ± 1.5 a	6920 ± 724 a
	20	94.8 ± 3.0	4.8 ± 0.4	47.6 ± 3.5 a	25.0 ± 2.0 a	6760 ± 579 a
	50	96.2 ± 3.3	5.0 ± 0.0	57.6 ± 6.0 a	26.4 ± 4.5 a	7784 ± 1184 a
	100	92.2 ± 6.1	5.0 ± 0.0	23.0 ± 6.5 b	14.4 ± 4.4 b	3304 ± 696 b
	F	1.867	1.680	10.479	7.595	10.370
	p-value	0.138	0.178	0.000 *	0.001 *	0.000 *
Interaction species × treatment						
	F	2.371	1.543	6.227	2.357	2.761
	p-value	0.59	0.201	0.000 *	0.06	0.033 *

Values are mean ± standard deviation, asterisk and different letters indicate statistically significant differences at *p*-value ≤ 0.05 among treatments in each species. F-value and *p*-value of the ANOVA test. MGT, mean germination time; SVI, seedling vigor index.

At 100% extract concentration, a reduction by 27% was observed in the *O. sativa* and *E. oryzoides* shoot length (*p*-values < 0.05) and by about 50% for their SVI values (*p*-values = 0.000). Roots decreased by 47% and 61% (*p*-values = 0.000), respectively.

Lastly, the significant interaction "species × *L. multiflorum* root extract treatment" with respect to root length and SVI (*p*-values < 0.05), thanks also to the effect of the 50% extract concentration on the roots of *E. oryzoides* (−17%), showed a greater inhibition of the growth of weed seedlings compared to that of rice.

The results of *L. multiflorum* root powder bioassay supported previous data on *E. oryzoides* showing that both used quantities (0.4 and 0.8 g/dm^2) significantly influenced only its root elongation (decrease between 36% and 38% compared to the control) and SVI (decrease between 29% and 30%). Contrastingly, the powdered roots showed no effect against *O. sativa*, for which all the values of the measured indices were comparable to those of controls (Table 6). On the basis of these results, the interaction "species × *L. multiflorum* root powder treatment" was significant, showing a greater reduction in *E. oryzoides* root length and SVI.

Table 6. Germination indices measured for *E. oryzoides* and *O. sativa* under the effect of different quantity of *L. multiflorum* powdered roots.

Species	Powdered Root Quantity (g/dm^2)	Germination (%)	MGT	Root Length (mm)	Shoot Length (mm)	SVI
E. oryzoides	0.00	100.0 ± 0.0	5.0 ± 0.1	90.6 ± 5.9 a	36.2 ± 1.7	12688 ± 647 a
	0.4	94.0 ± 5.5	5.0 ± 0.1	57.9 ± 9.7 b	37.3 ± 1.4	8984 ± 1389 b
	0.8	94.0 ± 5.5	5.1 ± 0.2	56.2 ± 21.9 b	39.2 ± 3.9	8872 ± 1965 b
	F	3.000	1.471	9.275	1.704	11.394
	p-value	0.088	0.268	0.004 *	0.223	0.002 *
O. sativa	0.00	92.5 ± 8.3	5.0 ± 0.0	63.2 ± 5.1	25.4 ± 1.3	8166 ± 346
	0.4	92.0 ± 8.4	5.0 ± 0.1	62.7 ± 12.6	31.8 ± 4.5	8639 ± 1427
	0.8	90.0 ± 7.1	5.1 ± 0.1	52.7 ± 8.4	29.5 ± 5.3	7478 ± 1697
	F	0.139	1.600	2.063	3.205	1.015
	p-value	0.872	0.242	0.170	0.077	0.391
Interaction species × treatment						
	F	0.467	0.286	4.843	1.724	6300
	p-value	0.632	0.754	0.017 *	0.200	0.006 *

Values are mean ± standard deviation, asterisk and different letters indicate statistically significant differences at *p*-value ≤ 0.05 among treatments in each species. F-value and *p*-value of the ANOVA test. MGT, mean germination time; SVI, seedling vigor index.

2.4. Seed Effects

The phytotoxic activity of *L. multiflorum* seeds was also assessed. Their aqueous extract impacted similarly on both target species that achieved growth values comparable to those of their controls for all considered indices (*p*-values > 0.05) (Table 7). The germination percentage of treated *E. oryzoides* and *O. sativa* was greater than 90%. MGT was the same as for untreated seeds while the root and shoot development showed insignificant differences as well as SVI values (*p*-values > 0.05).

Table 7. Germination indices measured for *E. oryzoides* and *O. sativa* under the effect of *L. multiflorum* seed extract.

Species	Seed Extract	Germination (%)	MGT	Root Length (mm)	Shoot Length (mm)	SVI
E. oryzoides	0%	96.7 ± 5.2	5.7 ± 0.4	42.0 ± 6.5	23.5 ± 3.0	6323 ± 785
	100%	91.7 ± 20.4	5.7 ± 0.1	50.8 ± 13.7	24.5 ± 7.2	6703 ± 1866
	F	0.338	0.037	2.034	0.086	0.212
	p-value	0.574	0.850	0.184	0.775	0.655
O. sativa	0%	90.0 ± 25.3	4.7 ± 0.1	51.4 ± 5.1	23.3 ± 5.1	5432 ± 2640
	100%	98.0 ± 12.1	4.7 ± 0.1	50.1 ± 3.6	21.8 ± 2.7	6040 ± 1308
	F	1.356	0.448	0.271	0.425	0.256
	p-value	0.271	0.519	0.614	0.529	0.624
Interaction species × treatment						
	F	1.640	0.146	2.301	0.387	0.024
	p-value	0.215	0.706	0.145	0.541	0.878

Values are mean ± standard deviation. F-value and *p*-value of the ANOVA test. MGT, mean germination time; SVI, seedling vigor index.

2.5. Polyphenol Content in L. multiflorum Extracts

Figure 1 shows the polyphenol content in the aqueous extracts of the different investigated *L. multiflorum* parts measured using the Folin-Ciocalteau reagent.

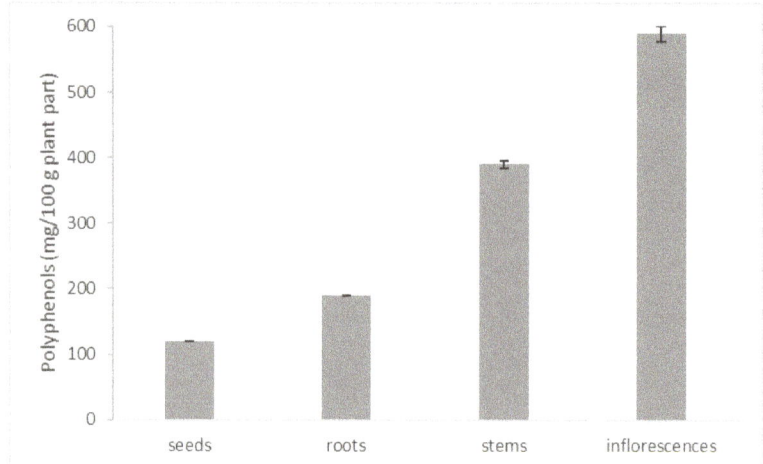

Figure 1. Total polyphenols detected in the aqueous extracts of the various *L. multiflorum* organs.

The highest content, equal to 590 mg GAE/100 g plant part, was identified in the inflorescences extract. Gradually lower quantities were found in stems (390 mg GAE/ 100 g), roots (190 mg GAE/ 100 g) and seeds (120 mg GAE/100 g), in accordance with the decreasing phytotoxic activity recorded for the various *L. multiflorum* parts against *E. oryzoides* and *O. sativa*.

3. Discussion

Different studies reported the allelopathic activity of some *Lolium* species including *L. multiflorum* [25–28]. Usually, it is treated more as a weed capable of undermining the crop rather than as a crop cultivated with a function in the weed control and, therefore, in the crop protection. For example, Lehoczky and co-workers [26] described the inhibitory effects of aqueous extract obtained from *L. multiflorum* shoots on some of the main grown crops such as *Hordeum vulgare* L., *Triticum aestivum* L. and *Zea mays* L. However, other authors investigated the impact of decaying residues from *L. multiforum* used as a cover crop on *O. sativa* seedling development obtaining opposite results due to both inhibitory and stimulating effects [29–31]. To the best of our knowledge, very few data refer to the effectiveness of *L. multiflorum* against the weed growth [32]. Moreover, the relationship between *L. multiflorum* and *E. oryzoides* was never investigated.

In this context, our results are particularly interesting and can be at the basis of the weed management strategies adopted by organic farmers who cultivated *L. multiflorum* before rice.

L. multiflorum showed a preferential action with impacts significantly different and more severe on *E. oryzoides* rather than on *O. sativa*. Both *L. multiflorum* treatments, i.e., aqueous extracts and powder, obtained from all the investigated organs—inflorescences, stems and roots—showed a significant inhibitory effect on the weed. In particular, the stem and inflorescence aqueous extracts and the inflorescence powder significantly affected both seed germination and seedling growth, while the root aqueous extract and the stem and root powder reduced the root length. The root development of *E. oryzoides* showed always a greater reduction than those of *O. sativa*. In addition, species-specific phytotoxic effects were evident for the inflorescence and stem extracts, also regarding the shoot development and seed germination.

Additionally, *O. sativa* suffered the inhibitory effect of the aqueous extracts from *L. multiflorum*. In particular, inflorescence and stem extracts were able to reduce both the seed germination and the seedling growth (i.e., root and shoot elongation), while the root extracts affected only the seedling growth. On the other hand, the powder treatments showed minor activity and only those obtained from inflorescence had significant effect, inhibiting the seed germination and the seedling growth.

Therefore, the data showed the existence of a phytotoxic activity by *L. multiflorum*, instead of its stimulating effects on rice, and, in general, a more marked action of the inflorescence, followed by stems and roots.

Finally, the seed aqueous extract was unable to affect *E. oryzoides* neither *O. sativa*. In both bioassays, all their growth parameters reached high values, similar to those of controls. The ineffectiveness of *L. multiflorum* seeds in influencing the development of other seeds could be attributed to the fact that the phytotoxic substances present in the cover crop are synthesized in a subsequent growth stage of the plant.

The preferential impact of *L. multiflorum* on the root development confirmed previous data documenting that the phytotoxic effect most observed in vegetative structures occurs on the root system [20]. Nevertheless, some studies on the relationships between species showed the different impact of the aqueous biomass extract on the measured variables. For example, Hoffman et al. [33] reported significant inhibition of root and shoot growth, without effect on germination, while Turk et al. [34] documented the decrease of germination and no reduction of the hypocotyl as well as Han et al. [35] recorded the inhibition of germination and root development but no effect on the shoots. On the other hand, the activity of phytotoxic compounds and their effects such as reduction in seed germination and seedling growth are caused by a variety of specific interactions and cannot been explained by just a single mode of action [28].

Lastly, polyphenols are a heterogeneous group of substances produced by the secondary metabolism of plants, where, in relation to chemical diversity, they play different roles. They can be simple low molecular weight compounds or complex structures conjugated with sugar moieties useful to plants for their structure, pigmentation, pollination, defense from predators and pathogens. Furthermore, their action as allelochemicals is known and investigated [36–39]. The different phytotoxicity of the investigated *L. multiflorum* organs could be partially related to the decreasing polyphenol values detected starting from the inflorescences.

Some allelopathic compounds were previously isolated from the aqueous leachates of decaying *L. multiflorum* residues. In particular, benzenepropanoic acid has proven to be effective in inhibiting the root and shoot growth of rice seedlings [29]. Other compounds such as caffeic acid, p-coumaric acid, ferulic acid and hydrocinamic acid were identified in the water fraction obtained from the fermentation of *L. multiflorum* shoots and roots. These phenolic acids seem to be responsible for the ability of the extract to reduce the shoot and root elongation in different rice cultivars [31]. Furthermore, the same type of extract was also able to affect the growth of two wheat cultivars [32].

In conclusion, the data obtained from our in vitro tests substantially confirmed the farmer empirical observations regarding the use of *L. multiflorum* as a cover crop, namely the corresponding reduction of *E. oryzoides* incidence, and explained the *O. sativa* poor density observed in certain fields under the same practice. In their opinion, *L. multiflorum* appears to negatively affect both the weed and rice development in the early growth stages, but *O. sativa* is less influenced than *E. oryzoides*, and on this thin difference it is possible to play for giving the crop a competitive advantage over the weed.

The farmers' empirical knowledge comes from their direct long-time experiences in managing complex agro-ecosystems or are drawn from the rural tradition, thus including and safeguarding a stock of precious knowledge often fragmented or lost. It could be not a coincidence that in the past century, when the local farms combined the rice production with livestock, the cultivation of forage species such as *L. multiflorum* in rotation with rice was a common practice. Hence, the study results validate the usefulness of the farmers' contribution in participatory research as a valuable guide for scientific inquiry and as a support for innovations in sustainable agriculture.

Moreover, *L. multiflorum* could be considered the starting point to formulate new plant-based and eco-friendly herbicides, functional to reduce the use of more dangerous synthetic compounds and the consequent environmental pressure due to agronomic practices in the rice area.

4. Materials and Methods

4.1. Plant Material

Seeds of *O. sativa* (cv. Rosa Marchetti), *E. oryzoides* and *L. multiflorum* were obtained from 'Terre di Lomellina' organic farm located in the northern Italy (GPS coordinates: 45°10′28.329′′N 8°35′44.198′′E). They were stored at 4 °C until use after surface sterilization with 1% sodium hypochlorite by shaking for 7 min and repeatedly rinsing with distilled water.

In the same farm, fresh plants of *L. multiflorum* were also collected. Their inflorescences, stems and roots were separately air-dried at room temperature (25 °C) in the shade and preserved in paper bags until extraction.

4.2. Aqueous Extract Bioassay

The aqueous extract of each powdered part—inflorescences, stems and roots—of *L. multiflorum* was prepared mixing a suitable amount with distilled water (1:10, *w*/*v*) and shaking it at room temperature for 24 h. Afterwards, the mixture was filtered through gauzes to remove residues and centrifuged at 4500 rpm for 30 min. The obtained extracts were used as such (100%) and diluted with distilled water to give final concentrations of 1%, 10%, 20% and 50%.

Otherwise, in order to simulate the leaching from seeds, 60 unsterilized seeds were placed in 30 mL of distilled water on an orbital shaker, at room temperature, for 24 h. Subsequently, the obtained extract was filtered before use.

Ten sterilized seeds of *E. oryzoides* and *O. sativa* were sown into each Petri dish (90 mm diameter) containing 2 filter papers and 4 mL of each extract or its dilution were added. The same volume of distilled water was used as a control (0%). Petri dishes prepared in a vertical laminar flow hood and sealed with parafilm were kept in a growth chamber (25 °C/16 h light and 18 °C/8 h dark) for seven days.

Concerning the inflorescence, stem and root extracts, five Petri dishes were realized for each combination of "species × *L. multiflorum* treatment", according with the following randomized block design: two species (*E. oryzoides* and *O. sativa*) × six levels of concentration (100%, 50%, 20%, 10%, 1%, 0%) × three *L. multiflorum* organs (inflorescences, stems and roots) × five replicates. A similar experimental design with five repetitions was followed for the seed extract, considering only two levels of concentration (100% and 0%) and one *L. multiflorum* organ (seed).

4.3. Plant Part Powder Bioassay

Different quantities (0.4 and 0.8 g/dm^2) of each powdered part—inflorescences, stems and roots—of *L. multiflorum* were spread on two filter papers in Petri dishes (90 mm diameter). Afterwards, ten sterilized seeds of *E. oryzoides* and *O. sativa*, respectively, were placed and soaked with 5 mL of distilled water. The same volume of distilled water was used in the control samples (0 g/dm^2 of powder). Petri dishes prepared in a vertical laminar flow hood and sealed with parafilm were kept in a growth chamber (25 °C/16 h light and 18 °C/8 h dark) for seven days. Five Petri dishes were realized for each combination of "species × *L. multiflorum* treatment" according with the following randomized block design: two species (*E. oryzoides* and *O. sativa*) × three levels of quantity (0.5 g, 0.25 g, 0 g) × three *L. multiflorum* organs (inflorescences, stems and roots) × five replicates.

4.4. Seedling Growth Parameter and Germination Indices

The number of germinated seeds in each Petri dish was recorded daily. At the seventh day, the length of their radicles and shoots was measured on graph paper under a stereomicroscope. The collected data were used to calculate the germination percentage, SVI [40] and MGT [41], respectively, by the following equations:

$$\text{Germination Percentage} = \frac{\text{Germinated Seed Number}}{\text{Seed Total Number}} \times 100 \qquad (1)$$

$$\text{SVI} = (\text{Mean Root Length} + \text{Mean Shoot Length}) \times \text{Germination Percentage} \qquad (2)$$

$$\text{MGT} = \frac{\sum D \times \text{Germinated Seed Number}}{\sum \text{Germinated Seed Number}}, \qquad (3)$$

where D is the number of days from the beginning of germination.

4.5. Determination of Polyphenolic Content

The total polyphenolic content of the aqueous extracts was determined colorimetrically by the Folin-Ciocalteau method described by Scalbert et al. [42] with slight modifications. Briefly, 0.5 mL of each extract was added to 2.5 mL of 10% Folin-Ciocalteau reagent, previously diluted with distilled water. After 3 min, 2 mL of 7.5% sodium carbonate solution was added. The mixture was incubated in the dark for 1 h at room temperature and its absorbance was measured at 765 nm using a UV-vis spectrophotometer (Jenway 7205). A calibration curve was prepared with gallic acid standard solution at various concentrations (10 to 100 mg/L). The results were expressed as mg gallic acid equivalent (GAE)/100 g dry plant part. All the measurements were taken in triplicate and the mean values were calculated.

4.6. Statistical Analysis

The data were analyzed, with the support of IBM SPSS software, through the analysis of variance carried out separately for each bioassay (i.e., extract and powder bioassays) and *L. multiflorum* organs (i.e., inflorescences, stems, roots, seeds). The germination indices (i.e., germination percentage, SVI, MGT, root length, shoot length) measured for the two species (i.e., *E. oryzicola* and *O. sativa*) under different treatments were taken into account as dependent variables.

The one-way ANOVA and the Turkey's-b post hoc test were performed in order to establish the significant effect (at $\alpha \leq 0.05$) of the treatments with *L. multiflorum* (i.e., the different levels of concentration or quantity in extract and powder bioassay, respectively), on the species, and describe the homogenous subsets.

Moreover, the two-way ANOVA was performed, considering as factors the treatments with *L. multiflorum* and the species, in order to highlight the significant interaction (at $\alpha \leq 0.05$) between "species × *L. multiflorum* treatments", and then highlighting the species-specific effects of the treatments and the different behaviors or susceptibility between the rice crop and the weed.

Author Contributions: Co-first authors, S.V. and F.O; co-last authors, S.B. and M.I. Conceptualization, S.B., M.I. and S.V.; methodology, F.O., M.I., V.V. and S.V.; validation, M.I. and S.V.; formal analysis, F.O. and V.V.; investigation, V.V. and S.V.; resources, S.B. and M.I.; data curation, F.O. and S.V.; writing—original draft preparation, F.O. and S.V.; writing—review and editing, F.O., M.I. and S.V.; visualization, F.O., V.V. and S.V.; supervision, M.I. and S.V.; project administration, M.I. and S.V.; funding acquisition, S.B. and M.I. All authors have read and agreed to the published version of the manuscript.

Funding: This research received no external funding.

Acknowledgments: This research was supported by "Risobiosystems Project" (Italian Ministry Mipaaf funds, for research and innovation in the organic rice sector), and by EcorNaturaSì s.p.a.. We gratefully thank Dr. Stefano Gomarasca for his help in plant identification and seed treatment; all the members of the multi-actor community "RisoBioVero", the farmers network "Noi Amici della Terra" and the farm "Terre di Lomellina di Rosalia Caimo Duc" for the efforts in promoting the networking, knowledge exchange and dissemination of best practices; the farm "Una Garlanda" as the early pioneer of the practice.

Conflicts of Interest: The authors declare no conflict of interest.

References

1. Kendig, A.; Williams, B.; Smith, C.W. Rice weed control. In *Rice Origin, History, Technology and Production*; Smith, C.W., Dilday, R.H., Eds.; John Wiley & Sons Inc.: New York, NY, USA, 2003; Volume 3.7, pp. 458–460.
2. Fischer, A.J.; Ateh, C.M.; Bayer, D.E.; Hill, J.E. Herbicide-resistant *Echinochloa oryzoides* and *E. phyllopogon* in California *Oryza sativa* fields. *Weed Sci.* **2000**, *48*, 225–230. [CrossRef]
3. Gibson, K.D.; Fischer, A.J.; Foin, T.C.; Hill, J.E. Implications of delayed *Echinochloa* spp. germination and duration of competition for integrated weed management in water-seeded rice. *Weed Res.* **2002**, *42*, 351–358. [CrossRef]
4. Labrada, R. Major weed problems in rice. In *Rice Information*; FAO: Rome, Italy, 2002; Volume 3.
5. Osuna, M.D.; Okada, M.; Ahmad, R.; Fischer, A.J.; Jasieniuk, M. Genetic diversity and spread of thiobencarb resistant early watergrass (*Echinochloa oryzoides*) in California. *Weed Sci.* **2011**, *59*, 195–201. [CrossRef]
6. Busi, R. Resistance to herbicides inhibiting the biosynthesis of very-long-chain fatty acids. *Pest Manag. Sci.* **2014**, *70*, 1378–1384. [CrossRef] [PubMed]
7. Directive 79/117/EE. Available online: https://eurlex.europa.eu/LexUriServ/LexUriServ.do?uri=CONSLEG:1979L0117:20040520:EN:PDF (accessed on 3 February 2020).
8. Directive 91/414/CEE. Available online: https://eur-lex.europa.eu/legal-content/EN/ALL/?uri=CELEX%3A31991L0414 (accessed on 3 February 2020).
9. Ministero Della Salute. Available online: http://www.salute.gov.it/portale/news/p3_2_1_1_1.jsp?lingua=italiano&menu=notizie&p=dalministero&id=3644 (accessed on 4 February 2020).
10. Hill, J.E.; Smith, R.J., Jr.; Bayer, D.E. Rice weed control: Current technology and emerging issues in temperate rice. *Aus. J. Exp. Agric.* **1994**, *34*, 1021–1029. [CrossRef]
11. Tran, D.V. World rice production: Main issues and technical possibilities. *Cah. Opt. Méditerr.* **1997**, *24*, 57–69.
12. Padel, S. Conversion to organic farming: A typical example of the diffusion of an innovation? *Sociol. Rural.* **2001**, *41*, 40–61. [CrossRef]
13. Reganold, J.P.; Wachter, J.M. Organic agriculture in the twenty-first century. *Nat. Plants* **2016**, *2*, 1–8. [CrossRef]
14. Delmotte, S.; Tittonell, P.; Mouret, J.C.; Hammond, R.; Lopez-Ridaura, S. On farm assessment of rice yield variability and productivity gaps between organic and conventional cropping systems under Mediterranean climate. *Eur. J. Agron.* **2001**, *35*, 223–236. [CrossRef]
15. Hazra, K.K.; Swain, D.K.; Bohra, A.; Singh, S.S.; Kumar, N.; Nath, C.P. Organic rice: Potential production strategies, challenges and prospects. *Org. Agric.* **2018**, *8*, 39–56. [CrossRef]
16. Teasdale, J.R.; Mangum, R.W.; Radhakrishnan, J.; Cavigelli, M.A. Weed seedbank dynamics in three organic farming crop rotations. *Agron. J.* **2004**, *96*, 1429–1435. [CrossRef]
17. Liebman, M.; Davis, A. Managing weeds in organic farming systems: An ecological approach. In *Organic Farming: The Ecological System*; Agronomy monograph 54; Francis, C., Ed.; American Society of Agronomy: Madison, WI, USA, 2009; pp. 173–196.
18. Clark, A. *Managing Cover Crops Profitably*, 3rd ed.; Sustainable Agriculture Research and Education (SARE) Program: Beltsville, MD, USA, 2007; pp. 32–33.
19. Tabaglio, V.; Marocco, A.; Schulz, M. Allelopathic cover crop of rye for integrated weed control in sustainable agroecosystems. *Ital. J. Agron.* **2013**, *8*, 35–40. [CrossRef]
20. Patni, B.; Chandra, H.; Mishra, A.P.; Guru, S.K.; Vitalini, S.; Iriti, M. Rice allelopathy in weed management. An integrated approach. *Cell. Mol. Biol.* **2018**, *64*, 84–93. [CrossRef] [PubMed]
21. Flamini, G. Natural herbicides as a safer and more environmentally friendly approach to weed control: A review of the literature since 2000. *Stud. Nat. Prod. Chem.* **2012**, *38*, 353–396.
22. Regulation EU 2018/848. Available online: https://eur-lex.europa.eu/legal-content/EN/TXT/?uri=uriserv%3AOJ.L_.2018.150.01.0001.01.ENG (accessed on 22 January 2020).
23. Orlando, F.; Alali, S.; Vaglia, V.; Pagliarino, E.; Bacenetti, J.; Bocchi, S. Participatory approach for developing knowledge on organic rice farming: Management strategies and productive performance. *Agric. Syst.* **2020**, *178*, 102739. [CrossRef]
24. Ispra. Available online: https://www.isprambiente.gov.it/files2018/pubblicazioni/rapporti/Rapporto_282_2018.pd (accessed on 22 January 2020).

25. Amini, R.; An, M.; Pratley, J.; Azimi, S. Allelopathic assessment of annual ryegrass (*Lolium rigidum*): Bioassays. *Allelopath. J.* **2009**, *24*, 67–76.
26. Lehoczky, E.; Nelima, M.O.; Szabó, R.; Szalai, A.; Nagy, P. Allelopathic effect of *Bromus* spp. and *Lolium* spp. shoot extracts on some crops. *Commun. Agric. Appl. Biol. Sci.* **2011**, *76*, 537–544.
27. Ferreira, M.I.; Reinhardt, C.F.; van der Rijst, M.; Marais, A.; Botha, A. Allelopathic root leachate effects of *Lolium multiflorum* x *L. perenne* on crops and the concomitant changes in metabolic potential of the soil microbial community as indicated by the Biolog Ecoplate™. *Int. J. Plant Soil Sci.* **2017**, *19*, 1–14. [CrossRef]
28. Favaretto, A.; Scheffer-Basso, S.M.; Perez, N.B. Allelopathy in Poaceae species present in Brazil. A review. *Agron. Sustain. Dev.* **2018**, *38*, 22. [CrossRef]
29. Li, G.; Zeng, R.S.; Li, H.; Yang, Z.; Xin, G.; Yuan, J.; Luo, Y. Allelopathic effects of decaying Italian ryegrass (*Lolium multiflorum* Lam.) residues on rice. *Allelopath. J.* **2008**, *22*, 15–23.
30. Li, G.X.; Li, H.J.; Yang, Z.; Xin, G.; Tang, X.R.; Yuan, J.G. The rhizosphere effects in "Italian ryegrass-rice" rotational system V. Evidences for the existence of rice stimulators in decaying products of Italian ryegrass residues. *Acta Sci. Nat. Univ. Sunyatseni* **2008**, *47*, 88–93.
31. Jang, S.J.; Kim, K.R.; Yun, Y.B.; Kim, S.S.; Kuk, Y.I. Inhibitory effects of Italian ryegrass (*Lolium multiflorum* Lam.) seedlings of rice (*Oryza sativa* L.). *Allelopath. J.* **2018**, *44*, 219–232. [CrossRef]
32. Jang, S.J.; Beom, Y.Y.; Kim, Y.J.; Kuk, Y.I. Effects of downy brome (*Bromus tectorum* L.) and Italian ryegrass (*Lolium multiflorum* Lam.) on growth inhibition of wheat and weeds. *Philipp. Agric. Sci.* **2018**, *101*, 20–27.
33. Hoffman, M.L.; Weston, L.A.; Snyder, J.C.; Regnier, E.E. Allelopathic influence of germinating seeds and seedlings of cover crops on weed species. *Weed Sci.* **1996**, *44*, 579–584. [CrossRef]
34. Turk, M.A.; Tawaha, A.M. Allelopathic effect of black mustard (*Brassica nigra* L.) on germination and growth of wild oat (*Avena fatua* L.). *Crop Prot.* **2003**, *22*, 673–677. [CrossRef]
35. Han, C.M.; Wu, N.; Li, W.; Pan, K.; Wang, J.C. Allelopathic effect of ginger on seed germination and seedling growth of soybean and chive. *Sci. Hortic.* **2008**, *116*, 330–336. [CrossRef]
36. Appiah, K.S.; Mardani, H.K.; Omari, R.A.; Eziah, V.Y.; Ofosu-Anim, J.; Onwona-Agyeman, S.; Amoatey, C.A.; Kawada, K.; Katsura, K.; Oikawa, Y.; et al. Involvement of carnosic acid in the phytotoxicity of *Rosmarinus officinalis* leaves. *Toxins* **2018**, *10*, 498. [CrossRef]
37. Scavo, A.; Rialb, C.; Molinillob, J.M.G.; Varelab, R.M.; Mauromicalea, G.; Maciasb, F.A. The extraction procedure improves the allelopathic activity of cardoon (*Cynara cardunculus* var. *altilis*) leaf allelochemicals. *Ind. Crops Prod.* **2019**, *128*, 479–487. [CrossRef]
38. Fiorentino, A.; Della Greca, M.; D'Abrosca, B.; Oriano, P.; Golino, A.; Izzo, A.; Monaco, P. Lignans, neolignans and sesquilignans from *Cestrum parqui* l'Her. *Biochem. Syst. Ecol.* **2007**, *35*, 392–396. [CrossRef]
39. Della Greca, M.; Fiorentino, A.; Monaco, P.; Previtera, L.; Zarrelli, A. Effusides I-V: 9, 10-dihydrophenanthrene glucosides from *Juncus effusus*. *Phytochemistry* **1995**, *40*, 533–535. [CrossRef]
40. Abdul-Baki, A.A.; Anderson, J.D. Vigor determination in soybean seed by multiple criteria. *Crop Sci.* **1973**, *13*, 630–633. [CrossRef]
41. Ellis, R.A.; Roberts, E.H. The quantification of ageing and survival in orthodox seeds. *Seed Sci. Technol.* **1981**, *9*, 373–409.
42. Scalbert, A.; Monties, B.; Janin, G. Tannins in wood: Comparison of different estimation methods. *J. Agric. Food Chem.* **1989**, *37*, 1324–1332. [CrossRef]

© 2020 by the authors. Licensee MDPI, Basel, Switzerland. This article is an open access article distributed under the terms and conditions of the Creative Commons Attribution (CC BY) license (http://creativecommons.org/licenses/by/4.0/).

Article

Additive Effect of Botanical Insecticide and Entomopathogenic Fungi on Pest Mortality and the Behavioral Response of Its Natural Enemy

G. Mandela Fernández-Grandon [1,*], Steven J. Harte [1], Jaspher Ewany [1], Daniel Bray [1] and Philip C. Stevenson [1,2]

1. Natural Resources Institute, University of Greenwich, Central Avenue, Chatham Maritime, Kent ME4 4TB, UK; s.j.harte@gre.ac.uk (S.J.H.); EwanyJaspher@yahoo.com (J.E.); d.bray@gre.ac.uk (D.B.); p.c.stevenson@gre.ac.uk (P.C.S.)
2. Royal Botanic Gardens, Kew, Richmond, Surrey TW9 3DS, UK
* Correspondence: m.fernandez-grandon@gre.ac.uk; Tel.: +44-0-1634-883057

Received: 25 December 2019; Accepted: 29 January 2020; Published: 1 February 2020

Abstract: Sustainable agricultural intensification employs alternatives to synthetic insecticides for pest management, but these are not always a direct replacement. Botanical insecticides, for example, have rapid knockdown but are highly labile and while biological pesticides are more persistent, they are slow acting. To mitigate these shortcomings, we combined the entomopathogenic fungus (EPF) *Metarhizium anisopliae* with pyrethrum and evaluated their efficacy against the bean aphid, *Aphis fabae*. To ascertain higher trophic effects, we presented these treatments to the parasitoid, *Aphidius colemani*, on an aphid infested plant in a Y-tube olfactometer and measured their preferences. Aphid mortality was significantly higher than controls when exposed to EPF or pyrethrum but was greater still when exposed to a combination of both treatments, indicating an additive effect. This highlights the potential for applications of pyrethrum at lower doses, or the use of less refined products with lower production costs to achieve control. While parasitoids were deterred by aphid infested plants treated with EPF, no preference was observed with the combination pesticide, which provides insight into the importance that both application technique and timing may play in the success of this new technology. These results indicate the potential for biorational pesticides that combine botanicals with EPF.

Keywords: biopesticide; organic pesticide; Y-tube olfactometer; pyrethrum; parasitoid; entomopathogenic fungi; leaf disc assay; insect behavior; survival analysis

---

1. Introduction

Pesticidal plant extracts are an important component of sustainable integrated pest management (IPM) and can offer an effective alternative to synthetic pesticides for management of pests, especially for smallholders [1]. Pesticidal plants typically have lower impact on higher trophic levels, including natural enemies of pests, so are better suited to sustainable production [2,3] and are locally available at a lower cost than synthetic chemicals [4]. The most important commercial botanical pesticides include pyrethrum and neem products [5]. Pyrethrum is a natural insecticide extracted from the flowers of *Chrysanthemum cinerariaefolium* and *Chrsanthemum cineum* [6,7] which has been used for controlling field, household, and storage pests, and parasites in livestock and humans [8–10]. A combination of awareness of the negative impacts of synthetic pesticides and increased pressure from regulatory authorities on permitted chemicals has seen the global demand for biopesticides grow over the past decade at an estimated 15% per annum [11–13], compared to 3% per annum for synthetic pesticides [11].

Natural pyrethrum contains six entomotoxic compounds: cinerin I and II, pyrethrin I and II and jasmolin I and II [14]. Pyrethrins enter the insect body via ingestion or contact, penetrate the epidermis and are distributed throughout the body in the haemolymph [7]. The insecticide disrupts the insect's peripheral and central nervous systems by causing repetitive discharges of nerves, resulting in paralysis [15]. Pyrethrins have a rapid "knockdown" effect preceding insect death [7,14] and insects usually die in a few minutes or hours following exposure to a fatal dose [8].

Pyrethrins influenced the development of some of the most widely used synthetic insecticides—the pyrethroids, including cypermethrin, permethrin, deltamethrin, fenvalerate and bifenthrin. The drive towards different synthetic pyrethroids was to increase their stability in the environment, providing effective control for longer. This has proven successful as pyrethroids are routinely used as agricultural insecticides with high adoption rates internationally [8].

The active components in pyrethrum are highly labile in ultraviolet (UV) light, non-persistent, and are less toxic to humans and the environment [9,16–18]. Although the lack of persistence has previously limited use of natural pyrethrum as an agricultural insecticide [8,15] it is less disruptive to IPM programs that include beneficial insects than conventional insecticides.

One of the emerging technologies as part of IPM is the use of entomopathogenic fungi (EPF) [19], which can be used to control a wide range of agricultural pests [20]. They have no negative effects on human health [21]. Entomopathogenic fungi are specific pathogens of insects that can infect their hosts through the external cuticle [20]. Rapid penetration and infection of a susceptible host occurs at high humidity [20], but spores can remain viable on the cuticle during unfavorable conditions and penetrate when humidity rises, even if only for a short time [22].

Upon successful penetration, the fungi develop, colonize the insect's internal organs and the insect eventually dies. It is after insect death that hyphae emerge, followed by spore formation and production of numerous conidia on the cadaver [20]. EPF, such as *Metarhizium anisopliae* (Metschnikoff) used in this study, can only complete their lifecycles and increase populations by producing numerous conidia after the death of infected hosts [23] which will disperse, infecting more insects and continue the propagation cycle.

There are successful synergies and compatibilities of EPF with different plant-based pesticides for improved pest control. For instance, the combination of sub-lethal doses of the EPF *Beauvaria bassiana* and neem extract increased mortality against whitefly, *Bemisia tabaci,* nymphs when neem insecticide was drenched with simultaneous application of *B. bassiana* on tomatoes plants [24,25]. *Beauvaria bassiana* (isolate PL63) was compatible with botanical extracts from *Trichilia catigua* leaves and effectively controlled insect pests in Brazil [26]. Another study by Shoukat et al. [27], revealed that both *M. anisopliae* and *B. bassiana* showed a synergistic effect when mixed with neem extract, *Azadirachta indica,* and increased mortality against 3rd instars of *Culex pipiens* in the field. These examples of the compatibility of EPF with pesticidal plant extracts suggests potential in combining EPF and pyrethrum to improve efficacy through the rapid knockdown of pyrethrum and the persistent control offered by EPF to control agricultural pests and combat pest resistance.

A benefit that may be realized through a combination biopesticide over synthetic pesticides could be reduced impact on beneficial insects in the environment through decreased exposure and greater specificity. EPF has been reported to be pathogenic to beneficial insects, including parasitoids. *B. bassiana* was found to infect and kill adult *Aphidius colemani* with infection rates ranging from 46.3% to 60% [28] and between 57.6 to 66% [29] under greenhouse environments. Shipp et al. [29] did not recommend the use of adult *A. colemani* together with *B. bassiana* for pest control in greenhouses. The exposure of *A. colemani* to different EPF strains such as *M. anisopliae* is still to be explored.

However, time of application of EPF has been manipulated to reduce parasitoid mortality and affect the use of EPF together with parasitoids in pest control. For instance, parasitoid *Aphidius matricariae* and *B. bassiana* (strain EUT116) were effective against the aphid, *Myzus persicae* when the fungus was applied 96 hours after the release of parasitoids [30]. *Beauvaria bassiana* (strain PL63) and the aphid parasitoid, *Diaeretiella rapae* were recommended against *M. persicae* [31]. Another study by Mohammed and Hatcher [32], revealed that introducing EPF *Lecanicillium muscarium* six days after releasing *A. colemani* was effective against *M. persicae* in a greenhouse environment. Based on these results, usage of selected EPF isolates and applying EPF after parasitism can reduce detrimental effects on parasitoids. However, to inform the effective use of EPF with the beneficial insects, it is important that we understand their interaction with the fungi.

In this study, we assessed the efficacy of pyrethrum, *Metarhizium anisopliae* EPF and the combination of both on mortality of the aphid pest, *Aphis fabae*, and how the aphid parasitoid, *Aphidius colemani*, responded to the odors associated with these compounds. The components were selected for this proof of principle because of the ready availability of pyrethrum in low-income areas and the known efficacy of this *M. anisopliae* strain against this highly problematic aphid pest. To gauge the potential in establishment of EPF on the aphid population, we also recorded incidence of visible fungal establishment with hyphae seen emerging from the cuticle and the number of offspring produced by the aphids following exposure. We found that both pyrethrum and the EPF led to a significant increase of mortality on *A. fabae* which was further enhanced when they were presented in combination. Visible fungal growth occurred more rapidly on aphids treated with the combination biopesticide, indicating establishment in population could be accelerated through the multimodal action. It was also shown that its associated parasitoid, *A. colemani*, preferentially selects plants that do not contain EPF when foraging using odor, however, the preference exhibited is absent in the combined treatment.

2. Results

2.1. Aphid Mortality Assay

2.1.1. Survival

No significant interaction was found between the application rate of EPF 0 CFU mL^{-1} (carrier oil only), 1×10^6 CFU mL^{-1} or 1×10^8 CFU mL^{-1} and pyrethrum presence (10 ppm pyrethrins) or absence (0 ppm) on aphid survival ($\chi^2 = 0.70$, df = 2, $p = 0.70$). However, significant independent effects of both pyrethrum ($\chi^2 = 6.56$, df = 1, $p = 0.01$) and EPF concentration ($\chi^2 = 16.8$, df = 2, $p = 0.001$) were found on aphid survival (Figure 1). Addition of pyrethrum led to a 40.5 h reduction in predicted mean aphid survival at 0 CFU mL^{-1} EPF (from 80.1 h to 39.6 h). At 1×10^6 CFU mL^{-1} survival was reduced by 29.2 h through addition of pyrethrum (from 67.3 h to 35.3 h), and by 10 h at 1×10^8 CFU mL^{-1} (from 19.7 h to 9.7 h). Survival was reduced significantly through addition of EPF compared to No EPF ($\chi^2 = 6.9$, df = 1, $p = 0.009$), and from 1×10^6 CFU mL^{-1} to 1×10^8 CFU mL^{-1} EPF ($\chi^2 = 9.9$, df = 1, $p = 0.002$).

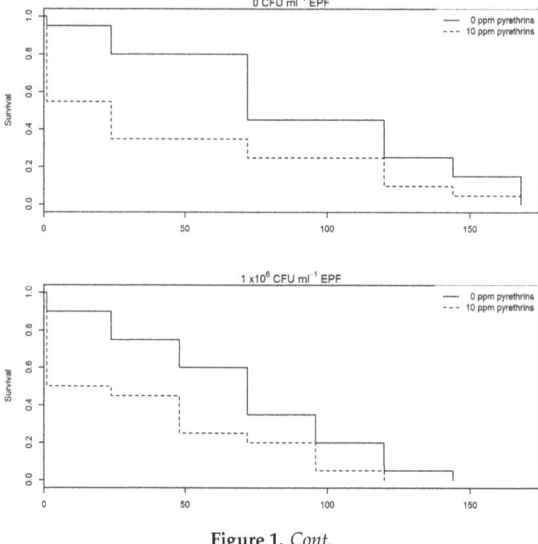

Figure 1. *Cont.*

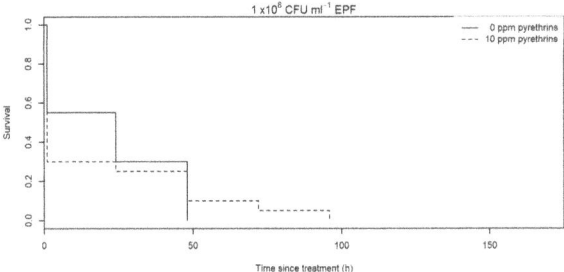

Figure 1. Survival of individual *A. fabae* exposed to *M. anisopliae* alone (solid line) or in combination with pyrethrum (dotted line). Entomopathogenic fungus (EPF) was tested at 0 colony forming units (CFU) mL^{-1} (carrier oil only, top graph), 1×10^6 CFU mL^{-1} (middle graph) and 1×10^8 CFU mL^{-1}.

2.1.2. Hyphal Growth on Insect Surface

Overall, aphids (n = 40) which had not been treated with EPF (0 CFU mL^{-1}) showed no hyphal growth up to 192 h after treatment. Aphids not treated with EPF were therefore excluded from further analysis of time until visible fungal growth. Increasing concentration of treatment from 1×10^6 CFU mL^{-1} to 1×10^8 CFU mL^{-1} significantly reduced time until hyphal growth ($\chi^2 = 10.74$, df = 1, $p = 0.001$). Addition of pyrethrum at 10 ppm pyrethrins also significantly reduced time until hyphae formation was observed ($\chi^2 = 10.74$, df = 1, $p < 0.001$). However, no interaction was found between EPF level and pyrethrum treatment on time until hyphal growth was observed ($\chi^2 = 2.37$, df = 1, $p = 0.12$). Addition of pyrethrum reduced predicted mean time until growth was seen by 84 h at 1×10^6 CFU mL^{-1} (from 226 h to 142 h) and 63 h at 1×10^8 CFU mL^{-1} (from 170 h to 107 h) (Figure 2).

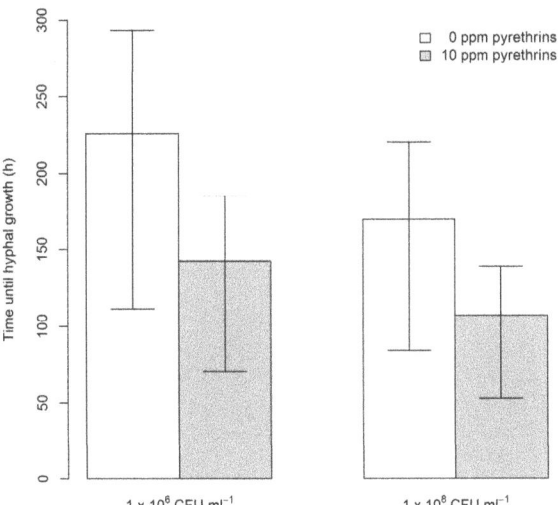

Figure 2. Predicted time (± upper and lower quantiles) until hyphal growth was observed on individual *A. fabae* exposed to *M. anisopliae* alone or in combination with pyrethrum. CFU: colony forming units.

2.1.3. Number of Offspring

Due to a very low offspring count in the first block of replicates, only aphids tested in the second block of treatments were used in the analysis. In this block, a significant interaction was found between EPF level (0 CFU mL^{-1} (carrier oil only), 1×10^6 CFU mL^{-1} or 1×10^8 CFU mL^{-1}) and pyrethrum presence (10 ppm pyrethrins) or absence (0 ppm) on total number of offspring produced by each aphid ($\chi^2 = 7.01$, df = 2, $p = 0.03$). At 0 CFU mL^{-1} and 1×10^6 CFU mL^{-1}, addition of pyrethrum resulted in significantly fewer offspring produced (Tukey's test, $p < 0.05$, Figure 3). However, no significant effect of pyrethrum was found on number of offspring produced by aphids exposed to 1×10^8 CFU mL^{-1} of EPF.

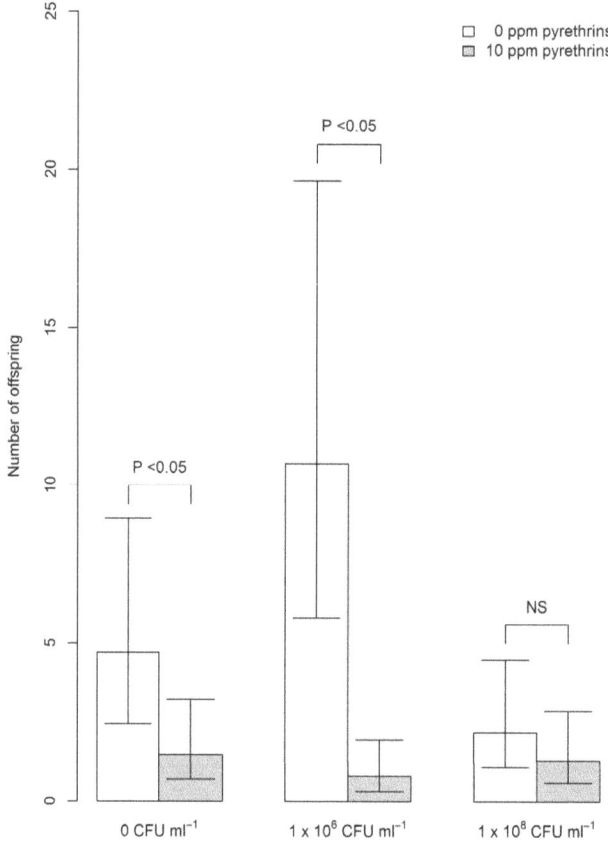

Figure 3. Predicted mean number (± 95% confidence intervals) of offspring produced by individual *A. fabae* exposed to *M. anisopliae* alone (white bars) or in combination with pyrethrum (10 ppm pyrethrins; grey bars). EPF was tested at 0 colony forming units (CFU) mL^{-1} (carrier oil only, left), 1×10^6 CFU mL^{-1} (middle) and 1×10^8 CFU mL^{-1} (right).

2.2. Parasitoid Dual-Choice Assays

When presented with the choice of aphid-infested or uninfested plant material through a dual-choice assay, more female parasitoids chose the aphid-infested material (exact binomial test, $n = 50$, $p = 0.0066$) (Figure 4). This served as a positive control to confirm that parasitoids would orientate towards *A. fabae* in the absence of visual cues. All other experimental treatments used only aphid-infested bean plants. When presented with a choice between two aphid-infected plants, one of which had been treated with pyrethrum, no significant difference was found between proportion of parasitoids choosing an untreated plant, and a plant which had been treated with pyrethrum (exact binomial test, $n = 50$, $p = 0.322$). However, significantly fewer parasitoids chose the EPF-treated compared to the untreated plant (exact binomial test, $n = 50$, $p < 0.001$). When the pyrethrum was presented alongside the EPF and compared to an untreated plant, the parasitoid displayed no preference (exact binomial test, $n = 50$, $p = 0.203$) (Figure 4). For all the replicates there was only one non-responder recorded for failing to make a choice in the allotted time (treatment containing pyrethrum + EPF), this individual was excluded from the analysis.

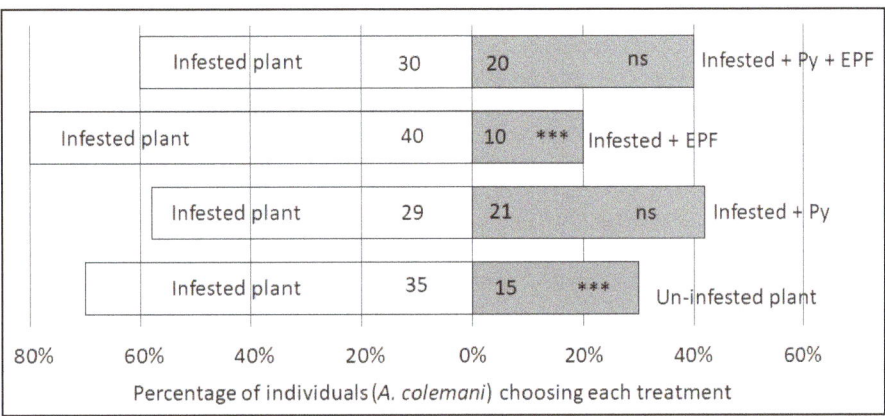

Figure 4. Parasitoid, *A. colemani*, responses to treatments in a Y-tube olfactometer. N = 50 for each treatment, with the number of individuals choosing each side indicated on the relevant bar. Where 'Infested' refers to aphid colonized plant, Py is treatment with pyrethrum and EPF is treatment with entomopathogenic fungi, *Metarhizium anisopliae* (ICIPE 62) *** Indicates $p < 0.01$, ns = not significant.

3. Discussion

Our working hypotheses for this study were:

1. Efficacy of pyrethrum and EPF would be enhanced when presented in combination.
2. The biopesticides would affect parasitoid plant/host preference.

Through the evaluation of aphid mortality, hyphae formation and offspring production after exposure to the biopesticides, we found that efficacy was enhanced when the components were presented in combination. This additive effect of combination was observed through reduced survival, more rapid formation of hyphae and reduced fecundity. Increased mortality was recorded as EPF concentration increased, though from a practical perspective, the level of mortality achieved with lower dosage may be sufficient in pest control programs and may even be preferable if it permits low-level persistence of the host for biocontrol agents. One of the difficulties in controlling aphid populations lies in their high rate of fecundity. The significant reduction in fecundity noted with exposure to the combination treatment could be critical in more effective control as it may curb the exponential rate of population growth. However, in this study only ten replicates were evaluated for offspring production in each treatment. This was due to only four offspring being produced in the initial block of replicates. It is not clear why this number was this low. To evaluate this further we suggest the experiment be repeated and assessment to include the intrinsic rate of increase and other population metrics to gain greater insight into how this is likely to affect population dynamics.

The increase in mortality may be due to the bimodal effects of the combined treatment as the immediate attack on the nervous system provided by the pyrethrum leave individuals more susceptible to infection from the EPF. This concurs with previous studies on pests that have shown additive or synergistic effects when EPF and pesticidal plant products are presented in combination [33–35]. In addition to the increased mortality with a combined biopesticide application, it was notable that the time lag before hyphae were observed was shortened in the lower dosage application (226 h to 142 h). This difference of 84 h could have a considerable impact on the viability of such a product and especially when considered over multiple generations of the pest. One of the most widely recognized drawbacks to EPF application is the slow-acting nature of the product [35–38]. Although it can offer sustained control, this relies on its establishment within the population through propagation via the host. A guiding principle behind the exploration of a combined biopesticide is for a product that overcomes the short-lived nature of pyrethrum and the slow-acting nature of the EPF. An accelerated rate of establishment, as indicated by this study, could be a critical advantage to such a product. However, we recognize that this is a laboratory evaluation of the interaction under controlled conditions and with limited replication. The real impact of this would need to be established through longer running trials which assess the formation of conidia from the host and population suppression over multiple generations in a more field-realistic situation.

The results of the mortality experiments show strong support for the development of combined biopesticides as a new tool for IPM. In addition to a direct increase in pest mortality, there are indications that propagation of the EPF may be occurring more rapidly and fecundity of the pest is being suppressed. Furthermore, it is important to note the pyrethrin dosages applied in the trials (10 ppm) were a fifth of that recommended for effective pest control in the field. This was used to allow any potential synergy or additive effect of combination to be recorded as it was identified in preliminary trials that mortality was too high at 50 ppm to determine these effects. The mortality observed at this low dosage of pyrethrins when presented in combination with the EPF has greater practical and commercial appeal for this technology. Refinement of pyrethrum remains a relatively expensive process and one limited by the technology available in an area. The use of less refined material could lead to lower cost production, a reduction of impact on non-target species and greater potential for formulation in and for lower income countries. We also identify that there are various shortcomings to the use of these biopesticides which a combination approach will need to overcome and will assess through field trials.

Evidence from the Y-tube olfactometer behavioral assays supported our second hypothesis that the behavior of the parasitoid would be affected by volatiles from treated materials. The control demonstrated that the parasitoid was able to detect *A. fabae* feeding on the bean plants and would move towards them, behaviors indicative of foraging. When EPF was applied as the only variable present, wasps showed a preference for the plant-aphid treatment without the EPF. The avoidance of EPF by natural enemies has previously been observed in studies looking at ladybirds, *Coccinella septempunctata*, and anthocorids, *Anthocoris nemorum* [39,40] respectively. The behavior possibly confers fitness benefits as the wasp may reduce its exposure to the fungal pathogen. However, studies on parasitoid *Cephalonomia tarsalis* showed no avoidance behavior in response to the EPF *B. bassiana* [41]. The avoidance behavior of parasitoids to EPF may be species dependent, which highlights the importance of study on commercially relevant organisms. It is interesting to note that EPF avoidance observed in this study was absent in the presence of pyrethrum which could indicate either an interaction between the components or the perception of the EPF-aphid-plant treatment.

Future research should consider other aspects of the behavior of beneficial insects and elucidate the mechanisms behind this observed preference for non-EPF treated material. Experiments could be performed to disentangle the interaction between plants, aphids and EPF to identify whether individual components or the interaction is responsible for the deterrence that was observed. Next steps in this direction would be to evaluate the direct impact of relevant EPF strains to the parasitoids, the odors involved in potential repellency from EPF and how these behaviors affect success in field settings.

Our findings align with what has been found previously that plant-based insecticides can be complemented by addition of EPF. The full potential of such a technology is still to be explored and different formulations should be investigated using different EPF and botanical components. It is also important that these experiments are taken out of the laboratory and into the field to assess their efficacy in highly variable field conditions. In future testing we also suggest that the impact on beneficial insects in the environment is considered as a high priority and should extend to include pollinators as well as natural enemies. Their susceptibility to the combined formulations should be assessed and the findings should inform the future use of this technology.

4. Materials and Methods

4.1. Insect Rearing

Aphids, *Aphis fabae*, were obtained from colonies at Harper Adams University. The population was maintained on potted broad bean plants at temperatures of 27 °C, on a 12 h, L:D cycle. Fresh plant material was introduced into the colony if alatae were seen to form with old material removed following 24 h.

Parasitoid species, *Aphidius colemani*, were used in laboratory bioassays. The parasitoids were obtained from Bioline Agro-Sciences Ltd, UK from a different rearing background to that used in experiments. It has previously been found that parasitoids 2 to 5 days old display greater fecundity and higher rates of parasitism [42–44]. For these experiments female *A. colemani* 3 to 5 days old were used to increase the likelihood of host seeking behavior.

4.2. Entomopathogenic Fungi

The commercially available strain of *Metarhizium anisopliae* isolate, ICIPE 62, in an oil emulsion was obtained from Real IPM Kenya (Madaraka, Thika, Kenya) and used for the experimental work. Dilutions were conducted as necessary to obtain the concentration of colony forming units (CFU) required for trials.

4.3. Preperation of Pyrethrum

Semi-refined pyrethrum product was supplied by Botanical Extracts EPZ Ltd (Twiga Crescent, Export Processing Zone, Athi River, Kenya) for use in the experimental work. The product was analyzed to ensure the correct dosage of active ingredient was used in experimental work. Pyrethrins were analyzed using an Agilent 1200 series HPLC system (Agilent Technologies, Santa Clara, United States) consisting of modular quaternary pump, degasser, auto-sampler, column oven and photodiode array detector. Separation of entomotoxic pyrethrum constituents was achieved on a Waters X-Select T3 column (250 × 4.6 mm, 3.5 µm) and a guard column with the same characteristics, all kept at 25 °C. The chromatographic conditions were: flow rate 1 mL min^{-1}, sample injection volume of 10 µL and mobile phases; A (100% H_2O), B (100% MeOH) and mobile phase C (1% formic acid). 27/68/5 (A/B/C) which was held for 2 mins, raised to 5/90/5 over 22 min (24 min total), followed by wash and re-equilibration steps. The photodiode array detection was conducted by scanning between 200 and 600 nm.

Individual compounds in each sample were identified by comparing their retention times and UV–Vis spectra with those of a standard pyrethrum sample purchased from Sigma (Gillingham, Dorset, UK). Quantitative determination of the target compounds in the extracts was performed using external calibration curves in the concentration range of 0.05 to 1 mg mL^{-1} (detection at 225 nm).

Total w/w% of the active compounds was 17%, comprised of cinerin I (0.5%), cinerin II (3.7%), pyrethrin I (0.5%), pyrethrin II (1.4%), jasmolin I (11%) and jasmolin II (1.5%). Retention times were 18.9, 11.4, 19.1, 11.8, 22.3 and 14.1 minutes respectively.

The pyrethrum extract was subsequently diluted with deionized H_2O, without the need for surfactants or adjuvants, to the required ppm values of pyrethrins, as specified in the mortality assay methods. All ppm values quoted in this section are for total pyrethrin content i.e., cinerin I & II, pyrethrin I & II and jasmolin I & II.

4.4. Mortality Assays

Leaf discs were removed from broad bean leaves using a sterilized 26 mm cork bore. Discs were then dipped into either negative control (H_2O), low concentration of EPF (1×10^6 CFU mL^{-1}), high concentration of EPF (1×10^8 CFU mL^{-1}), pyrethrum (10 ppm pyrethrins), low dose combination (EPF at 1×10^6 CFU mL^{-1} with 10 ppm pyrethrins) or full combination dose (EPF at 1×10^8 CFU mL^{-1} with 10 ppm pyrethrins). The concentration applied of pyrethrins is lower than the recommended 50 ppm for aphid control. A lower concentration of 10 ppm was selected from preliminary trials demonstrating that this elevated mortality but did not lead to 100% mortality, making it a viable candidate to assess interaction with the fungi.

During the preparation of leaf discs, a 1.5% agar solution (Oxoid Technical Agar No. 2) was prepared using distilled water. Once the solution was fully mixed, 10 ml was decanted into 29 mL pots (4.5 cm height × 4 cm diameter) to cool. Once the solution was viscous but not completely set, leaf discs were embedded into the agar ensuring the edges were sufficiently covered.

A single adult *A. fabae* was gently removed from plants using a fine paintbrush and placed onto the center of each leaf disc. A partially mesh lid was placed on each pot preventing aphid escape while avoiding a build-up of moisture. Pots were placed onto the trays in a Latin square design which was altered with each replicate to reduce positional bias. Treatments were maintained in a 26 °C room, on an 12:12 L:D light cycle.

All samples were monitored daily with counts made for offspring which were removed as they were found, mortality of the adults, and visible formation of hyphae. Monitoring was conducted at the same time on each day with the same experimenter conducting the observations to ensure consistency. A total of 20 replicates were performed for each treatment. The experiment was conducted across two blocks.

4.5. Parasitoid Choice Assays

Behavioral response of *A. colemani* to treatments of aphid-infested plants (positive control), uninfested plant (negative control), aphid-infested plant + pyrethrum, aphid-infested plant + EPF, and aphid-infested plant + pyrethrum + EPF was tested using a dual-choice assay.

Three days prior to each experimental day, bean plants were transferred from glasshouse to the bioassay room. Broad bean, *Vicia faba*, Dwarf Sutton variety (Kings seeds, Colchester, UK) were used for all experiments. All plants were used 3 weeks after germination with between 4 to 5 leaf pairs developed. Using a fine paintbrush, 50 aphids of mixed ages where transferred from the colony onto fresh bean plants and left for 72 h to establish plants in the 'Infested' treatments. This level is sufficient to induce production of defensive plant volatiles [45]. Uninfested control plants were maintained in the same condition with no direct contact with aphids. The pyrethrum and EPF suspension were placed in separate 1 L spray bottles and sprayed on the plants in a closed arena. All plant leaves were sprayed both on the top and the underside with pyrethrum or EPF to reflect application of these materials in a field environment. After administering the treatment, plants were left for 15 min before conducting the experiments.

In the treatment combining EPF and pyrethrum insecticide, equal amounts of each were sprayed separately. The order of spraying was alternated between trials. The sprayed plants were then placed in separate glass vessels to conduct the experiment. All the treatments were tested against the positive control to make four treatment sets per day.

A glass Y-tube olfactometer (stem 8 cm, arms 8 cm, internal diameter 1 cm and 120° angle between arms) was used to assay parasitoid response to plant treatments in the absence of visual stimuli. The Y-tube was connected to two, 3 L Kilner jars containing the positive control and the other contained either an experimental treatment or negative control. The Kilner jars used were modified with an airtight inlet and outlet fittings, allowing a continuous flow of charcoal filtered air at 200 mL/min through each vessel to the two arms of the Y-tube olfactometer via Teflon tubing.

Experiments were conducted during active foraging times for the parasitoid (4–9 h after scotophase). During the experiment, individual *A. colemani* were released using a 1 ml pipette head connected to the base of the Y-tube olfactometer and removed immediately after the parasitoid entered the tube. Each *A. colemani* was observed for a maximum of 5 min or until it travelled 6 cm up one of the Y-tube arms and remained there for 30 s. Wasps that did not enter the Y-tube after 5 min were recorded as non-responders.

All the four treatments were tested on each experimental day, with the order determined using a Latin square. The bioassay arena of the Y-tube blocked the entrance of light from all sides except from the direction of the Y-tube olfactometer arms. In each treatment, the Y-tube arms and odor sources were swapped after five parasitoids were tested to minimize directional bias or any bias of choice due to light in one arm. A different Y-tube was used in each experiment, and all Y-tubes were cleaned with 70% ethanol and left to dry before use. Each parasitoid was used only once, reflecting a true biological replicate. Fifty replicates were completed for each treatment.

4.6. Statistical Analyses

4.6.1. Survival

The first objective of statistical analysis was to determine whether application of pyrethrum and EPF would result in lower aphid survival than treatment with EPF alone. Differences between treatment groups were first visualized by plotting Kaplan–Meier survival estimates (Figure 1). A model with Weibull errors was used to test whether there was an interaction between EPF level (0 CFU mL^{-1}, 1×10^6 CFU mL^{-1} or 1×10^8 CFU mL^{-1}) and presence (10 ppm pyrethrins) or absence (0 ppm) of pyrethrum on aphid survival. The time interval in which the aphid died was entered as the dependent variable. Aphids which did not die during the experiment were recorded as censored cases. Independent effects of EPF level and pyrethrum presence or absence were then tested separately. Differences in survival between individual EPF levels were tested through model simplification (presence vs. absence of EPF, low (1×10^6 CFU mL^{-1}) vs high (1×10^8 CFU mL^{-1}) EPF. Significance of all effects was assessed through χ^2 test changes in residual deviance following deletion from the model [46]. All analyses were performed in R [47–49].

4.6.2. Visible Fungal Growth

A model similar to the survival analysis was applied with Weibull errors was used to determine whether application of pyrethrum would result in faster hyphae formation of EPF at the three levels of EPF tested. In this model, time interval in which hyphae were first observed was entered as the dependent variable. Replicates in which spores were not recorded were entered as censored cases. EPF level and pyrethrum treatment were entered as factors in the model. An interaction was included to determine whether the effect of pyrethrum on time until hyphae emergence varied with initial EPF concentration.

4.6.3. Number of Offspring

A generalized linear model with negative binomial areas [50] was used to determine whether application of pyrethrum and EPF had significant effects on aphid offspring production. Total number of offspring produced by each aphid across the experiment was entered as the dependent variable. EPF level (0 CFU mL^{-1}, 1×10^6 CFU mL^{-1} or 1×10^8 CFU mL^{-1}) and pyrethrum level (0 ppm pyrethrins or 10 ppm pyrethrins) were entered as factors in the model. An interaction term was included to test whether the effect of pyrethrum on number of offspring produced varied with EPF level applied. Significance of each term was assessed through likelihood ratio tests (χ^2) following deletion from the model. Differences between individual factor levels in numbers of offspring produced were assessed through Tukey's tests performed on estimated means [51]. Analysis was restricted to the second round of aphids tested, as none of the aphids in the first round produced offspring.

4.6.4. Parasitoid Dual-Choice Assays

Choice assays using the Y-tube olfactometer data was analyzed using two-sided exact binomial tests, with an assumption of a 50:50 distribution if the wasps were moving at random. Wasps which made no choice after 5 minutes were excluded from analysis. Tests were conducted using R (v 3.5.1, Vienna, Austria).

5. Conclusions

Here we show that a combination of the entomopathogenic fungi, *Metarhizium anisoplae*, and pyrethrum led to a higher rate of mortality in *Aphis fabae* than for the individual pesticides when tested alone. Thus, the combination of these two biopesticides was effective at killing the target pest more effectively than the individual components with no apparent contraindications, illustrating a novel approach to compensate for their individual shortcomings; the rapid breakdown of pyrethrum and slow activation of EPF. The development of fungi on the external cuticle was also observed earlier when EPF was presented with pyrethrum, which may be key to more rapid establishment in the population. A surprising finding was that when used alone, the EPF was repellent to the parasitoid *Aphidius colemani*, however, when presented in combination with the pyrethrum had no effect on foraging behavior of the natural enemy. Thus, the combination of EPF and pyrethrum may be better suited to use in an IPM system that included natural enemies, though timing considerations may be critical. Combinations of biopesticides that have different mechanisms of action have the potential to improve the efficacy of the individual components and reduce the build-up of resistance. The additive effect also suggests that there is potential for applications of pyrethrum at lower doses and so reducing effects at higher trophic levels or that less refined products could be used at lower production costs to achieve control. Studies of behavior provide insight into the importance that application technique and timing play in the effectiveness of this novel combination pesticide technology.

Author Contributions: Conceptualization, P.C.S. and G.M.F.-G.; methodology, G.M.F.-G and S.J.H.; formal analysis, D.B.; investigation, S.J.H., J.E. and G.M.F.-G.; writing—original draft preparation, G.M.F.-G. and J.E.; writing—review and editing, D.B., S.J.H., J.E. and P.C.S.; supervision, G.M.F.-G.; project administration, G.M.F-G.; funding acquisition, G.M.F.-G. and P.C.S. All authors have read and agreed to the published version of the manuscript.

Funding: Funding for this project was provided by BBSRC and InnovateUK through the UK-China Agritech Challenge to PCS and MFG as part of the 'Environmentally Benign Combination Biopesticides–Transforming Pest Control in Chinese and UK Agriculture' and initiated through a grant awarded to MFG by the Higher Education Innovation Fund (HEIF).

Acknowledgments: We are grateful to Tom Pope of Harper Adams for provision of the *Aphis fabae*, Botanical Extracts EPZ Ltd for supplying the pyrethrum, and Real IPM for suppling the Met62 used in these experiments. All these materials were provided without charge for use in the study.

Conflicts of Interest: The authors declare no conflict of interest.

References

1. Belmain, S.R.; Stevenson, P.C. Ethnobotanicals in Ghana: Reviving and modernising an age-old practise. *Pestic Outlook* **2001**, *12*, 233–238.
2. Tembo, Y.; Mkindi, A.G.; Mkenda, P.A.; Mpumi, N.; Mwanauta, R.; Stevenson, P.C.; Ndakidemi, P.A.; Belmain, S.R. Pesticidal Plant Extracts Improve Yield and Reduce Insect Pests on Legume Crops Without Harming Beneficial Arthropods. *Front. Plant Sci.* **2018**, *9*, 1425. [CrossRef] [PubMed]
3. Mkenda, P.; Mwanauta, R.; Stevenson, P.C.; Ndakidemi, P.; Mtei, K.; Belmain, S.R. Extracts from Field Margin Weeds Provide Economically Viable and Environmentally Benign Pest Control Compared to Synthetic Pesticides. *PLoS ONE* **2015**, *10*, 0143530. [CrossRef] [PubMed]
4. Deng, A.L.; Ogendo, J.O.; Owuor, G.; Bett, P.K.; Omolo, E.O.; Mugisha-Kamatenesi, M.; Mihale, J.M. Factors determining the use of botanical insect pest control methods by small-holder farmers in the Lake Victoria basin Kenya. *AJEST* **2009**, *3*, 108–115.
5. Grzywacz, D.; Stevenson, P.C.; Mushobozi, W.L.; Belmain, S.; Wilson, K. The use of indigenous ecological resources for pest control in Africa. *Food Secur.* **2014**, *6*, 71–86. [CrossRef]
6. Cox, C. Pyrethrum/pyrethrin insecticide fact sheet. *J. Pestic. Reform* **2002**, *22*, 14–20.

7. Schleier, J.J., III; Peterson, R.K.D. Pyrethrins and pyrethroid insecticides. In *Green Trends in Insect Control*. Lopez O.; Fernandez-Bolanos, J., Ed.; RSC: London, UK, 2011; Volume 3, pp. 94–131.
8. Casida, J.E. Pyrethrum flowers and pyrethroid insecticides. *Environ. Health Perspect.* **1980**, *34*, 189–202. [CrossRef]
9. Sola, P.; Mvumi, B.M.; Ogendo, J.O.; Mponda, O.; Kamanula, J.F.; Nyirenda, S.P.; Belmain, S.R.; Stevenson, P.C. Botanical pesticide production, trade and regulatory mechanisms in sub-Saharan Africa: Making a case for plant-based pesticidal products. *Food Secur.* **2014**, *6*, 369–384. [CrossRef]
10. Gallo, M.; Formato, A.; Ianniello, D.; Andolfi, A.; Conte, E.; Ciaravolo, M.; Varchetta, V.; Naviglio, D. Supercritical fluid extraction of pyrethrins from pyrethrum flowers (*Chrysanthemum cinerariifolium*) compared to traditional maceration and cyclic pressurization extraction. *J. Supercrit. Fluid* **2017**, *119*, 104–112. [CrossRef]
11. Chandler, D.; Bailey, A.; Tatchell, M.; Davidson, G.; Greaves, J.; Grant, W. The development, regulation and use of biopesticides for integrated pest management. *Philos. Trans. R. Soc. B* **2011**, *366*, 1987–1998. [CrossRef]
12. Marrone, P. The Market and Potential for Biopesticides Biopesticides. *ACS Symp. Ser. Am. Chem. Soc.* **2014**, *1172*, 245–258.
13. Damalas, C.A.; Koutroubas, S.D. Current Status and Recent Developments in Biopesticide Use. *Agriculture* **2018**, *8*, 13. [CrossRef]
14. Chen, M.; Du, Y.; Zhu, G.; Takamatsu, G.; Ihara, M.; Matsuda, K.; Zhorov, B.; Dong, K. Action of six pyrethrins purified from the botanical insecticide pyrethrum on cockroach sodium channels expressed in Xenopus oocytes. *Pestic. Biochem. Physiol.* **2018**, *151*, 82–89. [CrossRef] [PubMed]
15. Davies, T.G.E.; Field, L.M.; Usherwood, P.N.R.; Williamson, M.S. DDT, pyrethrins, pyrethroids and insect sodium channels. *IUBMB Life* **2007**, *59*, 151–162. [CrossRef] [PubMed]
16. Oruonye, E.D.; Okrikata, E. Sustainable use of plant protection products in Nigeria and challenges. *J. Plant Breed. Crop Sci.* **2010**, *2*, 267–272.
17. Mpumi, N.; Mtei, K.; Machunda, R.; Ndakidemi, P.A. The Toxicity, Persistence and Mode of Actions of Selected Botanical Pesticides in Africa against Insect Pests in Common Beans, *P. vulgaris*: A Review. *Am. J. Plant Sci.* **2016**, *7*, 138–151. [CrossRef]
18. Karani, A.O.; Ndakidemi, P.A.; Mbega, E.R. Botanical Pesticides in Management of Common Bean Pests: Importance and Possibilities for Adoption by Small-scale Farmers in Africa. *J. Appl. Life Sci.* **2017**, *12*, 1–10. [CrossRef]
19. Lacey, L.A.; Grzywacz, D.; Shapiro-Ilan, D.I.; Frutos, R.; Brownbridge, M.; Goettel, M.S. Insect pathogens as biological control agents: Back to the future. *J. Invertebr. Pathol.* **2015**, *132*, 1–41. [CrossRef]
20. Inglis, D.G.; Gottel, M.S.; Butt, T.M.; Strasser, H. Use of Hyphomycetous Fungi for Managing Insect Pests. In *Fungi as Biocontrol Agents: Progress, Problems and Potential*; Butt, T.M., Jackson, C., Magan, N., Eds.; CAB International: Wallingford, UK, 2001; pp. 23–69.
21. Sinha, K.K.; Choudhary, A.K.; Priyanka, K. Chapter 15-Entomopathogenic Fungi. In *Ecofriendly Pest Management for Food Security*; Omkar, Ed.; Academic Press: Cambridge, MA, USA, 2016; pp. 475–505.
22. Steinkraus, D.C. Factors affecting transmission of fungal pathogens of aphids. *J. Invertebr. Pathol.* **2006**, *92*, 1–41. [CrossRef]
23. Meyling, N.; Eilenberg, J. Ecology of the entomopathogenic fungi *Beauveria bassiana* and *Metarhizium anisopliae* in temperate agroecosystems: Potential for conservation biological control. *Biol. Control.* **2007**, *43*, 145–155. [CrossRef]
24. Islam, M.T.; Omar, D.; Latif, M.A.; Morshed, M.M. The integrated use of entomopathogenic fungus, *Beauveria bassiana* with botanical insecticide, neem against *Bemisia tabaci* on eggplant. *Afr. J. Microbiol. Res.* **2011**, *5*, 3409–3413.
25. Islam, M.T.; Omar, D. Combined effect of *Beauveria bassiana* with neem on virulence of insect in case of two application approaches. *J. Anim. Plant Sci.* **2012**, *22*, 77–82.
26. Ribeiro, L.P.; Blume, E.; Bogorni, P.C.; Dequech, S.T.B.; Brand, S.C.; Junges, E. Compatibility of *Beauveria bassiana* commercial isolate with botanical insecticides utilized in organic crops in southern Brazil. *Biol. Agric. Hortic.* **2012**, *28*, 223–240. [CrossRef]
27. Shoukat, R.F.; Freed, S.; Ahmad, K.W. Evaluation of binary mixtures of entomogenous fungi and botanicals on biological parameters of *Culex pipiens* (Diptera: Culicidae) under laboratory and field conditions. *Int. J. Mosq. Res.* **2016**, *3*, 17–24.

28. Ludwig, S.W.; Oetting, R.D. Susceptibility of Natural Enemies to Infection by *Beauveria bassiana* and Impact of Insecticides on *Ipheseius degenerans* (Acari: Phytoseiidae). *J. Agric. Urban Entomol.* **2001**, *18*, 169–178.
29. Shipp, J.L.; Zhang, Y.; Hunt, D.W.A.; Ferguson, G. Influence of Humidity and Greenhouse Microclimate on the Efficacy of *Beauveria bassiana* (Balsamo) for Control of Greenhouse Arthropod Pests. *Environ. Entomol.* **2003**, *32*, 1154–1163. [CrossRef]
30. Rashkia, M.; Kharazi-pakdel, A.; Allahyari, H.; van Alphen, J.J.M. Interactions among the entomopathogenic fungus, *Beauveria bassiana* (Ascomycota: Hypocreales), the parasitoid, *Aphidius matricariae* (Hymenoptera: Braconidae), and its host, *Myzus persicae* (Homoptera: Aphididae). *Biol. Control* **2009**, *50*, 324–328. [CrossRef]
31. Martins, I.C.F.; Silva, R.J.; Alencar, J.R.D.C.C.; Silva, K.P.; Cividanes, F.J.; Duarte, R.T.; Agostini, L.T.; Polanczyk, R.A. Interactions Between the Entomopathogenic Fungi *Beauveria bassiana* (Ascomycota: Hypocreales) and the Aphid Parasitoid *Diaeretiella rapae* (Hymenoptera: Braconidae) on *Myzus persicae* (Hemiptera: Aphididae). *J. Econ. Entomol.* **2014**, *107*, 933–938. [CrossRef] [PubMed]
32. Mohammed, M.M.; Hatcher, P.E. Combining entomopathogenic fungi and parasitoids to control the green peach aphid *Myzus persicae*. *Biol. Control* **2017**, *110*, 44–55. [CrossRef]
33. Ali, S.; Farooqi, M.A.; Sajjad, A. Compatibility of entomopathogenic fungi and botanical extracts against the wheat aphid, *Sitobion avenae* (Fab.) (Hemiptera: Aphididae). *Egypt. J. Biol. Pest Control* **2018**, *28*, 97. [CrossRef]
34. Otieno, J.A.; Pallmann, P.; Poehling, H.M. Additive and synergistic interactions amongst *Orius laevigatus* (Heteroptera: Anthocoridae), entomopathogens and azadirachtin for controlling western flower thrips (Thysanoptera: Thripidae). *BioControl* **2017**, *62*, 85–95. [CrossRef]
35. Johnson, D.J.; Goettel, M.S. Reduction of grasshopper populations following field application of the fungus *Beauveria bassiana*. *Biocontrol Sci. Technol.* **1992**, *3*, 165–175. [CrossRef]
36. Jaros-Su, J.; Groden, E.; Zhang, J. Effects of Selected Fungicides and the Timing of Fungicide Application on *Beauveria bassiana*-Induced Mortality of the Colorado Potato Beetle (Coleoptera: Chrysomelidae). *BioControl* **1999**, *15*, 259–269.
37. Benjamin, M.A.; Zhioua, E.; Ostfeld, R.S. Laboratory and Field Evaluation of the Entomopathogenic Fungus *Metarhizium anisopliae* (Deuteromycetes) for Controlling Questing Adult *Ixodes scapularis* (Acari: Ixodidae). *J. Med. Entomol.* **2002**, *39*, 723–728. [CrossRef] [PubMed]
38. Peng, G.; Wang, W.; Yin, Y.; Zeng, D.; Xia, Y. Field trials of *Metarhizium anisopliae* var. *acridum* (Ascomycota: Hypocreales) against oriental migratory locusts, *Locusta migratoria manilensis* (Meyen) in Northern China. *Crop. Prot.* **2008**, *27*, 1244–1250. [CrossRef]
39. Ormond, E.L.; Alison, P.M.T.; Pell, J.K.; Freeman, S.N.; Roy, H.E. Avoidance of a generalist entomopathogenic fungus by the ladybird, *Coccinella septempunctata*. *FEMS Microbiol. Ecol.* **2011**, *77*, 229–237. [CrossRef] [PubMed]
40. Meyling, N.V.; Pell, J.K. Detection and avoidance of an entomopathogenic fungus by a generalist insect predator. *Ecol. Entomol.* **2006**, *31*, 162–171. [CrossRef]
41. Lord, J.C. Response of the wasp, *Cephalonomia tarsalis* (Hymenoptera: Bethylidae) to *Beauveria bassiana* (Hyphomycetes: Moniliales) as free conidia or infection in its host, the saw-toothed grain beetle, *Oryzaephilus surinamensis* (Coleoptera: Silvanidae). *Biol. Control* **2001**, *21*, 300–304. [CrossRef]
42. Rajapakse, R.H.S. Effect of host age, parasitoid age, and temperature on interspecific competition between *Chelonus insularis* Cresso, *Cotesia marginiventris* Cresson and *Microplitis manilae* Ashmead. *Int. J. Trop. Insect Sci.* **1992**, *13*, 87–94. [CrossRef]
43. Silva-Torres, C.S.A.; Barros, R.; Torres, J.B. Effect of age, photoperiod and host availability on the parasitism behaviour of *Oomyzus sokolowskii* Kurdjumov (Hymenoptera: Eulophidae). *Neotrop. Entomol.* **2009**, *38*, 512–519. [CrossRef]
44. Pizzol, J.; Desneux, N.; Wajnberg, E.; Thiéry, D. Parasitoid and host egg ages have independent impact on various biological traits in a *Trichogramma* species. *J. Pest Sci.* **2012**, *85*, 489–496. [CrossRef]
45. Guerrieri, E.; Poppy, G.M.; Powell, W.; Tremblay, W.; Pennacchio, F. Induction and Systemic Release of Herbivore-Induced Plant Volatiles Mediating In-Flight Orientation of *Aphidius ervi*. *J. Chem. Ecol.* **1999**, *25*, 1247–1261. [CrossRef]
46. Crawley, M.J. *The R Book*, 2nd ed.; Wiley: Chichester, UK, 2013; pp. 869–892.
47. R Core Team. *R: A Language and Environment for Statistical Computing*; R Foundation for Statistical Computing: Vienna, Austria, 2018.

48. Therneau, T.M.; Grambsch, P.M. *Modeling Survival Data: Extending the Cox Model.*; Springer: New York, NY, USA, 2000; pp. 1–350.
49. Therneau, T. A Package for Survival Analysis in S. version 2.38. 2015. Available online: https://CRAN.R-project.org/package=survival. (accessed on 25 December 2019).
50. Venables, W.N.; Ripley, B.D. *Modern Applied Statistics with S*, 4th ed.; Springer: New York, NY, USA, 2002.
51. Length, R. emmeans: Estimated Marginal Means, aka Least-Squares Means. R package version 1.3.3. 2019. Available online: https://cran.r-project.org/web/packages/emmeans/index.html (accessed on 25 December 2019).

© 2020 by the authors. Licensee MDPI, Basel, Switzerland. This article is an open access article distributed under the terms and conditions of the Creative Commons Attribution (CC BY) license (http://creativecommons.org/licenses/by/4.0/).

Article

Extracts of Common Pesticidal Plants Increase Plant Growth and Yield in Common Bean Plants

Angela G. Mkindi [1], Yolice L. B. Tembo [2], Ernest R. Mbega [1], Amy K. Smith [3,4], Iain W. Farrell [3], Patrick A. Ndakidemi [1], Philip C. Stevenson [3,5] and Steven R. Belmain [5,*]

[1] Department of Sustainable Agriculture, Biodiversity and Ecosystems Management, Centre for Research, Agricultural Advancement, Teaching Excellence and Sustainability (CREATES), The Nelson Mandela African Institution of Science and Technology, Box 447 Arusha, Tanzania; angela.mkindi@nm-aist.ac.tz (A.G.M.); ernest.mbega@nm-aist.ac.tz (E.R.M.); patrick.ndakidemi@nm-aist.ac.tz (P.A.N.)
[2] Department of Crop and Soil Sciences, Lilongwe University of Agriculture and Natural Resources, Bunda, Malawi; ytembo@bunda.luanar.mw
[3] Royal Botanic Gardens, Kew, Richmond, Surrey TW9 3DS, UK; AmyKendall.Smith@kew.org (A.K.S.); I.Farrell@kew.org (I.W.F.); P.C.Stevenson@greenwich.ac.uk (P.C.S.)
[4] Faculty of Biological Sciences, University of Leeds, Leeds LS2 9JT, UK
[5] Natural Resources Institute, University of Greenwich, Central Avenue, Chatham Maritime, Kent ME4 4TB, UK
* Correspondence: S.R.Belmain@greenwich.ac.uk; Tel.: +44-1634-883761

Received: 10 December 2019; Accepted: 20 January 2020; Published: 23 January 2020

Abstract: Common bean (*Phaseolus vulgaris*) is an important food and cash crop in many countries. Bean crop yields in sub-Saharan Africa are on average 50% lower than the global average, which is largely due to severe problems with pests and diseases as well as poor soil fertility exacerbated by low-input smallholder production systems. Recent on-farm research in eastern Africa has shown that commonly available plants with pesticidal properties can successfully manage arthropod pests. However, reducing common bean yield gaps still requires further sustainable solutions to other crop provisioning services such as soil fertility and plant nutrition. Smallholder farmers using pesticidal plants have claimed that the application of pesticidal plant extracts boosts plant growth, potentially through working as a foliar fertiliser. Thus, the aims of the research presented here were to determine whether plant growth and yield could be enhanced and which metabolic processes were induced through the application of plant extracts commonly used for pest control in eastern Africa. Extracts from *Tephrosia vogelii* and *Tithonia diversifolia* were prepared at a concentration of 10% w/v and applied to potted bean plants in a pest-free screen house as foliar sprays as well as directly to the soil around bean plants to evaluate their contribution to growth, yield and potential changes in primary or secondary metabolites. Outcomes of this study showed that the plant extracts significantly increased chlorophyll content, the number of pods per plant and overall seed yield. Other increases in metabolites were observed, including of rutin, phenylalanine and tryptophan. The plant extracts had a similar effect to a commercially available foliar fertiliser whilst the application as a foliar spray was better than applying the extract to the soil. These results suggest that pesticidal plant extracts can help overcome multiple limitations in crop provisioning services, enhancing plant nutrition in addition to their established uses for crop pest management.

Keywords: induced systemic response; foliar fertiliser; rutin; tryptophan; phenylalanine; botanicals

1. Introduction

Common bean (*Phaseolus vulgaris*) is a strategic crop in low- and middle-income countries, known for its economic and nutritional benefits [1,2]. Tanzania is among the top 20 producers of common bean in the world [3]. However, bean productivity is generally stagnant across much of Africa due to a number of suboptimal provisioning services such as poor soil fertility and pest damage that are limiting potential yields [4,5]. Although chemical fertilizers can dramatically increase bean yields, they are largely unaffordable and unavailable to most smallholder farmers [6] and contribute to reduced soil stability [7,8], pollution [9] and carbon footprint [10]. Natural soil fertility enhancement through the use of manure, composts, green mulches, cover crops and crop rotation are not widely used by smallholder farmers, arguably due to high labour costs and poor local knowledge [11–14].

Sustainable technologies for pest management in legume crops often relies on the breeding of resistant varieties [15]. However, the use of pesticidal plant extracts in smallholder farming systems is also an established agro-ecologically sustainable pest control method [16–20]. Although the economics and cost-benefits of smallholder use of crude plant extracts for pest management are certainly favourable in many situations [19], uptake and promotion of pesticidal plants could be further facilitated by increased evidence on potential multiple benefits of their use [21], making their use even more attractive to smallholder farmers. For example, recent research has shown that the impact of pesticidal plants on beneficial arthropods such as pollinators and predators, is much less than that observed when using synthetic pesticides [18]. Research has also demonstrated that other potential benefits to smallholder use of pesticidal plants could be through direct effects on plant vigour by functioning as a green fertiliser or through the provision of additional nutrition and inducing systemic plant responses [22,23]. Very often plants used as pesticides have multiple uses such as providing fruits, seeds, fibre, timber or in traditional medicines [24–26]. Alternative uses can also include use as green mulches and cover crops to improve the soil fertility, where previous research points particularly to the use of *Tephrosia vogelii* and *Tithonia diversifolia* [27–30]. This study, therefore, sought to evaluate the contribution of extracts from *T. vogelii* and *T. diversifolia* on the growth, yield and metabolism of common beans. Evidence from this study could validate farmer observations and increase the perceived value of using such extracts, thus encouraging wider uptake in smallholder farming systems.

2. Results and Discussions

2.1. Growth and Yield of Common Beans in Response to the Application of Treatments

Extracts were applied to the leaves through foliar spraying or directly to soil as a soil drench in order to compare the effects on bean plant growth and yield. Significant variation in the growth of common beans was observed according to treatments where *T. vogelii* extracts resulted in significantly higher plant height, number of leaves and branches, leaf area, stem width and leaf greenness. However, water, water and soap and synthetic pesticide treatments were significantly lower in terms of plant height number of leaves, number of branches per plant, leaf area, stem width and leaf greenness (Table S1).

Yield was measured using the number of pods per plant and seed yield per plant (Table 1). Significantly higher numbers of pods and seeds were recorded in the *T. vogelii* treatment, followed by *T. diversifolia* and the foliar fertiliser for pods per plant and seed yield per plant. The control treatments (water, water and soap and synthetic pesticide) recorded significantly lower numbers for pods per plant and seed yield. Both number of pods per plant and seeds per pod showed a significant variation with respect to the method of application with higher values recorded for the number of seeds per pod and seed yield per plant when treated by foliar spray compared to when the treatments were applied to the soil for pod number and seed yield.

Table 1. Effects of foliar fertiliser, synthetic and plant pesticide treatments and application method on the yield of common beans.

Treatment Applied	Number of Pods Per Plants	Seed Yield/Plant (g)
Foliar Fertiliser	3.1 ± 0.26 b	2.7 ± 0.33 b
Synthetic pesticide	2.1 ± 0.24 c	1.3 ± 0.19 c
Tephrosia vogelii	4.1 ± 0.23 a	3.8 ± 0.23 a
Tithonia diversifolia	3.1 ± 0.31 b	3.3 ± 0.23 b
Water	1.9 ± 0.23 c	1.5 ± 0.16 c
Water and soap	1.6 ± 0.22 c	1.7 ± 0.11 c
Method of Application		
Foliar spray	2.9 ± 0.21 a	2.7 ± 0.20 a
Soil drenching	2.4 ± 0.16 b	2.1 ± 0.16 b
2-Way ANOVA (F-Statistics)		
Treatment	15.2 ***	29.0 ***
Treatment method	6.7 *	14.8 ***
Treatment * Treatment method	2.0 *	3.1 *

The values presented are means ± SE. *, *** = significant at $p \leq 0.05$, $p \leq 0.001$ respectively. Means followed by the same letter in a column are not significantly different.

As the effect was much more pronounced when applied to the leaves compared to the soil, our data suggest that the plant extracts contribute to plant nutrition as a foliar fertiliser, which may be particularly useful in smallholder farming systems where soils are often degraded. Furthermore, these data suggest that previous reports on the use of these pesticidal plants in crop protection [17,18,31] have maintained crop yield not only by fighting pests, but by functioning as a foliar fertiliser. Contribution to growth and yield is likely to be related to the addition of nitrogen [32] where *T. diversifolia* [33,34] and *T. vogelii* [35–37] are known to produce nitrogen-rich green biomass.

2.2. Effect of Treatments and Application Method on Common Bean Metabolite Production

Analysis of chlorophyll content, flavonoids and anthocyanins indicated that the *T. vogelii* treatment resulted in significantly higher chlorophyll concentration, followed by the foliar fertiliser and *T. diversifolia* (Table 2). Lower chlorophyll content was observed in water, water and soap and the synthetic pesticide. Flavonoid content was highest in *T. diversifolia* treated plants, followed by the foliar fertiliser and *T. vogelii*, and these were significantly different from the water and water and soap treatments. Pereira et al. [38] reported that chlorophyll content could enhance photosynthesis rates, which ultimately influences plant vigour. No significant variation was observed in anthocyanin content across treatments or modes of application suggesting that the influence of treatments on plant metabolism was specific.

As expected, the commercial foliar fertiliser had a significant effect on metabolite production. The effect of *T. diversifolia* on chlorophyll content was supported by previous research by Oke et al. [39]. Leaf samples were further analysed to identify the contribution of treatments on the amounts of specific metabolites including primary metabolites (phenylalanine and tryptophan) and the secondary metabolite, rutin. An analysis of variance showed that these metabolites were higher when exposed to the foliar spray method of application in comparison with soil drenching (Table 3).

Table 2. Effect of treatment on the presence of key metabolite groups in common bean.

Treatments	Chlorophylls (mg/L)	Flavonoids (Abs g DM^{-1})	Anthocyanins (Abs g DM^{-1})
Foliar fertiliser	19.3 ± 1.84 b	2.8 ± 0.28 ab	0.1 ± 0.01 a
Synthetic pesticide	13.7 ± 0.74 c	2.4 ± 0.14 bcd	0.1 ± 0.00 a
Tephrosia vogelii	24.6 ± 1.29 a	2.7 ± 0.23 abc	0.1 ± 0.01 a
Tithonia diversifolia	18.9 ± 0.89 b	3.0 ± 0.16 a	0.1 ± 0.01 a
Water	12.7 ± 0.53 c	2.1 ± 0.17 d	0.1 ± 0.03 a
Water and soap	14.0 ± 0.49 c	2.2 ± 0.15 cd	0.1 ± 0.02 a
Method of Application			
Soil drench	15.9 ± 0.89 b	2.5 ± 0.12 a	0.1 ± 0.01 a
Foliar spray	18.5 ± 1.14 a	2.6 ± 0.13 a	0.1 ± 0.01 a
2-Way ANOVA (F-Statistics)			
Treatment	27.8 ***	3.4 *	0.6ns
Method of application	12.7 **	0.5ns	0.4ns
Treatment * Method of application	3.0 *	1.3ns	0.3ns

The values presented are means ± SE. *, **, *** = significant at $p \leq 0.05$, $p \leq 0.01$, $p \leq 0.001$ respectively, ns = not significant. Means followed by the same letter in a column are not significantly different.

Table 3. Two-way Analysis of Variance on the influence of mode of application on the relative abundance (mg/g dry weight) of phenylalanine, tryptophan and rutin.

Method of Application	Phenylalanine	Tryptophan	Rutin
Foliar spray	43608.3 ± 4557.06 a	45478.3 ± 5450.15 a	15093.8 ± 1675.05 a
Soil drench	26209.9 ± 2127.52 b	26805.8 ± 2566.88 b	9342.5 ± 895.06 b
Two-way ANOVA (F-statistics)	13.4 ***	10.3 **	12.8 ***

The values presented are means ± SE. **, *** = significant at $p \leq 0.01$, $p \leq 0.001$ respectively. Means followed by the same letter in a column are not significantly different.

Overall, the foliar application was more effective in inducing changes, regardless of treatment (Figure 1). Foliar application was effective because it facilitated direct contact between the solution applied and the leaf surface where adsorption takes place [40,41], whereas application to the soil is indirect [36]. From this study, the production of amino acids induced by *T. diversifolia* and *T. vogelii* was similar to that observed with Neem (*Azadirachta indica*) where similar metabolic changes were reported by Sharma [42]. Similarly, Neem extracts applied to tomatoes have been observed to increase the abundance of several flavonoids through the jasmonate pathway [22].

Primary and secondary metabolites in plants can contribute to the development and growth of crop plants [22] as well as contribute to plant defence mechanisms [43]. Flavonoids are known to help a plant relate with other organisms and the environment thereby responding to biotic and abiotic stress [44,45]. Their contribution to growth is explained by their effect on auxin transport, shoot growth, root development and nitrogen fixing processes in legumes [46–49]. Examples of flavonoids in bean plants are kaempferol, quercetin [50,51], and rutin [52]. Flavonoids are also reported to mediate plant resistance to herbivores [53] thus, their increased occurrence could enhance defence against antagonists. Amino acids such as phenylalanine and tryptophan are known to contribute to plant growth and metabolism such as auxin biosynthesis in the rhizosphere [54], growth and nodulation [55]. Hence, applications that increase such metabolites in common beans could be beneficial to provide sustainable production techniques for bean resistance to pests, growth and yield as reported for ginger (*Zingiber officinale*) [56].

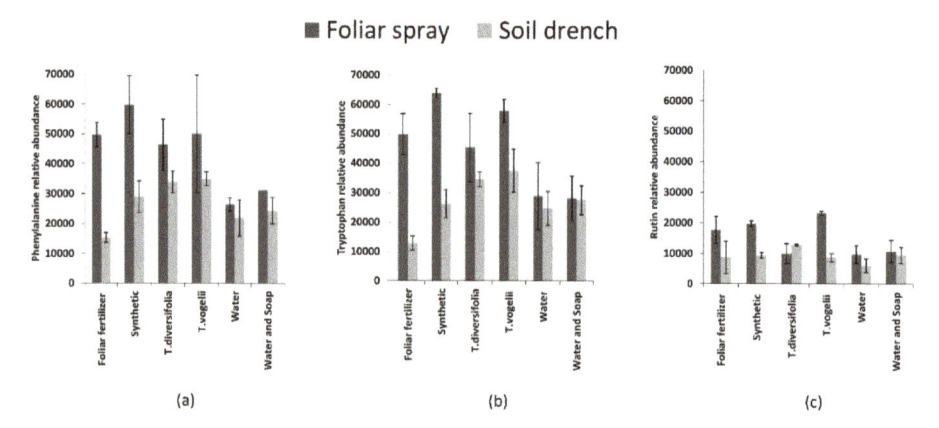

Figure 1. Relative abundance (mg/g dry weight) of (**a**) phenylalanine, (**b**) tryptophan and (**c**) rutin in common bean plants when exposed to different experimental treatments.

2.3. Correlations Between Bean Plant Growth Yield Parameters and Common Bean Metabolites

Three principal components (PC1, PC2 and PC3) were retained to explain 87.2% variance of the dependent variables (Table S1). The criteria for selection were based on a cumulative variance of 70% and an eigenvalue greater than one. The first principal component accounted for a total variance of 57.37%, while the second and third components explained 18.3% and 8.7% of the total variance, respectively. PCA observations of the treatments and their modes of application indicated the plant extracts applied to the bean plant or the soil were grouped together, implying that their contribution to bean growth was related (Figure 2a). Regardless of the plant extract species, application to the leaves had a negative relation with application to the soil. *T. vogelii* (Foliar spray) and water (Soil drench) were the treatments showing the highest and lowest influence, respectively. Furthermore, applying water had a low effect on the bean crop development regardless of the method of application.

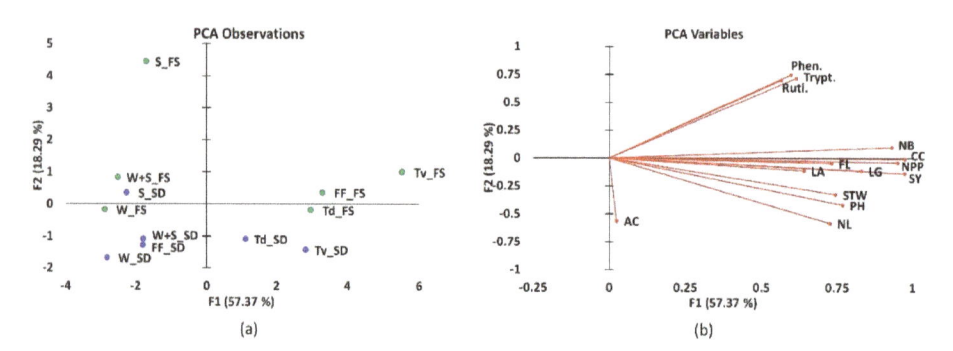

Figure 2. Two-dimensional principal component analysis (PCA) of (**a**) treatments applied using foliar spray and soil drench methods. Green marks indicate the treatments applied using foliar spray (FS) while blue marks indicate the application by soil drench (SD) where Tv = *T. vogelii*; Td = *T. diversifolia*; FF = foliar spray; W = water only; W + S = water and soap; S = synthetic; and (**b**) the covariance among all growth and metabolite parameters where CC = Chlorophyll content; FL = Flavonoids; AN = Anthocyanins; PH = Plant height; NL = Number of leaves; NB = Number of branches; LA = Leaf area; SW = Stem width; LG = Leaf greenness; NPP = Number of pods per plant; and SY = seed yield/plant.

Anthocyanin content correlated with the second principal component, which was different from the rest of the variables that all correlated with the first principal component (Figure 2b). This difference is likely to be based on the fact that anthocyanin values were minimal across all the treatments, with no significant difference observed in influencing bean development across the treatments. The first principal component's interpretation showed that yield parameters (number of pods per plant and seed yield per plant) and chlorophyll content explained more of the variation describing effects of the treatments. The number of branches showed a positive correlation with key metabolites, e.g., rutin (0.61), phenylalanine (0.58) and tryptophan (0.63). The PCA correlation matrix, eigenvalues, factor loadings, and factor scores at $p = 0.05$ can be found in Tables S2–S6.

3. Materials and Methods

3.1. Bean Rearing and Plant Material Preparation

The experiment was carried out in a controlled pest-free glass house at the Nelson Mandela African Institution of Science and Technology, Arusha, Tanzania (Latitude 3°24′ S Longitude 36°47′ E, elevation of 1168 masl with a mean annual rainfall of 1200 mm, mean maximum temperature of 21.7 °C and mean minimum temperature of 13.6 °C). Each treatment unit consisted of eight bean plants. Common bean seeds used for the experiment were of Lyamungo 90 variety, purchased from the Seliani Agricultural Research Institute. Two seeds were planted in each pot, later thinned to one plant per each pot, using 2-litre volume pots containing standard potting compost. All pots were arranged in a complete randomized block design on a bench in the glasshouse, providing even lighting, ventilation, temperature (25 ± 5 °C) and equal amounts of water per pot.

Pesticidal plant materials (*T. vogelii* and *T. diversifolia*) were collected from Lyamungo field areas, dried under the shade and ground into fine powder using previously reported methods and locations [18]. *T. vogelii* and *T. diversifolia* are among a large group of insecticidal plants that have been used for decades for pest control [17–19,57]. Positive controls included synthetic pesticide (Karate, lambda cyhalothrin) and a commercial foliar fertiliser (BioForce, an organic extract from seaweeds and blue green algae) which were applied according to instructions provided on the label. Pesticidal plant powders were extracted in soapy water (0.1% soap) to produce an extract solution of 10% (*w/v*) following previously reported methods [18]. Negative control treatments were with plain water, and water with 0.1% soap.

All treatments were applied in two different methods, either as a foliar spray using a hand sprayer or directly to the soil with a small watering can, ensuring equal amounts were applied to each plant. The treatments were applied fortnightly from the first week after plant germination until the time of bean flowering, i.e., a total of four treatment applications.

3.2. Collection of Growth Parameters Data and Leaf Samples for Chemical Analysis

Growth parameters and samples for chlorophyll content and bean leaf chemistry analysis were collected before bean flowering. Yield parameters were collected close to the maturity of the beans and the total yield collected after final bean harvesting. The growth parameters that were measured included plant height, number of leaves, number of branches, main stem width, leaf area and leaf greenness. Leaf greenness was scored using a scale of 1–5 where 1 was regarded as low greenness and 5 as high greenness using a leaf colour chart as previously reported [58]. Leaf area was determined from the direct measurements of length as a distance between the base and apex of the leaflet, and the width between positions of the leaflets. Leaf area was then calculated using the formula described by Bhatt [59]

$$LA = 11.98 + 0.06 \, L \times W \tag{1}$$

where LA = Leaf area', L = leaf length and W = leaf width.

Plant leaf samples were harvested three days after spraying the beans. Harvesting was done at the vegetative stage, just before flowering. Four plants from each treatment were randomly selected from each plant. The leaves were thoroughly washed with distilled water. Two leaves from each plant were placed in a desiccator with silica gel, desiccated and prepared for phytochemical analysis. The other two leaves collected from each plant were used for spectrophotometric analysis described below.

3.3. Spectrophotometric Analysis of Key Metabolite Groups in Bean Leaves

3.3.1. Chlorophyll Content Analysis

Chlorophyll concentration was determined through the extraction of chlorophyll from the third leaf of the growing tip of each plant using Dimethyl Sulphoxide (DMSO) as described by Hiscox, 1980 [60]. This involved placing 100 mg of the middle portion of the leaf in a 15 mL vial containing 7 mL DMSO and incubating at 65 °C for 24 h after which the leaves were completely transparent signifying chlorophyll extraction. The extracted liquid was transferred to graduated tubes and made up to a total volume of 10 mL with DMSO and then kept at 4 °C waiting for analysis. To determine the chlorophyll content, 300 microliters of the sample were transferred into an 86-well plate, where the absorbance at 645 nm and 663 nm were read using a spectrophotometer (Synergy, Multi-mode reader, Biotek Instrument Inc. Winooski, VT, USA) against DMSO as a blank. Chlorophyll levels in milligrams per litre (mg/l) were then calculated using the formula described by Arnon [61].

$$\text{Total Chl} = 20.2 \times D645 \text{ nm} + 8.02 \times D663 \text{ nm} \tag{2}$$

where Chl = Chlorophyll, D = the Absorbance value at the respective wavelengths obtained from the spectrophotometer.

3.3.2. Anthocyanins and Flavonoids Analysis

Flavonoids and anthocyanins in bean plant leaves were determined using the method described by Makoi et al. [62]. Dried and ground bean leaves were used, where 0.1 g of the plant powder was extracted in 10 mL acidified methanol, made at a ratio of 79:20:1 of MeOH:H_2O:HCl. The extract was incubated for 72 h in darkness for auto extraction and then filtered through a filter paper (Whatman #2). Absorbance of the clear supernatant was measured at 300, 530, and 657 nm in a spectrophotometer (Synergy, Multi-mode reader, Biotek Instrument Inc. Winooski, VT, USA) against acidified methanol as a standard. Flavonoid concentration was obtained from the measured absorption at 300 nm and expressed in Abs g DM^{-1}.

$$\text{Abs g}^{-1} \text{ DM} = \text{Abs300} \tag{3}$$

Anthocyanins were measured by using the formula described by Lindoo and Caldwell [63].

$$\text{Abs g}^{-1} \text{ DM} = \text{Abs530} - 1/3 \times \text{Abs657} \tag{4}$$

where Abs = Absorption readings recorded from the spectrophotometer. The resulting concentration was expressed as Abs g DM^{-1}.

3.4. HPLC Detection of Primary and Secondary Metabolites

Desiccated beans leaves were powdered using an electric coffee grinder, and 50 mg of the powder was extracted in methanol (1 mL) and left to stand for 24 h at room temperature before chemical analysis. Extracts were transferred to Eppendorf tubes and centrifuged for 20 min at 500 rpm. From this 300 µL supernatant was transferred into HPLC glass vials for separation. The sample analyses were performed by Liquid Chromatography-Electrospray Ionization Mass Spectroscopy (LC-ESIMS) and UV spectroscopy using a Thermo Fisher Velos Pro LC-MS. Aliquots of extract were injected directly onto a Phenomenex (Macclesfield, Cheshire, UK) Luna C18(2) columns (150 × 3.0 mm i.d., 5 um particle size) and the compounds were eluted using methanol (A), water (B) and acetonitrile containing 1% formic acid (C) with A = 0%, B = 90% at T = 0 min; A = 90%, B = 0% at T = 20 min and held for 10 min with C at 10% throughout the analyses. Column temperature was 30 °C with flow rate = 0.5 mL min^{-1}.

High resolution MS spectra were used to provide additional data for compound identification and were recorded for a subset of samples using a Thermo LTQ-Orbitrap XL mass spectrometer (Waltham, MA, United States) with compound separation on an Accela LC system.

3.5. Statistical Analysis

The experiment was conducted following the completely randomised block design with eight replications to assess yield and growth of common beans and four replications to assess the metabolites. Effects of treatments and their interactions observed were subjected to Analysis of Variance. The means of treatments and interactions were compared using the least significant difference (LSD) test at a significant level of $p \leq 0.05$. Principal Component Analysis (PCA) was performed to explain potential covariance between bean plant growth, yield parameters and common bean metabolites. All the analyses were done using XLSTAT version 2019.2.2.59614 (Addinsoft (2019). XLSTAT statistical and data analysis solution. Boston, MA, USA. https://www.xlstat.com).

4. Conclusions

In this study, foliar sprays of the pesticidal plants *T. vogelii* and *T. diversifolia* enhanced common bean growth, yield and induced essential metabolites known for facilitating plant growth. Thus, their use helps to reduce the need for both synthetic pesticides and fertilisers by sustainably reducing arthropod pests whilst increasing plant nutrition. As soil fertility and crop pests are considered two of the main problems contributing to the yield gaps of smallholder farmers, using botanical extracts for crop production can help farmers move towards more sustainable agro-ecological approaches to crop production, tackling two problems at the same time. Pesticidal plants such as *T. vogelii* and *T. diversifolia* can be obtained cheaply in many African countries. *T. vogelii* can easily be propagated, although it should not be cultivated near large bodies of water as the rotenoid compounds can be harmful to fish. *T. diversifolia* is widely growing in roadsides and field margins and is considered invasive in some parts of Africa, therefore, care is also needed when cultivating this plant to keep it under control. Other commonly used pesticidal plant species may also have beneficial impacts on crop growth, where further validation is recommended.

Supplementary Materials: The following are available online at http://www.mdpi.com/2223-7747/9/2/149/s1, Table S1: Effects of foliar fertiliser, synthetic pesticides and botanical plants extract on common beans growth. The values presented are means ± SE. *, **, *** = significant at $p \leq 0.05$, $p \leq 0.01$, $p \leq 0.001$ respectively, ns = not significant. Means followed by the same letter in a column are not significantly different. Tables S2–S6: The correlation matrix used, eigenvalues, factor loadings, and factor scores.

Author Contributions: Conceptualization, A.G.M., P.C.S., E.R.M., S.R.B., Y.L.B.T., and P.A.N.; methodology, A.G.M., P.C.S., I.W.F. and S.R.B.; software, P.C.S.; validation, P.C.S.; formal analysis, A.G.M., P.C.S. and A.K.S., and I.W.F.; investigation, A.G.M., resources, P.C.S. and S.R.B.; data curation, P.C.S., I.W.F. and A.G.M.; writing—original draft preparation, A.G.M.; writing—review and editing, P.A.N., E.R.M., P.C.S. and S.R.B.; visualization, P.C.S.; supervision, P.A.N., S.R.B. and P.C.S.; project administration, S.R.B.; funding acquisition, S.R.B. and P.A.N. All authors have read and agreed to the published version of the manuscript.

Funding: This research was funded by grants from the McKnight foundation to SRB Grant No: 17-070 and the World Bank to PAN Grant No.5799-TZ.

Conflicts of Interest: The authors declare no conflict of interest.

References

1. Ndakidemi, P.A.; Dakora, F.D.; Nkonya, E.M.; Ringo, D.; Mansoor, H. Yield and economic benefits of common bean (*Phaseolus vulgaris*) and soybean (*Glycine max*) inoculation in northern Tanzania. *Aust. J. Exp. Agric.* **2006**, *46*, 571. [CrossRef]
2. Dakora, F.D.; Keya, S.O. Contribution of legume nitrogen fixation to sustainable agriculture in sub-Saharan Africa. *Soil Biol. Biochem.* **1997**, *29*, 809–817. [CrossRef]
3. Hillocks, R.J.; Madata, C.S.; Chirwa, R.; Minja, E.M.; Msolla, S. Phaseolus bean improvement in Tanzania, 1959–2005. *Euphytica* **2006**, *150*, 215–231. [CrossRef]

4. Bucheyeki, T.L.; Mmbaga, T.E. On-Farm Evaluation of Beans Varieties for Adaptation and Adoption in Kigoma Region in Tanzania. *ISRN Agron.* **2013**, *2013*, 1–5. [CrossRef]
5. Laizer, H.C.; Chacha, M.N.; Ndakidemi, P.A. Farmers' Knowledge, Perceptions and Practices in Managing Weeds and Insect Pests of Common Bean in Northern Tanzania. *Sustainability* **2019**, *11*, 4076. [CrossRef]
6. Katungi, E.; Farrow, A.; Chianu, J.; Sperling, L.; Beebe, S. Common bean in Eastern and Southern Africa: A situation and outlook analysis. *Int. Cent. Trop. Agric.* **2009**, *61*, 1–44.
7. Blanco-Canqui, H.; Schlegel, A.J. Implications of inorganic fertilization of irrigated corn on soil properties: Lessons learned after 50 years. *J. Environ. Qual.* **2013**, *42*, 861–871. [CrossRef]
8. Xin, X.; Zhang, J.; Zhu, A.; Zhang, C. Effects of long-term (23 years) mineral fertilizer and compost application on physical properties of fluvo-aquic soil in the North China Plain. *Soil Tillage Res.* **2016**, *156*, 166–172. [CrossRef]
9. Joshi, R.; Singh, J.; Vig, A.P. Vermicompost as an effective organic fertilizer and biocontrol agent: Effect on growth, yield and quality of plants. *Rev. Environ. Sci. Biotechnol.* **2014**, *14*, 137–159. [CrossRef]
10. Hillier, J.; Hawes, C.; Squire, G.; Hilton, A.; Wale, S.; Smith, P. The carbon footprints of food crop production. *Int. J. Agric. Sustain.* **2009**, *7*, 107–118. [CrossRef]
11. Jagisso, Y.; Aune, J.; Angassa, A. Unlocking the Agricultural Potential of Manure in Agropastoral Systems: Traditional Beliefs Hindering Its Use in Southern Ethiopia. *Agriculture* **2019**, *9*, 45. [CrossRef]
12. Cai, T.; Steinfield, C.; Chiwasa, H.; Ganunga, T. Understanding Malawian farmers' slow adoption of composting: Stories about composting using a participatory video approach. *L. Degrad. Dev.* **2019**, *30*, 1336–1344. [CrossRef]
13. Arlauskiene, A.; Jablonskyte-Rasce, D.; Slepetiene, A. Effect of legume and legume-festulolium mixture and their mulches on cereal yield and soil quality in organic farming. *Arch. Agron. Soil Sci.* **2019**, 1–6. [CrossRef]
14. Mtyobile, M.; Muzangwa, L.; Mnkeni, P.N.S. Tillage and crop rotation effects on soil carbon and selected soil physical properties in a Haplic Cambisol in Eastern Cape, South Africa. *Soil Water Res.* **2019**, *15*, 47–54. [CrossRef]
15. Rodríguez-González, A.; Casquero, P.A.; Cardoza, R.E.; Gutiérrez, S. Effect of trichodiene synthase encoding gene expression in Trichoderma strains on their effectiveness in the control of *Acanthoscelides obtectus*. *J. Stored Prod. Res.* **2019**, *83*, 275–280. [CrossRef]
16. Dougoud, J.; Toepfer, S.; Bateman, M.; Jenner, W.H. Efficacy of homemade botanical insecticides based on traditional knowledge. A review. *Agron. Sustain. Dev.* **2019**, *39*, 37. [CrossRef]
17. Mkindi, A.; Mpumi, N.; Tembo, Y.; Stevenson, P.C.P.C.; Ndakidemi, P.A.P.A.; Mtei, K.; Machunda, R.; Belmain, S.R.S.R. Invasive weeds with pesticidal properties as potential new crops. *Ind. Crops Prod.* **2017**, *110*, 113–121. [CrossRef]
18. Tembo, Y.; Mkindi, A.G.; Mkenda, P.A.; Mpumi, N.; Mwanauta, R.; Stevenson, P.C.; Ndakidemi, P.A.; Belmain, S.R. Pesticidal Plant Extracts Improve Yield and Reduce Insect Pests on Legume Crops Without Harming Beneficial Arthropods. *Front. Plant Sci.* **2018**, *9*, 1425. [CrossRef]
19. Mkenda, P.; Mwanauta, R.; Stevenson, P.C.; Ndakidemi, P.; Mtei, K.; Belmain, S.R. Extracts from field margin weeds provide economically viable and environmentally benign pest control compared to synthetic pesticides. *PLoS ONE* **2015**, *10*, e0143530. [CrossRef]
20. Rodríguez-González, Á.; Álvarez-García, S.; González-López, Ó.; Da Silva, F.; Casquero, P.A. Insecticidal Properties of *Ocimum basilicum* and *Cymbopogon winterianus* against *Acanthoscelides obtectus*, Insect Pest of the Common Bean (*Phaseolus vulgaris*, L.). *Insects* **2019**, *10*, 151. [CrossRef]
21. Rojht, H.; Košir, I.J.; Trdan, S. Chemical analysis of three herbal extracts and observation of their activity against adults of *Acanthoscelides obtectus* and *Leptinotarsa decemlineata* using a video tracking system. *J. Plant Dis. Prot.* **2012**, *119*, 59–67. [CrossRef]
22. Pretali, L.; Bernardo, L.; Butterfield, T.S.; Trevisan, M.; Lucini, L. Botanical and biological pesticides elicit a similar Induced Systemic Response in tomato (*Solanum lycopersicum*) secondary metabolism. *Phytochemistry* **2016**, *130*, 56–63. [CrossRef] [PubMed]
23. Siah, A.; Magnin-Robert, M.; Randoux, B.; Choma, C.; Rivière, C.; Halama, P.; Reignault, P. Natural Agents Inducing Plant. Resistance Against Pests and Diseases. In *Natural Antimicrobial Agents*; Springer: Berlin, Germany, 2018; Volume 19, pp. 121–159.

24. Haruna, Y.; Kwanashie, H.O.; Anuka, J.A.; Atawodi, S.E.; Hussaini, I.M. In vivo Anti-malarial Activity of Methanol Root Extract of *Securidaca longepedunculata* in Mice Infected with *Plasmodium berghei*. *Int. J. Mod. Biol. Med.* **2013**, *3*, 7–16.
25. Ngadze, R.T.; Linnemann, A.R.; Nyanga, L.K.; Fogliano, V.; Verkerk, R. Local processing and nutritional composition of indigenous fruits: The case of monkey orange (*Strychnos* spp.) from Southern Africa. *Food Rev. Int.* **2017**, *33*, 123–142. [CrossRef]
26. Isman, M.B.; Gunning, P.J.; Spollen, K.M. *Tropical Timber Species as Sources of Botanical Insecticides*; ACS: Washington, DC, USA, 1997; pp. 27–37.
27. Jama, B.; Palm, C.A.; Buresh, R.J.; Niang, A.; Gachengo, C.; Nziguheba, G.; Amadalo, B. *Tithonia diversifolia* as a green manure for soil fertility improvement in western Kenya: A review. *Agrofor. Syst.* **2000**, *49*, 201–221. [CrossRef]
28. Anjarwalla, P.; Belmain, S.R.; Sola, P.; Jamnadass, R.; Stevenson, P.C. *Handbook on Pesticidal Plants*; World Agroforestry Centre: Nairobi, Kenya, 2016; ISBN 978-92-9059-397-3.
29. Nyende, P.; Delve, R.J. Farmer participatory evaluation of legume cover crop and biomass transfer technologies for soil fertility improvement using farmer criteria, preference ranking and logit regression analysis. *Exp. Agric.* **2004**, *40*, 77–88. [CrossRef]
30. Desaeger, J.; Rao, M.R. The potential of mixed covers of Sesbania, *Tephrosia* and *Crotalaria* to minimise nematode problems on subsequent crops. *Field Crops Res.* **2001**, *70*, 111–125. [CrossRef]
31. Kayange, C.D.M.; Njera, D.; Nyirenda, S.P.; Mwamlima, L. Effectiveness of *Tephrosia vogelii* and *Tephrosia candida* Extracts against Common Bean Aphid (*Aphis fabae*) in Malawi. *Adv. Agric.* **2019**, *2019*, 1–6. [CrossRef]
32. Mafongoya, P.L.; Chintu, R.; Chirwa, T.S.; Matibini, J.; Chikale, S. Tephrosia species and provenances for improved fallows in southern Africa. *Agrofor. Syst.* **2003**, *59*, 279–288. [CrossRef]
33. Endris, S. Combined application of phosphorus fertilizer with *Tithonia* biomass improves grain yield and agronomic phosphorus use efficiency of hybrid maize. *Int. J. Agron.* **2019**, *2019*, 9. [CrossRef]
34. Pavela, R.; Dall'Acqua, S.; Sut, S.; Baldan, V.; Ngahang Kamte, S.L.; Biapa Nya, P.C.; Cappellacci, L.; Petrelli, R.; Nicoletti, M.; Canale, A.; et al. Oviposition inhibitory activity of the Mexican sunflower *Tithonia diversifolia* (Asteraceae) polar extracts against the two-spotted spider mite *Tetranychus urticae* (Tetranychidae). *Physiol. Mol. Plant Pathol.* **2018**, *101*, 85–92. [CrossRef]
35. Munthali, M.G.; Gachene, C.K.K.; Sileshi, G.W.; Karanja, N.K. Amendment of *Tephrosia* Improved Fallows with Inorganic Fertilizers Improves Soil Chemical Properties, N Uptake, and Maize Yield in Malawi. *Int. J. Agron.* **2014**, *2014*, 9. [CrossRef]
36. Rutunga, V.; Karanja, N.K.; Gachene, C.K.K. Six month-duration *Tephrosia vogelii* Hook.f. and *Tithonia diversifolia* (Hemsl.) a gray planted-fallows for improving maize production in Kenya. *Biotechnol. Agron. Soc. Environ.* **2008**, *12*, 267–278.
37. Snapp, S.S.; Rohrbach, D.D.; Simtowe, F.; Freeman, H.A. Sustainable soil management options for Malawi: Can smallholder farmers grow more legumes? *Agric. Ecosyst. Environ.* **2002**, *91*, 159–174. [CrossRef]
38. Pereira, L.D.M.; Pereira, E.D.M.; Revolti, L.T.M.; Zingaretti, S.M.; Môro, G.V. Seed quality, chlorophyll content index and leaf nitrogen levels in maize inoculated with *Azospirillum brasilense*. *Rev. Cienc. Agron.* **2015**, *46*, 630–637. [CrossRef]
39. OkeE, S.O.; Awowoyin, A.V.; Oseni, S.R.; Adediwura, E.L. Effects of Aqueous Shoot Extract of *Tithonia diversifolia* on the Growth of Seedlings of *Monodora tenuifolia* (Benth.), *Dialium guineense* (Willd.) and *Hildegardia barteri* (Mast.) Kosterm. *Not. Sci. Biol.* **2011**, *3*, 64–70. [CrossRef]
40. Fageria, N.K.; Filho, M.P.B.; Moreira, A.; Guimarães, C.M. Foliar fertilization of crop plants. *J. Plant Nutr.* **2009**, *32*, 1044–1064. [CrossRef]
41. Wang, D.; Deng, X.; Wang, B.; Zhang, N.; Zhu, C.; Jiao, Z.; Li, R.; Shen, Q. Effects of foliar application of amino acid liquid fertilizers, with or without *Bacillus amyloliquefaciens* SQR9, on cowpea yield and leaf microbiota. *PLoS ONE* **2019**, *14*, e0222048. [CrossRef]
42. Paul, P.; Sharma, P. *Azadirachta indica* leaf extract induces resistance in barley against leaf stripe disease. *Physiol. Mol. Plant Pathol.* **2002**, *61*, 3–13. [CrossRef]
43. Bohinc, T.; Ban, G.; Ban, D.; Trdan, S. Glucosinolates in plant protection strategies: A review. *Arch. Biol. Sci.* **2012**, *64*, 821–828. [CrossRef]
44. Khalid, M.; Bilal, M.; Huang, D.F. Role of flavonoids in plant interactions with the environment and against human pathogens—A review. *J. Integr. Agric.* **2019**, *18*, 211–230. [CrossRef]

45. Mierziak, J.; Kostyn, K.; Kulma, A. Flavonoids as important molecules of plant interactions with the environment. *Molecules* **2014**, *19*, 16240–16265. [CrossRef] [PubMed]
46. Buer, C.S.; Djordjevic, M.A. Architectural phenotypes in the transparent testa mutants of *Arabidopsis thaliana*. *J. Exp. Bot.* **2009**, *60*, 751–763. [CrossRef] [PubMed]
47. Buer, C.S.; Imin, N.; Djordjevic, M.A. Flavonoids: New roles for old molecules. *J. Integr. Plant. Biol.* **2010**, *52*, 98–111. [CrossRef]
48. Singla, P.; Garg, N. Plant flavonoids: Key players in signaling, establishment, and regulation of rhizobial and mycorrhizal endosymbioses. In *Mycorrhiza—Function, Diversity, State of the Art*, 4th ed.; Springer: Berlin, Germany, 2017; pp. 133–176. ISBN 9783319530642.
49. Nagata, M.; Yamamoto, N.; Miyamoto, T.; Shimomura, A.; Arima, S.; Hirsch, A.M.; Suzuki, A. Enhanced hyphal growth of arbuscular mycorrhizae by root exudates derived from high R/FR treated *Lotus japonicus*. *Plant Signal. Behav.* **2016**, *11*, e1187356. [CrossRef] [PubMed]
50. Dinelli, G.; Bonetti, A.; Minelli, M.; Marotti, I.; Catizone, P.; Mazzanti, A. Content of flavonols in Italian bean (*Phaseolus vulgaris* L.) ecotypes. *Food Chem.* **2006**, *99*, 105–114. [CrossRef]
51. Hu, Y.; Cheng, Z.; Heller, L.I.; Krasnoff, S.B.; Glahn, R.P.; Welch, R.M. Kaempferol in red and pinto bean seed (*Phaseolus vulgaris* L.) coats inhibits iron bioavailability using an in vitro digestion/human Caco-2 cell model. *J. Agric. Food Chem.* **2006**, *54*, 9254–9261. [CrossRef]
52. Gomez, J.D.; Vital, C.E.; Oliveira, M.G.A.; Ramos, H.J.O. Broad range flavonoid profiling by LC/MS of soybean genotypes contrasting for resistance to *Anticarsia gemmatalis* (Lepidoptera: Noctuidae). *PLoS ONE* **2018**, *13*, e0205010. [CrossRef]
53. Stevenson, P.C.; Anderson, J.C.; Blaney, W.M.; Simmonds, M.S.J. Developmental inhibition of *Spodoptera litura* (Fab.) larvae by a novel caffeoylquinic acid from the wild groundnut, *Arachis paraguariensis* (Chod et Hassl.). *J. Chem. Ecol.* **1993**, *19*, 2917–2933. [CrossRef]
54. Qureshi, M.A.; Iqbal, A.; Akhtar, N.; Shakir, M.A.; Khan, A. Co-inoculation of phosphate solubilizing bacteria and rhizobia in the presence of L-tryptophan for the promotion of mash bean (*Vigna mungo* L.). *Soil Environ.* **2012**, *31*, 47–54.
55. Hussain, M.I.; Akhtar, M.J.; Asghar, H.N.; Ahmad, M. Growth, nodulation and yield of mash bean (*Vigna mungo* L.) as affected by Rhizobium inoculation and soil applied L-tryptophan. *Soil Environ.* **2011**, *30*, 13–17.
56. Ghasemzadeh, A.; Jaafar, H.Z.E.; Rahmat, A. Elevated Carbon Dioxide Increases Contents of Flavonoids and Phenolic Compounds, and Antioxidant Activities in Malaysian Young Ginger (*Zingiber officinale* Roscoe.) Varieties. *Molecules* **2010**, *15*, 7907–7922. [CrossRef] [PubMed]
57. Grzywacz, D.; Stevenson, P.C.; Mushobozi, W.L.; Belmain, S.R.; Wilson, K. The use of indigenous ecological resources for pest control in Africa. *Food Secur.* **2014**, *6*, 71–86. [CrossRef]
58. Haripriya Anand, M.; Byju, G. Chlorophyll meter and leaf colour chart to estimate chlorophyll content, leaf colour, and yield of cassava. *Photosynthetica* **2008**, *46*, 511–516. [CrossRef]
59. Bhatt, M.; Chanda, S.V. Prediction of leaf area in *Phaseolus vulgaris* by non-destructive method. *Bulg. J. Plant Physiol.* **2003**, *29*, 96–100.
60. Hiscox, J.D.; Israelstam, G.F. A method for the extraction of chlorophyll from leaf tissue without maceration. *Can. J. Bot.* **1980**, *58*, 1332–1334. [CrossRef]
61. Arnon, D.I. Copper enzymes in isolated chloroplasts. Polyphenoloxidase in *Beta vulgaris*. *Plant Physiol.* **1949**, *24*, 1. [CrossRef]
62. Makoi, J.H.J.R.; Chimphango, S.B.M.; Dakora, F.D. Photosynthesis, water-use efficiency and $\delta 13C$ of five cowpea genotypes grown in mixed culture and at different densities with sorghum. *Photosynthetica* **2010**, *48*, 143–155. [CrossRef]
63. Lindoo, S.J.; Caldwell, M.M. Ultraviolet-B Radiation-induced Inhibition of Leaf Expansion and Promotion of Anthocyanin Production: Lack of Involvement of the Low Irradiance Phytochrome system. *Plant Physiol.* **1978**, *61*, 278–282. [CrossRef]

© 2020 by the authors. Licensee MDPI, Basel, Switzerland. This article is an open access article distributed under the terms and conditions of the Creative Commons Attribution (CC BY) license (http://creativecommons.org/licenses/by/4.0/).

Article

Fumigant Toxicity in *Myzus persicae* Sulzer (Hemiptera: Aphididae): Controlled Release of *(E)*-anethole from Microspheres

María J. Pascual-Villalobos [1,*], Manuel Cantó-Tejero [1], Pedro Guirao [2] and María D. López [3]

[1] Instituto Murciano de Investigación y Desarrollo Agrario y Alimentario (IMIDA), C/Mayor S/N La Alberca, 30150 Murcia, Spain; manuel.canto@carm.es
[2] Departamento de Producción Vegetal y Microbiología, Universidad Miguel Hernández, Escuela Politécnica Superior de Orihuela, Carretera de Beniel Km. 3.2, 03312 Orihuela, Alicante, Spain; pedro.guirao@umh.es
[3] Departamento de Producción Vegetal, Facultad de Agronomía, Universidad de Concepción, Campus Chillán, Avenida Vicente Méndez 595, P.O. Box 537, Chillán 3812120, Chile; lolalopezbelchi@gmail.com
* Correspondence: mjesus.pascual@carm.es

Received: 18 December 2019; Accepted: 16 January 2020; Published: 18 January 2020

Abstract: *(E)*-anethole is a phenylpropanoid that is the main compound found in the essential oils (EOs) of anise and fennel seeds, and either fumigant or direct contact activity of this compound has been demonstrated against aphids and stored product pests. In this work, solid microspheres were prepared by three methods—oil emulsion entrapment, spray-drying, and complexed with β-cyclodextrin. Fumigation activity of each microsphere preparation was tested against the green peach aphid, *Myzus persicae* Sulzer (Hemiptera: Aphididae), on pepper leaves. The best insecticidal activity was with *(E)*-anethole encapsulated in oil emulsion beads and introduced to aphids as a vapour over 24 h, with an LC_{50} of 0.415 µL/L compared to 0.336 µL/L of vapors from free *(E)*-anethole. Scanning electron microscopy of the beads revealed a compact surface with low porosity that produced a controlled release of the bioactive for more than 21 d, whilst most of the volatile was evaporated within two days if applied unformulated. Spray drying gave spherical particles with the greatest encapsulated yield (73%) of 6.15 g of *(E)*-anethole incorporated per 100 g of powder. Further work will be done on improving the formulation methods and testing the solid microspheres in all aphid stages scaling up the experimental assay. It is foreseen that nanotechnology will play a role in future developments of low risk plant protection products.

Keywords: encapsulation; essential oils; botanical active substances; insecticidal activity; aphids; anise; fennel; oil emulsion entrapment; spray drying

---

1. Introduction

(E)-anethole [trans-1-methoxy-4-(C1-propenyl) benzene] is an aromatic ether synthesized by some plants. This phenylpropanoid is the main compound in the essential oil of umbelifers such as anise (*Pimpinella anisum* L.) or fennel (*Foeniculum vulgare* Miller) but it is also present in other plant families-Schisandraceae-*Illicium verum* Hook. f, -Rutaceae-*Clausena anisata* (Willd) Hook f ex Benth,-Backhousiaceae-*Backhousia anisata* Vickery and -Magnoliaceae-*Magnolia salicifolia* (Sieb et Zucc) Maxim. [1–5].

The Apiaceae (formerly Umbelliferae) family comprises vegetables (celery-*Apium graveolens* L., parsley-*Petroselinum sativum* L., coriander-*Coriandrum sativum* L.), herbs, and spices (anise, fennel, cumin-*Cuminum cyminum* L.). Aniseeds have long been used to make schnapps like the popular French pastis, a beverage distilled from anise, liquorice, and fennel seeds macerate.

Fumigant toxicity of anise and cumin essential oils has been reported against the cotton aphid (*Aphis gossypii* Glover (Hemiptera: Aphididae) [6]. Vapours of anise essential oil (EO) or its main compound *(E)*-anethole were toxic (LD_{90} = 0.18 µL/cm^2 or 0.14 µL/cm^2 respectively) to the bird cherry-oat aphid (*Rhopalosiphum padi* L., Hemiptera: Aphididae) in a laboratory bioassay within small air-tight dishes (2.2 × 2.2 × 1 cm^3), according to reference [1].

A blend of *(E)*-anethole, limonene, and fenchone at 880 ppm was toxic (100% mortality) against *Rhyzopertha dominica* (F.) (Coleoptera: Bostrichidae), a pest of stored cereals (using a fumigant bioassay 1µL/vial of 15 mL at 30 °C in the dark), as reported in reference [7]. Another phenylpropanoid, estragole (also present in fennel EO) and fenchone were more active against *Sitophilus oryzae* L. (Coleoptera: Curculionidae) and *Callosobruchus chinensis* Fab. (Coleoptera: Bruchidae) than *(E)*-anethole [8]. *(E)*-anethole in combination with 1,8-cineole (1:1) was the best regarding fumigant toxicity on the red flour beetle adults, *Tribolium castaneum* Herbst (Coleoptera: Tenebrionidae), and it was also observed that heating enhanced the toxicity [9].

Other references in the literature [10–12] point out at direct contact activity of the substances against aphids and stored products pests (*Ephestia kuehniella* Zeller, Lepidoptera: Pyralidae). Fennel EO (with 67.5% of *(E)*-anethole and 25.5% of fenchone) was more active in *M. persicae* than anise (93% *(E)*-anethole), contact LD_{50} = 0.06 or 0.43%, respectively, by spraying on aphid-infested cabbage plants [13].

Solid nanoparticles of monoterpenes (carvacrol, thymol, eugenol) have been obtained using chitosan, β-cyclodextrin, zeine, modified starch, or polyethylen glycol (PEG) as encapsulating agents [14,15]. In previous works, beads of linalool were made by an oil emulsion entrapment method using starch, the encapsulation yield obtained was 86% and the time to release half of the bioactive exceeded 70 days [16]. Other authors prepared nanoparticles of *(E)*-anethole by emulsification and nanoprecipitation using a biodegradable polymer accepted for clinical drug delivery—polylactic-co-glycolic acid (PLGA) [17]—and after an initial burst release the activity against Gram+ bacteria lasted for more than four days. Another reference explains the encapsulation of *(E)*-anethole in liposomes, that are vesicles in which an aqueous phase is enclosed by a membrane of phospholipids; in this case, the liposomes were stable at 4 °C and provided a controlled release of *(E)*-anethole [18]. An enhancement of the antiaflatoxigenic efficacy of *I. verum* EO by nanoencapsulation in gel or lyophilized chitosan nanoparticles has also been reported [2].

Our work focusses on the use of encapsulated EOs as a fumigant system against insect pests in closed environments. For instance, solid formulations, prepared by emulsification of coriander and basil EOs in alginate and glycerol and dripping into a calcium chloride solution, were tested inside funnel traps and were as effective as the insecticide dichlorvos as killing agents for adults of the Indianmeal moth (*Plodia interpunctella* Hübner, Lepidoptera: Pyralidae) adults lured [19].

It is hypothesized that fumigant activity of plant volatiles could be exploited to control phytophagous insects of vegetables grown in greenhouses but this idea has not yet reached commercial development due to the volatility and low stability of these compounds. The objective of our work was to formulate *(E)*-anethole as solid microparticles (by oil emulsion entrapment, spray drying or molecular inclusion) and test the potential of the vapour released as aphicide on pepper leaves. Experiments were implemented with the green peach aphid, *Myzus persicae* Sulzer (Hemiptera: Aphididae), one of the main pests worldwide attacking fruit trees and vegetables and causing direct damage and transmission of virus diseases.

2. Results

2.1. Encapsulation of (E)-Anethole

The formulations prepared turned out to be within the micrometric range from 1.7 µm to 4 µm particle size. Spray drying (SD) gave the greatest encapsulation yield (73%) and loading (6.15 g/100 g of powder) of *(E)*-anethole in the capsules, although, by oil emulsion entrapment (OEE), the amount of bioactive loaded was quite similar (see Table 1).

Table 1. Physico-chemical parameters in the microspheres.

Formulation Method [1]	Dry Sphere Size (μm)	Loading (g Monoterpene 100 g^{-1} Powder)	Encapsulation Yield (%)
SD	4.00 a	6.15 a	73 a
OEE	1.70 b	5.20 a	26 b
MI	3.52 a	1.33 b	14 c

[1] SD = Spray-drying, OEE = oil-emulsion-entrapment, MI = molecular inclusion (see Section 4). Samples were prepared three times and then bulked. Different letters in the same column mean significant differences at ($p \leq 0.05$).

The plate in Figure 1A shows the *(E)*-anethole/β-cyclodextrin inclusion complex (MI) in the form of irregular particles, therefore this method is less suitable to encapsulate the bioactive product. On the other hand, spray drying (inlet air temperature of 100 °C) of an emulsion with maltodextrin (SD) produced spherical particles of all sizes pilled up due to the strong attraction to each other (Figure 1B). Finally, SEM micrographs (C) and (D) in Figure 1 represent the dry calcium alginate beads (OEE) and reveal a compact surface with low porosity achieved after using glycerol, the surfactant and a high percentage of sodium alginate (4%).

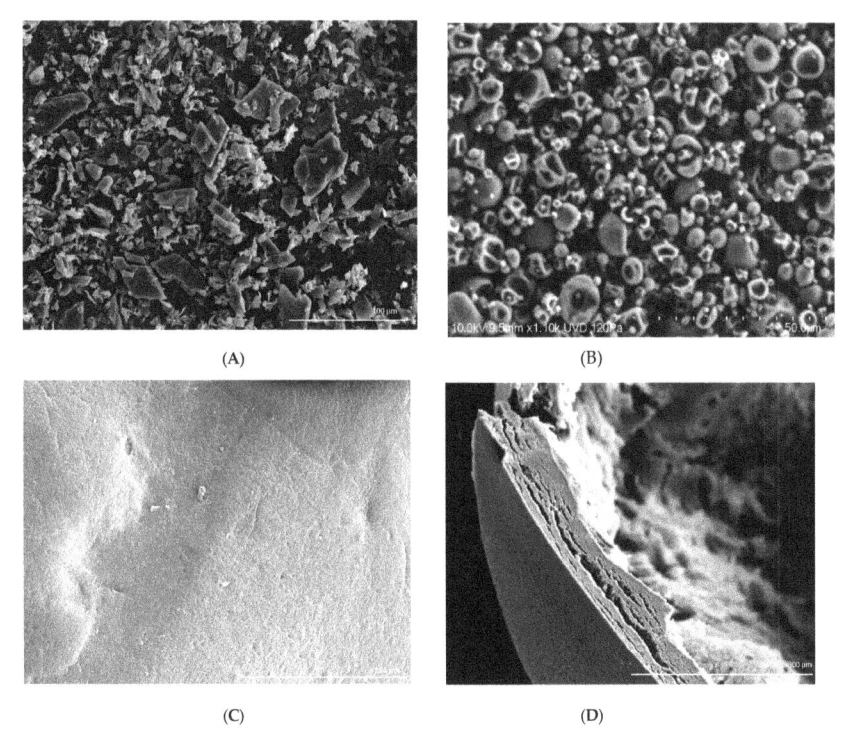

Figure 1. Scanning Electron Microscopy (SEM) micrographs of nanoparticles obtained by (**A**) spray drying (SD), (**B**) molecular inclusion (MI) with × 100-fold magnification, (**C**) oil emulsion entrapment (OEE) with × 100-fold magnification, and (**D**) oil emulsion entrapment (OEE) with × 190-fold magnification.

2.2. Fumigant Activity and Controlled Release of (E)-Anethole

Our formulations have good aphicidal potential (Table 2). Free *(E)*-anethole vapours were fast and entered the aphids giving the lowest LC$_{50}$ (0.336 μL/L) after 24 h. Meanwhile OEE formulation exhibited a LC$_{50}$ = 0.415 µL/L followed in activity by the SD preparation. Results of vapour toxicity apply just to the experimental conditions used (2.5 L dessicators plus two pepper leaves and 20 apterous *M. persicae* females in each leaf). Overall, the encapsulated *(E)*-anethole had a LC$_{90}$ from 0.78 to 3.38 µL/L after 24 h exposure to the aphids.

Table 2. Lethal Concentrations [1] of vapours of *(E)*-anethole (µL/L air) to *Myzus persicae* Sulzer (Hemiptera: Aphididae), pink clone, after 24 h.

Formulation Method [2]	LC$_{50}$	95% CI	LC$_{90}$	95% CI	χ^2
SD	1.292	1.169–1.476	3.383	2.706–4.305	0.487 [ns]
OEE	0.415	0.416–0.468	0.780	0.675–0.832	23.850 *
Free *(E)*-anethole	0.336	0.306–0.369	1.043	0.867–1.255	8.572 [ns]

[1] Probit analysis fitting lethal concentration 50 (LC$_{50}$) and 90 (LC$_{90}$) and confidence intervals. χ^2 non-significant (n.s.) or significant (*) at 0.1%. [2] SD = spray drying, OEE = oil emulsion entrapment (see Section 4).

The results are presented in more detail in Figure 2. The graph shows the dose response of the formulations including the molecular inclusion complexes (MI) for which the lethal concentrations could not be computed due to the very low mortality values obtained (this is why this treatment is not included in Table 2). The regression line of free *(E)*-anethole intercepts the probit = 5 line (that represents LC$_{50}$) first, indicating the greatest effect at a low concentration, while the OEE formulation intercepts the probit = 6.28 (that represents LC$_{90}$) first, indicating more effectivity at high doses (Figure 2). Overall a similar response of the preparations SD, MI, and free *(E)*-anethole is observed due to the parallel regression lines; what changes is the amount of product required to produce the same mortality.

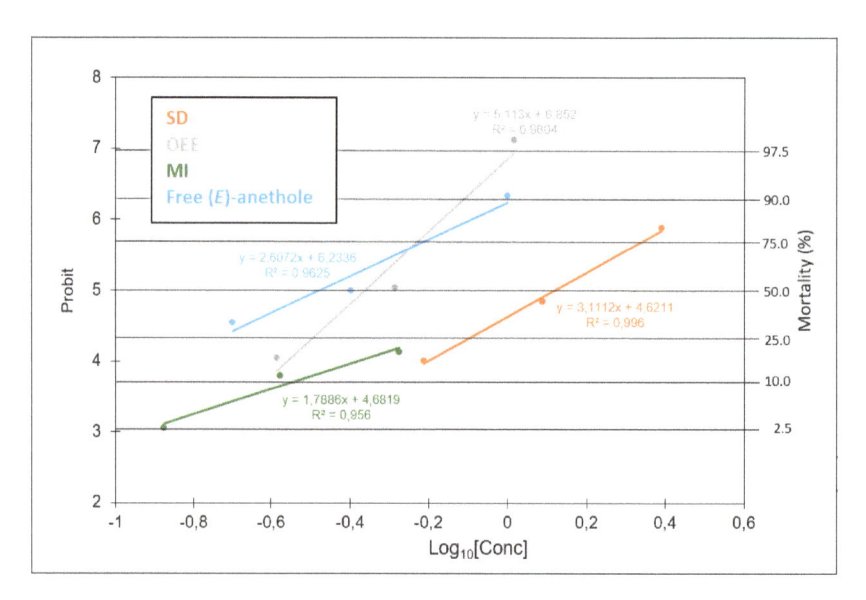

Figure 2. Regression lines of probit analysis for mortality against *Myzus persicae*. OEE = oil emulsion entrapment, SD = spray drying, MI = molecular inclusion, and free *(E)*-anethole.

In Figure 3, we can see that the OEE formulate was quite close in toxicity to free *(E)*-anethole after 24 h, but presumably, the former would have had effects beyond the short period of observation if evaluated. In this context, MI complexes hardly produced mortality in the short term.

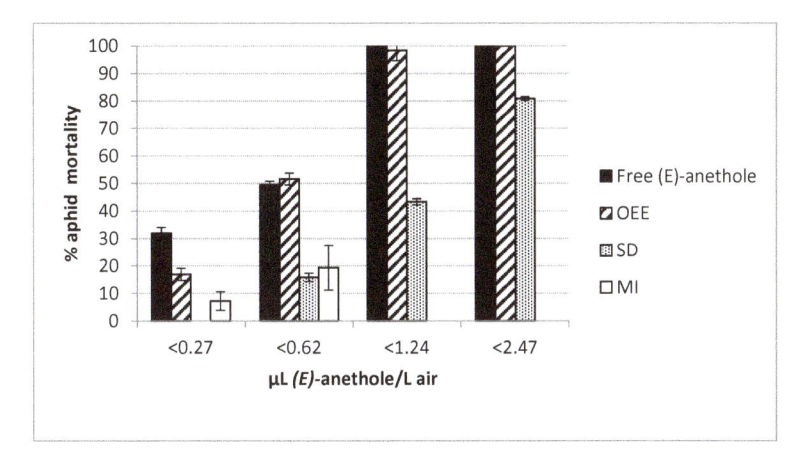

Figure 3. Mortality (%) in *Myzus persicae* Sulzer (Hemiptera: Aphididae) after exposure (24 h at 25 °C) to vapours (μl/L air) of *(E)*-anethole released from microspheres (OEE = oil emulsion entrapment, SD = spray drying, MI = molecular inclusion) or free *(E)*-anethole. Percentages of mortality refer to total number of insects tested in the six replications per dose and formulation ($n = 240$).

Such results are explained by different paces at which *(E)*-anethole is released from the microspheres (Figure 4). At 15 °C, there are statistically significant differences among all treatments (Figure 4A), and at 40 °C, there are statistically significant differences between free *(E)*-anethole and MI but not between OEE and SD (Figure 4B). The formulation slows down the availability of the toxic vapours in comparison with the free *(E)*-anethole particularly under the conditions of the fumigant bioassay (25 °C and mortality recorded after 24 h). It is foreseen, however, that the toxic vapours would last several weeks further.

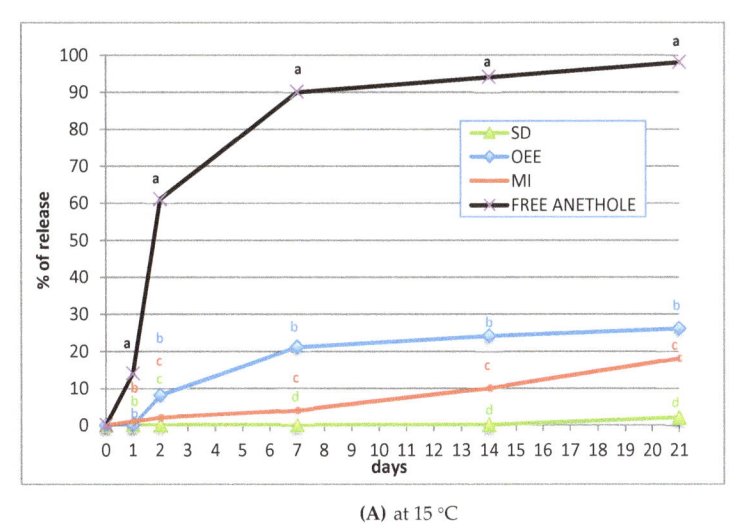

(**A**) at 15 °C

Figure 4. *Cont.*

(B) at 40 °C

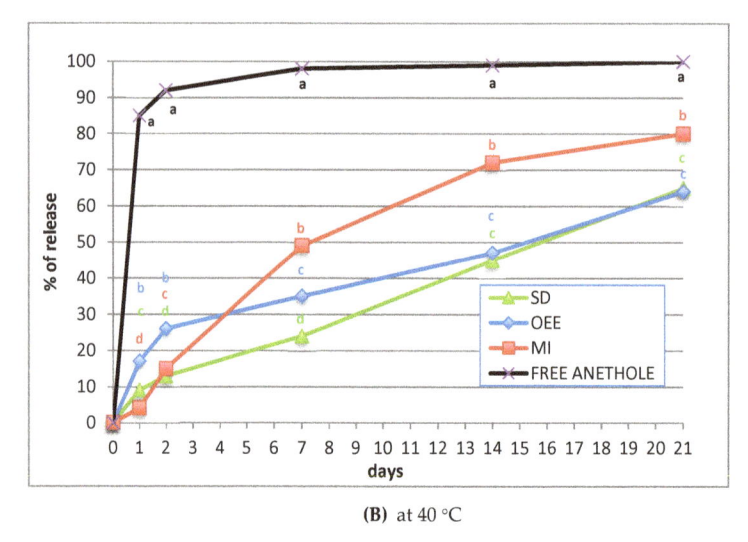

Figure 4. Release of free *(E)*-anethole and controlled release from formulated microspheres for 21 d (**A**) at 15 °C and (**B**) at 40 °C. Mean values in the same day with the same letter do not differ significantly ($p > 0.05$) using Duncan's test.

3. Discussion

Plants are a good natural source of *(E)*-anethole, fennel variations in the Iranian genotypes accounted for 1.2–88.4% of the EO whilst in anise 78.6–96% are common [1,20–22]. The mode of entry of the bioactive volatile in the insects is possibly via the respiratory system by inhalation [8,23] but its mode of action remains unclear. Some publications refer to greater activity when mixtures of volatiles for example limonene and fenchone [7,13] or 1,8-cineole [9] are applied together with *(E)*-anethole. Greater insecticidal fumigant activity against *Trichoplusia ni* Hübner (Lepidoptera: Noctuidae) of lemongrass or thyme EOs or the binary mixture of the two main compounds often had better action than pure compounds [24].

Therefore it is worthwhile to study further the fumigant effect of *(E)*-anethole in binary mixtures with monoterpenoids against *M. persicae* in all insect stages and expand the period of study (to several days) to provide new data on the advantages of a controlled release to be applied in pest control into a greenhouse.

The bioassay was done inside air-tight desiccators. Mortality was recorded after 24 h, and once opened, the concentration of the volatiles inside the desiccators could change; this was the main reason why we decided to take just one observation. Another reason was to be sure the leaves inside the desiccator were healthy enough for the aphids to remain alive, but for those affected by the insecticidal effects of anethole. The bioassay has to be improved for longer periods of observation.

If we compare our results with those of the literature, there is an agreement in the fumigant effect of EOs containing *(E)*-anethole. The lethal doses varied depending on the insect pest and the volume of the chamber used in the assays. The LD_{50} of fennel EO was 10.3 µL/L in *Brevicoryne brassicae* L. (Hemiptera: Aphidae), whereas 2 µL/L of cumin or origanum EOs (with carvacrol, *(E)*-anethole and pulegone in the oil) has been reported to cause 100% mortality in *A. gossypii* [25,26]. Our results of LC_{50} range from 0.3 to 1.47 µL/L of *(E)*-anethole (free or encapsulated) against *M. persicae*. Repellency was reported for vegetable aphid pests such as *M. persicae*, *A. gossypii*, and *Macrosiphum euphorbiae* Thomas (Hemiptera: Aphididae) in our previous works with values of RD_{50} = 0.07–0.09 µL/cm^2 for anise and RD_{50} = 0.04–0.08 µL/cm^2 for *(E)*-anethole [27,28], and the pure compound was more repellent for the pink clone of *M. persicae* and *A. gossypii*. The efficacy of anise EO by contact applications was greater against early nymphal instars (first and second nymphs), LD_{50} = 0.003% v/v, than to late nymphal instars (third and fourth nymphs), LD_{50} = 0.017% v/v, of apterous aphids [29]. Newly emerged adults of *T. castaneum* were highly susceptible to vapours of *(E)*-anethole in comparison with sclerotized older beetles in which concentrations at least of 20 µL/L were required to produce toxic effects [9]. Therefore, soft-bodied suckling pests such as aphids might be more susceptible to fumigation by EOs than stored product beetle pests.

Encapsulation offers clear advantages for a bioactive volatile—in addition to avoid releasing the product all at once—like protection against environmental conditions (light, temperature, oxygen, etc.). Further work will be done on improving the formulation methods described in this paper where encapsulation yields have ranged from 14 to 73%; of the three methods tested, OEE and SD are more promising. It would be of practical use the release of just the required amount of active that causes high insect mortality (previously calculated for each stage of development) for a prolonged period of time. Other authors have obtained loadings of 13%, particle size < 180 nm and bactericidal activity prolonged for more than 4 d from PLGA *(E)*-anethole nanoparticles [17]. Similarly, PEG nanoparticles of geranium and bergamot EOs slowed the release of the volatiles down and the residual contact activity against cockroaches was improved [30]. Polymer based nanoencapsulation of EO is considered for plant protection products and the type of polymers consist mainly of polysaccharides (chitosan, alginate and starch), polyesters (PEG) or biodegradable materials such as gum arabic or lecithins.

Plant essential oils are available raw materials, for example: anise EO is obtained from anise fruits at a yield of 2–6% and its market price is 7–9 €/Kg. We propose that botanicals coming from plants that have been used as foods or condiments be considered as safe plant protection products. In fact, the European Food Safety Authority (EFSA) regards them as Low Risk Active Substances (LRAS).

All classes of controlled release systems could be considered as new formulations for insecticide applications: nanocapsules or microcapsules with polymers, cyclodextrin complexes, solid-lipid nanoparticles, nanoemulsions or microemulsions, liposomes, and nanogels.

Nanotechnology is an area under development in plant protection and scaling up the experiments is important to be able to extrapolate the results to applications in agricultural production systems.

4. Materials and Methods

4.1. Materials

(E)-anethole (99%), calcium chloride, β-cyclodextrin (98%), maltodextrin and sodium alginate were obtained from Sigma-Aldrich, whereas glycerol (99.5% pure) was obtained from Labogros, France. Analytical grade solvents and surfactants (Tween 80) were from Sigma-Aldrich.

4.2. Microsphere Preparation

4.2.1. Beads of (E)-Anethole by Oil-Emulsion-Entrapment (OEE)

Beads were formed by dripping an alginate solution (containing a dispersion of *(E)*-anethole, glycerol and Tween 80) into a calcium solution. Diffusion of the calcium in alginate droplets led to their gelification. The preparation of the internal phase was carried out as follows: *(E)*-anethole (20 mL) was dispersed in glycerol (20 mL) and Tween 80 (20 mL). The blend was dispersed in 20 mL of alginate (40 g/L). This dispersion was dripped into a calcium chloride solution (10 g/100 mL). Beads were filtered with a wire mesh and finally were dried overnight at room temperature (15 °C). Samples were prepared three times and then bulked.

4.2.2. Preparation of β-Cyclodextrin/(E)-Anethole Molecular Inclusion (MI) Complexes

A chemical precipitation method was used to prepare β-cyclodextrin/*(E)*-anethole complexes. β-cyclodextrin (5 g) was dissolved in 300 mL of water at 55 °C for half an hour, 30 mL of *(E)*-anethole were added slowly to the suspension of β-cyclodextrin. The blend was frozen overnight at −20 °C.

The precipitated *(E)*-anethole/cyclodextrin complex was recovered by lyophilization (24 h) and filtration. Samples were prepared three times and then bulked.

4.2.3. Spray Drying of (E)-Anethole (SD)

An emulsion of *(E)*-anethole was mixed with 50 mL of maltodextrin (10%, *w/v*) then it was stirred at 300 rpm for 2 h. The emulsion was fed into a laboratory scale dryer (Mini Spray Dryer-B290, BÜCHI, Flawil, Switzerland) at room temperature with a flow rate of 4 mL min^{-1}. The inlet and outlet temperatures were maintained at 100 °C and 60 °C, respectively. The dried powder was collected and stored in an opaque, air-tight container at 4 °C for further analysis.

4.3. Encapsulation Yield and Loading

The amount of *(E)*-anethole into the dry microspheres was determined by GC/MS analysis as follows: 0.5 g of powder was dispersed in 8 mL of distilled water and 4 mL of hexane in 15 mL glass vials. Vials were heated and stirred in a hot plate at 60 °C for 30 min. The organic phase containing *(E)*-anethole was decanted, and the aqueous phase was exhaustively extracted with hexane four times (4 × 4 mL). These four phases were combined. The hexane was removed using a nitrogen stream. The quantitative analysis of *(E)*-anethole was carried out using a model 5890 Series II equipped with a DB-Waters 30 m × 0.32 mm capillary column coated with a polyethylene glycol film (1 µm thickness) and an Agilent model 5972 inert mass spectrometry (MS) detector (Agilent, Palo Alto, CA, USA). The initial oven temperature was held at 60 °C for 1 min. Afterwards, it was increased by 3 °C/min to 225 °C, with injector at 250 °C, column head pressure at 8.00 psi, helium carrier gas, flow rate of 2.6 mL/min, and splitless with 2 µL of sample injected. The content of *(E)*-anethole was calculated according to the area of the chromatographic peak and using linear regression. Prior to the quantification of monoterpene, the surface *(E)*-anethole in the formulation was washed.

Encapsulation yield is defined as the ratio between the quantities of *(E)*-anethole in the capsules versus the initial amount of *(E)*-anethole. Loading is defined as the quantity of *(E)*-anethole per 100 grams of dry microcapsules. SAS version 8.0 for windows (SAS Institute, Inc., USA) was used to compare mean values of the formulations by a Tukey test at $p < 0.05$.

4.4. Controlled Release of (E)-anethole through Different Matrix Blends

One gram of dry sample was placed into the vials without sealing. These vials were maintained in dry conditions at 15 °C and 40 °C in growth chambers (MLR-350H, Sanyo, Japan), and weight loss was monitored in an analytical balance as a function of time for 21 d. As a control, 1 g of *(E)*-anethole was set in a vial to study the weight loss for the same period of time. Data from three replications were recorded in this assay. Data were statistically analyzed by analysis of variance (ANOVA) using SPSS (PASW Statistic 18). Duncan's multiple tests were applied for the calculation of the significant differences among the controlled release of the blends at the 5% level ($p = 0.05$).

4.5. Scanning Electron Microscopy (SEM) Analysis

Microspheres were evaluated with SEM JEOL 6100 (SAI, Universidad de Murcia, Spain). The samples were mounted (both entire structures and cross sections) on specimen stubs with double sided adhesive tape. The specimens were coated with gold and examined at an accelerating voltage of 15 kV and a working distance of 20 mm. Topographical images were collected by an image capture system used for an X-ray detector (INCA, Oxford, UK) at a magnification of 370× and 1000×. The mean particle diameter, pore diameter, number of pores per unit area and the presence of pores were recorded. SAS version 8.0 for windows (SAS Institute, Inc., Cary, NC, USA) was used to compare mean values of the formulations by a Tukey test at $p < 0.05$.

4.6. Fumigation Bioassay

M. persicae, the green peach aphid, from a laboratory culture (pink clone), maintained at a constant temperature of 25 °C and a photoperiod of 16:8 h (light:dark) on pepper plants, was used for the insecticidal experiments. The experimental unit consisted of two pepper leaves, with twenty apterous females each, placed inside a 2.5 L air-tight desiccator. Each dose was replicated six times, with 240 insects per dose. The leaf petiole was into an Eppendorf tube with water. The products, either an amount of powder of the microspheres formulations (range 0–0.1 g/L air), to obtain a dose response, or the pure free *(E)*-anethole -pipetted onto a 2.1 cm^2 filter paper disk- (range 0–1 µl/L air), were placed in an unlid petri dish without direct contact with the insects. Therefore, only the volatile toxic effects were evaluated. The desiccators were maintained at 25 °C and 16:8 photoperiod for 24 h and aphid mortality was recorded. Controls were prepared exactly the same but the application of the products. Number of alive and dead aphids was recorded after 24 h. Probit analysis was performed to obtain LC_{50} and LC_{90}.

5. Conclusions

Spray drying of an emulsion of *(E)*-anethole with maltodextrin gave spherical particles with the greatest encapsulation yield and loading but the beads of *(E)*-anethole by oil-emulsion entrapment had better fumigant activity against *M. persicae*. Most of the free *(E)*-anethole vapours were available within 2 d of application whilst the preparations prolonged the release period for several weeks and required at least one week to release 20% of the bioactive depending on the temperature and the formulation, for instance *(E)*-anethole complexed with β-cyclodextrin required temperatures over 25 °C to release the product. Therefore, future experiments should expand the observation period and take into account the susceptibility of earlier nymphal instars to prove advantages of the practical use of *(E)*-anethole encapsulated in the form of microspheres.

Author Contributions: All authors have read and agreed to the published version of the manuscript. Conceptualization, M.J.P.-V.; methodology and analysis, M.D.L., M.C.-T., and P.G.; writing—original draft preparation M.J.P.-V.; writing—review and editing, P.G., M.C.-T., M.D.L., and M.J.P.-V.; funding acquisition, M.J.P.-V. and P.G.

Funding: The authors with to thank the funding received through the research projects FEDER 1420-19 and INIA RTA2017-00022. Manuel Cantó-Tejero was awarded with a grant (INIA CPD2016-0092) for a predoctoral contract at IMIDA, Murcia, Spain.

Conflicts of Interest: The authors declare no conflict of interest. The funders had no role in the design of the study; in the collection, analyses, or interpretation of data; in the writing of the manuscript, or in the decision to publish the results.

References

1. Pascual-Villalobos, M.J.; Cantó-Tejero, M.; Vallejo, R.; Guirao, P.; Rodríguez-Rojo, S.; Cocero, M.J. Use of nanoemulsions of plant essential oils as aphid repellents. *Ind. Crops Prod.* **2017**, *110*, 45–57. [CrossRef]
2. Dwivedy, A.K.; Singh, V.K.; Prakash, B.; Dubey, N.K. Nanoencapsulated *Illicium verum* Hook f. essential oil as an effective novel plant-based preservative against aflatoxin B_1 production and free radical generation. *Food Chem. Toxicol.* **2018**, *111*, 102–113. [CrossRef] [PubMed]
3. Addae-Mensah, I.; Asomaining, W.A.; Oteng-Yeboah, A.; Garneau, F.X.; Gagnon, H.; Jean, F.I.; Moudachirou, M.; Koumagic, K.H. (E)-anethole as a major essential oil constituent of *Clausena anisata*. *J. Essent. Oil Res.* **1996**, *8*, 513–516. [CrossRef]
4. Blewitt, M.; Southwell, I.A. *Backhousia anisata* Vickery an alternative source of (E)-anethole. *J. Essent. Oil Res.* **2000**, *12*, 445–454. [CrossRef]
5. Opdyke, D.L.J. *Monographs in Fragance Raw Materials*; Pergamon Press: Oxford, UK, 1979; p. 92.
6. Isman, M.B. Plant essential oils for pest and disease management. *Crop. Prot.* **2000**, *19*, 603–608. [CrossRef]
7. López, M.D.; Jordán, M.J.; Pascual-Villalobos, M.J. Toxic compounds in essential oils of coriander, caraway and basil active against stored rice pests. *J. Agric. Food Chem.* **2008**, *44*, 273–278. [CrossRef]

8. Kim, D.-H.-; Ahn, Y.-J. Contact and fumigant activity of constituents of *Foeniculum vulgare* fruits against three coleopteran stored-product insects. *Pest. Manag. Sci.* **2001**, *57*, 301–306. [CrossRef]
9. Koul, O.; Singh, G.; Singh, R.; Singh, J. Mortality and reproductive performance of *Tribolium castaneum* exposed to anethole vapours at high temperature. *Biopestic. Int.* **2007**, *3*, 126–137.
10. Zarrad, K.; Pascual-Villalobos, M.J. Testing liquid formulations of essential oils against aphid pests. In Proceedings of the IX Congreso Nacional de Entomología Aplicada, XV Jornadas Científicas de la SEEA, Valencia, Spain, 19–23 October 2015; p. 234.
11. Pascual-Villalobos, M.J.; Guirao, P.; Díaz-Baños, F.G.; Cantó-Tejero, M.; Villora, G. Oil in water nanoemulsions of botanical active substances. In *Nano-Biopesticides Today and Future Perspectives*; Koul, O., Ed.; Academic Press, Elsevier Inc.: London, UK, 2019; Volume 9, pp. 223–248.
12. Pascual-Villalobos, M.J.; Zarrad, K.; Castañé, C.; Riudavets, J. Liquid formulations of monoterpenoids as space and structural treatments of store rooms. In Proceedings of the 11th International Working Conference on Stored Product Protection, Chang Mai, Thailand, 24–28 November 2014; Arthur, F.H., Kengkouponich, R., Chayaprosert, W., Suthisut, D., Eds.; 2015; pp. 1061–1070.
13. Benelli, G.; Pavela, R.; Petrelli, R.; Capellaci, L.; Canale, A.; Senthil-Nathan, S.; Maggi, F. Not just popular spices! Essential oils from *Cuminum cyminum* and *Pimpinella anisum* are toxic to insect pests and vectors without affecting non-target invertebrates. *Ind. Crops Prod.* **2018**, *124*, 236–243. [CrossRef]
14. De Oliveira, J.L.; Campos, E.E.; Bakshi, M.; Abhilash, P.C.; Fernández Fraceto, L. Application of nanotechnology for the encapsulation of botanical insecticides for sustainable agriculture: Prospects and promises. *Biotechnol. Adv.* **2014**, *32*, 1550–1561. [CrossRef]
15. Mishra, P.; Seenivasan, R.; Mukherjee, A.; Chandrasekaran, N. Polymer/layered silicate nanocomposites as matrix for bioinsecticide formulations. In *Nano-Biopesticides Today and Future Perspectives*; Koul, O., Ed.; Academic Press, Elsevier Inc.: London, UK, 2019; Volume 6, pp. 161–178.
16. López, M.D.; Maudhuit, A.; Pascual-Villalobos, M.J.; Poncelet, D. Development of formulations to improve the controlled release of linalool to be applied as an insecticide. *J. Agric. Food Chem.* **2012**, *60*, 1188–1192. [CrossRef] [PubMed]
17. Manesh, M.E.; Ghaedi, Z.; Asemi, M.; Khanavi, M.; Manayi, A.; Jamalifar, H.; Atyavi, F.; Dinarvand, R. Study of antimicrobial activity of anethole and carvone loaded PLGA nanoparticles. *J. Pharm. Res.* **2013**, *7*, 290–295.
18. Gharib, R.; Greige-Gerges, H.; Jraij, A.; Auezova, L.; Charcoset, C. Preparation of drug-in-cyclodextrin-in-liposomes at a large scale using a membrane contactor: Application to trans-anethole. *Carbohydr. Polym.* **2016**, *154*, 276–286. [CrossRef] [PubMed]
19. Pascual-Villalobos, M.J.; López, M.D.; Castañé, C.; Soler, A.; Riudavets, J. Encapsulated essential oils as an alternative to insecticides in funnel traps. *J. Econ. Entomol.* **2015**, *108*, 2117–2120. [CrossRef]
20. Telci, I.; Dermitas, I.; Schin, A. Variations in plant properties and EO composition of sweet fennel (*Foeniculum vulgare* Mill.) fruits during stages of maturity. *Ind. Crops Prod.* **2015**, *30*, 126–130. [CrossRef]
21. Bahmani, K.; Darbandi, A.I.; Ramshini, H.A.; Moradi, N.; Akbari, A. Agro-morphological and phytochemical diversity of various Iranian fennel landraces. *Ind. Crops Prod.* **2015**, *77*, 282–294. [CrossRef]
22. Arslan, N.; Gürbüz, B.; Sarihan, E.O. Variations in EO content and composition in Turkish anise (*Pimpinella anisum* L.) populations. *Turk. J. Agric. For.* **2004**, *28*, 173–178.
23. Hamraoui, A.; Regnault-Roger, C. Comparaison des activités insecticides des monoterpènes sur deux espèces d'insectes ravageurs des cultures: *Ceratitis capitata* et *Rhopalosiphum padi*. *Acta Bot. Gallica* **1997**, *144*, 414–417. [CrossRef]
24. Tak, J.-H.; Jovel, E.; Isman, M.B. Contact, fumigant, and cytotoxic activities of thyme and lemongrass essential oils against larvae and an ovarian cell line of the cabbage looper, *Trichoplusia ni*. *J. Pest. Sci.* **2016**, *89*, 183–193. [CrossRef]
25. Jahan, F.; Abbasipour, H.; Hasanshahi, G. Fumigant toxicity and nymph production deterrence effect of five essential oils on adults of the cabbage aphid, *Brevicoryne brassicae* L. (Hemiptera: Aphidae). *J. Essent. Oil Bear. Plants* **2016**, *19*, 140–147. [CrossRef]
26. Tunc, I.; Sahinkaya, S. Sensitivity of two greenhouse pests to vapours of essential oils. *Entomol. Exp. Appl.* **1998**, *86*, 183–187. [CrossRef]

27. Cantó-Tejero, M.; Pascual-Villalobos, M.J.; Guirao, P. Estudio comparativo de la actividad repelente de aceites esenciales en varias de las principales especies de pulgón de pimiento. In Proceedings of the X Congreso Nacional de Entomología Aplicada, XVI Jornadas Científicas de la SEEA, Logroño, Spain, 20–24 October 2017; p. 127.
28. Cantó-Tejero, M.; Guirao, P.; Pascual-Villalobos, M.J.; Marcos-García, M.A. Aceites esenciales como repelentes frente a los pulgones *Myzus persicae* Sulzer y *Macrosiphum euphorbiae* (Thomas) (Hemiptera: Aphididae), Efectos sobre sus enemigos naturales *Sphaerophoria ruepellii* (Wiedemann) (Diptera: Syrphidae) y *Aphidius colemani* Viereck (Hymenoptera: Braconidae). In Proceedings of the XI Congreso Nacional de Entomología Aplicada, XVII Jornadas Científicas de la SEEA, Madrid, Spain, 4–8 November 2019; p. 80.
29. Al-Antary, T.M.; Belghasem, I.H.; Araj, S.A. Toxicity of anise oil against the green peach aphid *Myzus persicae* Sulzer using four solvents (Homoptera: Aphididae). *Fresenius Environ. Bull.* **2017**, *36*, 3705–3710.
30. González, J.O.W.; Stefanazzi, N.; Murray, A.P.; Ferrero, A.A.; Fernández-Band, B. Novel nanoinsecticides based on essential oils to control the German cockroach. *J. Pest. Sci.* **2015**, *88*, 393–404. [CrossRef]

© 2020 by the authors. Licensee MDPI, Basel, Switzerland. This article is an open access article distributed under the terms and conditions of the Creative Commons Attribution (CC BY) license (http://creativecommons.org/licenses/by/4.0/).

Article

Bioactivity of Common Pesticidal Plants on Fall Armyworm Larvae (*Spodoptera frugiperda*)

Kelita Phambala [1], Yolice Tembo [1], Trust Kasambala [1], Vernon H. Kabambe [1], Philip C. Stevenson [2,3] and Steven R. Belmain [2,*]

[1] Department of Crop and Soil Sciences, Lilongwe University of Agriculture and Natural Resources, Lilongwe Box 219, Malawi; kelitaphambala@yahoo.com (K.P.); ytembo@bunda.luanar.mw (Y.T.); tdonga@luanar.ac.mw (T.K.); kabambev@gmail.com (V.H.K.)
[2] Natural Resources Institute, University of Greenwich, Central Avenue, Chatham Maritime, Kent ME4 4TB, UK; p.c.stevenson@gre.ac.uk
[3] Biological Chemistry and In Vitro Research, Royal Botanic Gardens, Richmond TW9 3AB, UK
* Correspondence: s.r.belmain@gre.ac.uk; Tel.: +44-1634883761

Received: 17 December 2019; Accepted: 11 January 2020; Published: 15 January 2020

Abstract: The fall armyworm (FAW), *Spodoptera frugiperda* (Lepidoptera: Noctuidae) is a recent invasive pest species that has successfully established across sub-Saharan Africa where it continues to disrupt agriculture, particularly smallholder cereal production. Management of FAW in its native range in the Americas has led to the development of resistance to many commercial pesticides before its arrival in Africa. Pesticide use may therefore be ineffective for FAW control in Africa, so new and more sustainable approaches to pest management are required that can help reduce the impact of this exotic pest. Pesticidal plants provide an effective and established approach to pest management in African smallholder farming and recent research has shown that their use can be cost-beneficial and sustainable. In order to optimize the use of botanical extracts for FAW control, we initially screened ten commonly used plant species. In laboratory trials, contact toxicity and feeding bioassays showed differential effects. Some plant species had little to no effect when compared to untreated controls; thus, only the five most promising plant species were selected for more detailed study. In contact toxicity tests, the highest larval mortality was obtained from *Nicotiana tabacum* (66%) and *Lippia javanica* (66%). Similarly, in a feeding bioassay *L. javanica* (62%) and *N. tabacum* (60%) exhibited high larval mortality at the highest concentration evaluated (10% *w/v*). Feeding deterrence was evaluated using glass-fibre discs treated with plant extracts, which showed that *Cymbopogon citratus* (36%) and *Azadirachta indica* (20%) were the most potent feeding deterrents among the pesticidal plants evaluated. In a screenhouse experiment where living maize plants infested with fall armyworm larvae were treated with plant extracts, *N. tabacum* and *L. javanica* were the most potent species at reducing foliar damage compared to the untreated control whilst the synthetic pesticide chlorpyrifos was the most effective in reducing fall armyworm foliar damage. Further field trial evaluation is recommended, particularly involving smallholder maize fields to assess effectiveness across a range of contexts.

Keywords: botanical pesticide; pesticidal plant; pest management; invasive species; agro-ecological intensification; sustainable agriculture

---

1. Introduction

The fall armyworm, *Spodoptera frugiperda* (J.E. Smith) (Lepidoptera: Noctuidae) (FAW) is a polyphagous pest that is invasive and now widely established across sub-Saharan Africa. Although similar to the native African armyworm, *Spodoptera exempta* (Walker), FAW is more likely to persist year-round once established, attacking a much wider range of cereals as well as more than 100 other horticultural crops [1]. Native to North and South America, FAW was first reported in West and Central Africa in 2016 [2] and is now reported in at least 44 African countries [3]. FAW is a heavy foliage feeder that can cause 100% loss of cereal crops [4]. In the absence of effective control methods, potential maize yield losses caused by FAW have been estimated between 8.3 and 20.6 million metric tons per year in just 12 maize-producing countries in Africa. This represents a range of 21–53% of annual maize production averaged over a three-year period. The value of these losses was estimated at between US $2481 million and US $6187 million [5].

Current armyworm control relies on the use of synthetic pesticides; however, widespread over-use and misuse in the Americas have resulted in considerable problems with insecticide resistance particularly among the carbamates, pyrethroids and organophosphates [6] on which many African farmers rely. As African farmers have a long history of using plants with pesticidal properties [7–11], options for developing botanical biopesticides for FAW control may be more realistic than in other regions [12]. Recent research has evaluated several abundant pesticidal plant species, confirming that their use in smallholder farming can result in comparable yield to that when using commercial synthetics, without the severe environmental damage often associated with synthetic compounds [13–15]. With a need to develop new, effective and agro-ecologically sustainable methods for controlling FAW in Africa, we set out to screen some of the more promising pesticidal plant species where considerable knowledge already exists on their abundance, phytochemistry and safe use. The specific objective of the research presented here was to evaluate potential effects of pesticidal plants on the larval stage, assessing direct toxicity as well as post-ingestive toxicity and feeding deterrence. Finally, the most promising pesticidal plant extracts were evaluated in controlled trials using FAW-infested maize plants to determine whether the plant extracts reduced foliar damage under cropping conditions.

2. Results and Discussion

2.1. Contact Toxicity and Feeding Bioassays with Ten Pesticidal Plant Species

Water extracts (10% w/v) of ten common pesticidal plants which are regularly used by smallholder farmers showed variable effects on larval mortality (Figure 1). *Tephrosia vogelii* Hook.f. and *Lantana camara* L. showed very low mortality (<10%) in both feeding and contact toxicity bioassays which was surprising since previous research on both of these plant species demonstrated high and consistent efficacy against a range of pest species using the same extraction methods and plant sources and the same biologically active phytochemicals [13–17]. Low mortality (<40%) was also observed with *Vernonia amygdalina* Delile followed by *Aloe vera* (L.) Burm.f. and *Trichilia emetica* Vahl, despite evidence of efficacy against other target insect pests [11,18–21]. The most effective plant species were *Azadirachta indica* A. Juss., *Cymbopogon citratus* (DC.) Stapf, *Lippia javanica* (Burm.f.) Spreng., *Nicotiana tabacum* L. and *Ocimum basilicum* L. which caused at least 50% mortality through at least one bioassay [22]. Although some plant species had an effect through one application method only (*A. indica* and *L. javanica*), overall, the application method led to comparable effects for most plant species, which is verified through statistical analyses (Table S1). Other research on the evaluation of botanicals against FAW in Ethiopia [1], showed *N. tabacum* to cause 50% mortality to 3rd instars after 72 h exposure, which is considerably lower than the mortality observed in our bioassay. Further, this work found that *A. indica* and six other plant species were more effective than *N. tabacum* with mortality rates of 75–98%.

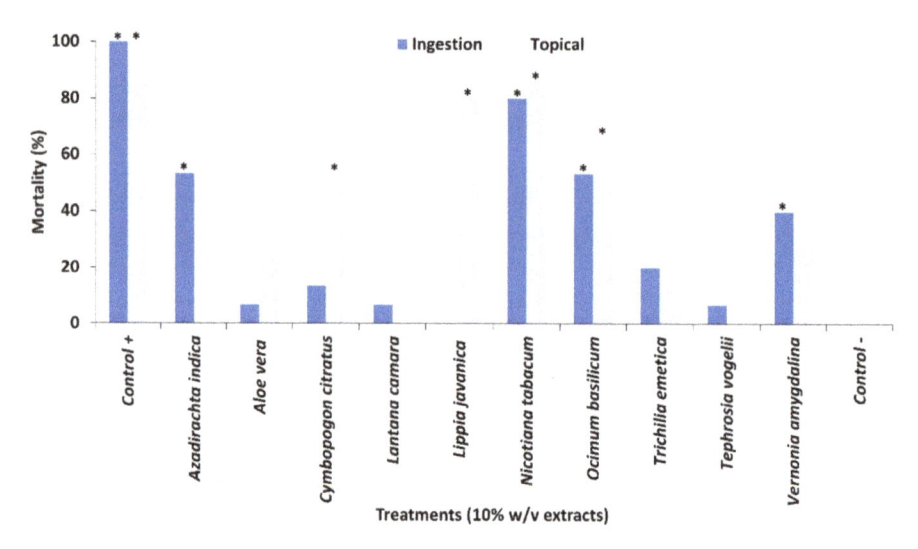

Figure 1. Mortality of 2nd instars when exposed to extracts of pesticidal plant either topically applied through a contact toxicity bioassay or through ingestion in a feeding bioassay. Treatments different from the untreated control ($p < 0.05$) are indicated by *. Significant differences are presented in Table S1.

2.2. Contact Toxicity and Feeding Bioassays with Five Pesticidal Plant Species

Plant extracts applied to glass fibre discs showed that the five most active plant species from the previous bioassay caused significantly greater mortality in comparison to the untreated control (Figure 2a). At the highest concentration (10%), *L. javanica* (62%) and *N. tabacum* (60%) exhibited high larval mortality. The lowest mortality was observed from *C. citratus* (16%) and *O. basilicum* (26%). Although there was a slight dose dependent effect, this was not significant for any of the plant extracts ($p < 0.05$; Table S1). Some differences in efficacy were observed in comparison to the first trial where extracts were prepared in water. Mortality trends between water and methanol extracts were similar with the exception of *L. javanica* where no mortality was observed in the feeding bioassay when applying the water extracts to discs. The reasons for this difference are not clear but may be caused by differences in methodology, in which water extract was presented on maize leaves to reflect farmer practices, whereas the methanol extract was applied to glass fibre discs to more easily assess potential effects of extract concentration. However, the differences are more likely due to the extraction efficiency of the different solvents where hydrophobic compounds with bioactivity were more efficiently extracted by methanol than water [11,13,16,23]. The lack of a clear dose effect as well as differences in mortality between the trials using water or methanol could also be caused by differences in larval feeding rates through feeding deterrence behaviours.

The topical application of plant extracts to FAW larvae showed a strong dose response for four out of the five plant species (Figure 2b; Table S1), whereas mortality from *O. basilicum* (4%) was not significantly different from the untreated control ($p < 0.5$). As expected, the 10% concentration exhibited high larval mortality of 50–66% for *N. tabacum* (66%), *L. javanica* (66%), *A. indica* (60%) and *C. citratus* (50%). However, the positive control of chlorpyrifos was superior to all plant extracts, causing nearly 100% mortality. Although these extracts were made using methanol, the mortality rates observed were not significantly different from the application of water extracts in the first trial ($p < 0.05$).

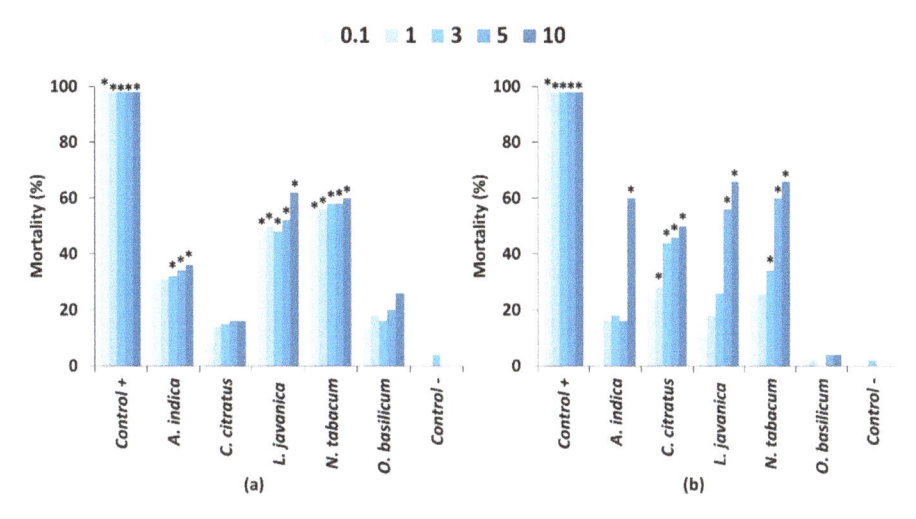

Figure 2. Mortality of fall armyworm 2nd instars seven days after exposure to five different concentrations of pesticidal plant extracts when applied through (**a**) a feeding bioassay and (**b**) a contact toxicity bioassay. Treatments differing significantly from the untreated control ($p < 0.05$) are indicated by *. Significant differences are presented in Table S1.

All of the five plant extracts used in this trial showed some degree of deterrence (Figure 3). The most potent feeding deterrents were *C. citratus* and *A. indica*. Although mortality appears to be relatively low with these two species, feeding deterrent compounds could help to reduce crop damage. In agreement with our observations, other studies demonstrated the deterrent effects of several *Cymbogon* species as well as *A. indica*. In a binary choice test, the essential oils isolated from *C. nardus*, *C. flexuosus* and *C. martini* exhibited strong antifeedant activity against *Acharia fusca* and *Euprosterna elaeasa* [24]. The antifeedant properties of *A. indica* are well established, particularly for a range of lepidopteran pests [25–27].

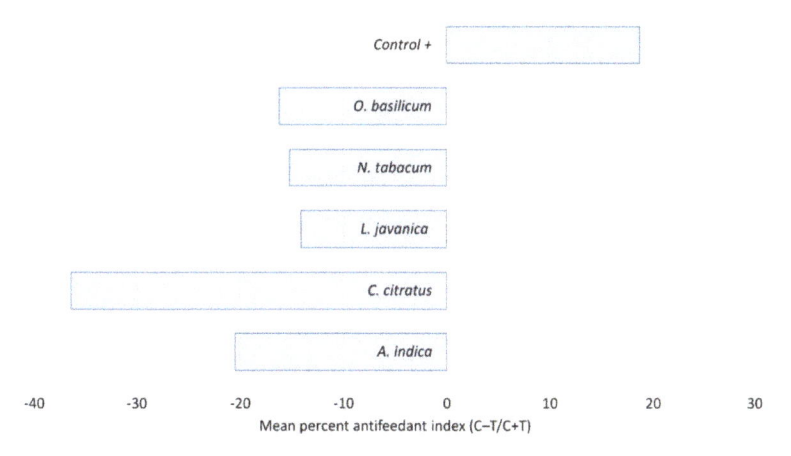

Figure 3. Antifeedant activity of five plant species extracts fed to fall armyworm 2nd instars, showing percent of feeding deterrence after 48 h. C = control; T = treated.

Trials evaluating the three most promising pesticidal plant species for their ability to control FAW larvae on living maize plants showed significant differences in effect among the treatments (Figure 4).

High foliar damage was observed in the negative controls (untreated, water and water plus 0.1% soap) with mean leaf damage scores of 6.5, 6.3 and 5.4, respectively. The lowest foliar damage score was observed in *N. tabacum* treatment (4.6); however, the slightly higher scores for *L. javanica* (5) and *O. basilicum* (5.2) were not significantly different from *N. tabacum*. The synthetic pesticide, chlorpyrifos, was the most effective in reducing FAW foliar damage with a mean score of 1.8.

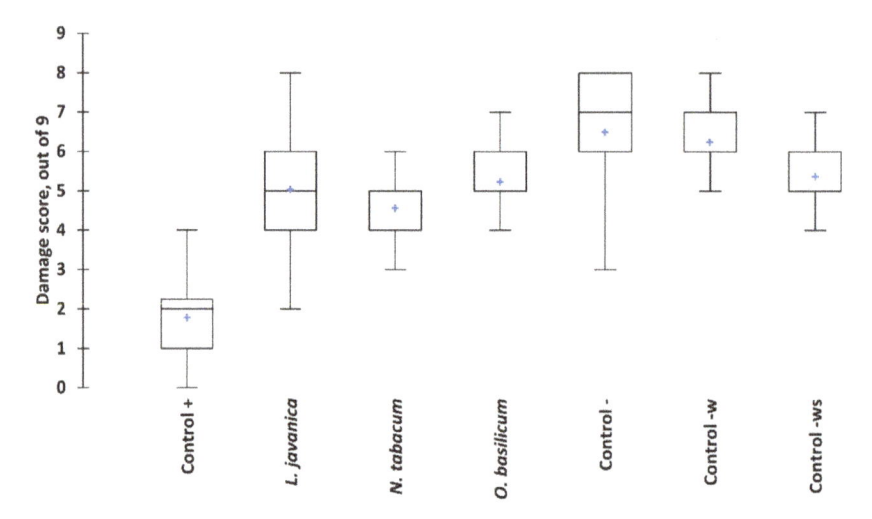

Figure 4. Fall armyworm damage to maize plants when exposed to different treatments over eight weeks. Boxes represent mean and 95% confidence intervals, tails are max. and min. values, blue crosses are median values. Significant differences are presented in Table S1.

The observed reduction in foliar damage may be attributed to a combination of toxicity, repellent and antifeedant effects of the plant extracts, with similar results observed from other studies [1]. The plant extracts did not reduce FAW damage as much as the synthetic pesticide chlorpyrifos, but most other studies on the use of pesticidal plants show similar lower mortality and damage rates when using natural pesticides in comparison to synthetic pesticides [13,15]. As most crops can compensate for some limited pest damage, further studies are required to determine whether these pesticidal plant treatments are able to maintain yield at comparable levels to synthetic pesticide use, which has been reported for a number of legume crops [14], cabbages [28,29] and sorghum [30].

3. Materials and Methods

3.1. Fall Armyworm Rearing

FAW larvae were collected from maize fields around Mitundu, Lilongwe District, Malawi (latitude 14°11' S longitude 33°46' E, elevation of 1100 metres above sea level (m a.s.l.). To establish a large colony the larvae were initially reared on portions of young maize leaves; however, once established, larvae were reared on an established artificial diet. The diet was composed of maize leaf powder, common bean powder, brewer's yeast, ascorbic acid, sorbic acid, methyl-*p*-hydroxybenzoate, vitamin E capsules, sucrose, formaldehyde and agar [31].

Neonates and 2nd instars were reared in 500 mL plastic containers containing young maize leaves, which were renewed daily. At the 2nd instar, the larvae were transferred to individual plastic containers (100 mL) to reduce cannibalism until pre-pupal stage and fed on an artificial diet which was changed every week until pupation. The pupae were harvested and transferred in Petri dishes lined with tissue paper and placed in adult emergence cages (mosquito netting around an 18 × 18 × 18 cm frame). After adult emergence, the moths were fed on honey from a honey-soaked wad of cotton wool in a container placed in each cage. Eggs laid on filter paper in the cages were removed daily and were disinfected by dipping them in 10% formaldehyde for 15 min. The eggs were then rinsed thoroughly with distilled water and dried on filter paper. Thereafter, eggs were placed in small containers until they hatched to repeat the rearing process [31].

3.2. Plant Material Collection and Extract Preparation

Initial plant screening was carried out with ten pesticidal plant species: *A. indica*, *O. basilicum*, *N. tabacum*, *C. citratus*, *T. vogelii*, *A. vera*, *L. camara*, *T. emetica*, *V. amygdalina* and *L. javanica*. Considerable phytochemical and efficacy knowledge was recorded by our group and others on all the material used of these species [10,15,17,18,20,24,32–37] which were sourced from the same locations with samples from four locations around Mitundu, Lilongwe District, Malawi combined to control for potential chemical variation across space [37,38]. The leaves of all plant species were collected from the wild from these known locations. Plant materials were shade dried, ground to a fine powder and kept in cool dark conditions until required. To produce 10% *w/v* extracts, 100 g of each plant powder were extracted in 1 L of water containing 0.1% soap for 24 h at room temperature. Thereafter the extracts were filtered and used immediately in bioassays. This trial included two control treatments, a positive control of chlorpyrifos and a negative control of water plus 0.1% soap.

Based on these bioassay results, five of the ten plant species were selected for further research: *A. indica*, *O. basilicum*, *N. tabacum*, *C. citratus* and *L. javanica*. In order to improve the extraction efficiency, the extracts were prepared by weighing 300 g of plant powder into 1.5 L of methanol for 24 h. Extracts were then filtered and placed in a fume hood for 24 h to allow the methanol to evaporate, leaving behind the extracted residue. Dried residue was weighed on an analytical scale and then resolubilized in acetone to make five concentrations of 10%, 5%, 3%, 1% and 0.1% *w/v* [39]. This trial included two control treatments, a positive control of chlorpyrifos and a negative control of acetone only.

The evaluation of FAW survival on living maize plants was carried out using the three best plant species: *O. basilicum*, *N. tabacum* and *L. javanica*. Aqueous extracts were prepared at 10% *w/v* using the extraction process described above in the first screening experiment. Four control treatments were also used in this study: chlorpyrifos as a positive control and three negative controls of no treatment application, water only and water with 0.1% soap.

3.3. Bioassay Methods

Contact toxicity was performed by means of topical application where 10 µL of the extract were applied topically on the bodies of the larvae using a 20 µL pipette. The larvae were then individually placed in plastic bottle tops containing plain artificial diet and covered with foil paper. Chlorpyrifos 48 EC was used as positive control using the manufacturer recommended rate of 20 mL of chlorpyrifos in 40 L of water while acetone or water was used as the negative control. Each replicate contained ten 2nd instars, with five replicates per each treatment and concentration evaluated. Final mortality data were recorded seven days afterwards by counting the number of dead larvae, with mortality data corrected using Abbott's formula [40].

The initial feeding bioassay screening ten plant species was carried out by dipping portions of young maize leaves into each extract, waiting one hour for the extract to dry and then placing five 2nd instars on the treated leaves to feed, three replicates per treatment. Treated maize leaves were replaced daily for seven days, with mortality data collected and corrected with Abbott's formula.

The subsequent feeding bioassay screening the shortlisted five plant species was carried out using previously reported methods [41]. Aliquots (100 µL) of 0.05 M sucrose in acetone were applied to individual glass-fibre discs (Whatman 2.1 cm diameter) and left to dry before aliquots (100 µL) of the plant extracts in acetone were applied to each disc. Once dry, the discs were weighed with an analytical balance, placed in individual containers and one 2nd instar was introduced with 10 larvae per treatment. Two control treatments were used: sucrose only and chlorpyrifos. After 48 h the remainder of the disc not eaten by the larvae was weighed and any living larvae were transferred to plain diet in individual containers. Mortality data were collected seven days from the start of the trial with mortality data corrected using Abbott's formula. A feeding deterrence index was calculated from the weights of control (C) and treated (T) discs using the following formula: Feeding deterrence = $(C - T)/(C + T) \times 100$ [42].

The final experiment evaluating FAW damage to living maize plants was carried out using maize variety sc403 planted in pots maintained in a screenhouse. Five seeds were planted per pot and were later thinned to three. Basal dressing fertiliser of 23:21:0 + 4 s was a pplied at seven days after planting at a rate of 100 kg/ha while a topdressing fertilizer of urea was a pplied at four weeks after planting at a rate of 159 kg/ha. All agronomic practices including watering and hand weeding were consistent across all maize plant pots. Maize plants were infested with 2nd instars at twenty days after seedling emergence. Each plant was infested with five larvae and the larvae were spaced at different leaf nodes to avoid cannibalism. Artificial infestation was done early in the morning to avoid exposing the larvae to harsh environments [31]. After infestation, plant pots were caged individually in cages of size 1.8 m × 0.6 m × 0.6 m to prevent the movement of larvae from one treatment to another. The experiment was laid out in a randomized complete block design replicated ten times where the cages acted as blocks. Hand-held plastic sprayer bottles were used to apply the treatments, ensuring consistent coverage of each plant. Treatments were first applied 48 h after infestation to allow the larvae to settle down and establish [43]. Subsequent applications were carried out at seven-day intervals. Foliar damage data were collected at an interval of seven days beginning from the first day after spraying. Using published methods [44], FAW foliar damage severity was recorded on an individual plant basis using a scale of 0–9 where 0 means no visible leaf damage, 1 = only pin-hole damage to the leaves, 2 = pin-hole and shot-hole damage to leaves, 3 = small elongated lesions (5–10 mm) on 1–3 leaves, 4 = midsized lesions (10–30 mm) on 4–7 leaves, 5 = large elongated lesions (>30 mm) or small portions eaten on 3–5 leaves, 6 = elongated lesions (>30 mm) and large portions eaten on 3–5 leaves, 7 = elongated lesions (>30 cm) and 50% of leaf eaten, 8 = elongated lesions (30 cm) and large portions eaten on 70% of leaves and 9 = most leaves have long lesions.

3.4. Data Analysis

Experiments were carried out following completely randomised block designs. Effects of treatments and their interactions observed were subjected to two-way analysis of variance. The means of treatments and interactions were compared using Tukey's honest significant difference (HSD) test at the 95% confidence interval. All the analyses were done using XLSTAT version 2019.2.2.59614 (Addinsoft, 2019); XLSTAT statistical and data analysis solution (Boston, MA, USA). Statistical outputs are provided in Table S1.

4. Conclusions

Recommendations from this research suggest that some relatively safe pesticidal plant species could provide an agro-ecologically sustainable pest management option for the exotic invasive FAW in Africa. Out of the original ten candidate plant species evaluated, four of these merit further investigation: *Azadirachta indica*, *Ocimum basilicum*, *Cymbopogon citratus* and *Lippia javanica*. These four plant species are cosmopolitan and frequently cultivated, so sustainable supplies for large-scale production would be feasible. Although *Tephrosia vogelii* did not show significant efficacy in our trials, further research should be recommended to confirm these results as *T. vogelii* is being recommended for fall armyworm control due to is known efficacy against a range of pests. Considerable knowledge on the chemistry of these plants is reported. Furthermore, *O. basilicum*, *C. citratus* and *L. javanica* are consumed as spices and teas, whilst *Azadirachta indica* has well-established low mammalian toxicity. Despite considerable work on its biopesticidal effects, *Nicotiana tabacum* is arguably the plant species with the highest vertebrate toxicity, well-known for the effects of nicotine and related alkaloids. However, despite potential human toxicity dangers, *N. tabacum* is being pursued as one of several potential botanical options for FAW control in several African countries, and thus merits further investigation regarding its safe use and non-target effects. Evidence from our work and previous research repeatedly shows that pesticidal plants do not cause mortality rates comparable to synthetic pesticides. However, the trade-off between lower mortality for lower environmental persistence needs to be seriously considered, particularly as there is growing evidence that less toxic natural pesticides can help facilitate natural pest regulation whilst not significantly sacrificing crop yield. The next step in evaluating the use of pesticidal plants for FAW control in Africa will require systematic trials under farmer field conditions that can assess their cost-benefits and impact on crop damage and yield in comparison to commercial synthetic pesticides.

Supplementary Materials: The following are available online at http://www.mdpi.com/2223-7747/9/1/112/s1, Table S1: Statistical analyses of fall armyworm bioassays evaluating pesticidal plants using two-way analysis of variance followed by Tukey's HSD test to separate the means. Means followed by the same letter in an experiment are not significantly different from each other.

Author Contributions: Conceptualization, Y.T., P.C.S. and S.R.B.; Methodology, K.P., Y.T., P.C.S. and S.R.B.; Software, K.P. and S.R.B.; Validation, K.P., Y.T., T.K., V.H.K., P.C.S. and S.R.B.; Formal analysis, K.P. and S.R.B.; Investigation, K.P., Y.T., T.K., V.H.K., P.C.S. and S.R.B.; Resources, K.P., Y.T., V.H.K., P.C.S. and S.R.B.; Data curation, K.P. and S.R.B.; Writing—Original draft preparation, K.P.; Writing—Review and editing, K.P., Y.T., T.K., V.H.K., P.C.S. and S.R.B.; Visualization, K.P. and S.R.B.; Supervision, Y.T., T.K., V.H.K., P.C.S. and S.R.B.; Project administration, Y.T. and S.R.B.; Funding acquisition, P.C.S. and S.R.B. All authors have read and agreed to the published version of the manuscript.

Funding: This research was funded by the McKnight Foundation, grant number 17-070.

Acknowledgments: The authors are grateful for the contributions of Aubrey Kaphukusi, Ipiana Kayira and Vincent Kalasanthenga for their technical laboratory and field assistance.

Conflicts of Interest: The authors declare no conflict of interest.

References

1. Sisay, B.; Tefera, T.; Wakgari, M.; Ayalew, G.; Mendesil, E. The Efficacy of Selected Synthetic Insecticides and Botanicals against Fall Armyworm, *Spodoptera frugiperda*, in Maize. *Insects* **2019**, *10*, 45. [CrossRef]
2. Goergen, G.; Kumar, P.L.; Sankung, S.B.; Togola, A.; Tamò, M. First Report of Outbreaks of the Fall Armyworm *Spodoptera frugiperda* (J E Smith) (Lepidoptera, Noctuidae), a New Alien Invasive Pest in West and Central Africa. *PLoS ONE* **2016**, *11*, e0165632. [CrossRef] [PubMed]
3. Rwomushana, I.; Bateman, M.; Beale, T.; Beseh, P.; Cameron, K.; Chiluba, M.; Clottey, V.; Davis, T.; Day, R.; Early, R.; et al. *Fall Armyworm: Impacts and Implications for Africa*; Evidence Note Update, October 2018; Centre for Agriculture and Bioscience International: Wallingford, UK, 2018.
4. Sisay, B.; Simiyu, J.; Mendesil, E.; Likhayo, P.; Ayalew, G.; Mohamed, S.; Subramanian, S.; Tefera, T. Fall Armyworm, *Spodoptera frugiperda* Infestations in East Africa: Assessment of Damage and Parasitism. *Insects* **2019**, *10*, 195. [CrossRef] [PubMed]
5. Day, R.; Abrahams, P.; Bateman, M.; Beale, T.; Clottey, V.; Cock, M.; Colmenarez, Y.; Corniani, N.; Early, R.; Godwin, J.; et al. Fall armyworm: Impacts and implications for Africa. *Outlooks Pest Manag.* **2017**, *28*, 196–201. [CrossRef]
6. Yu, S.J. Insecticide resistance in the fall armyworm, *Spodoptera frugiperda* (Smith, J.E.). *Pestic. Biochem. Physiol.* **1991**, *39*, 84–91. [CrossRef]
7. Belmain, S.R.; Stevenson, P.C. Ethnobotanicals in Ghana: Revising and modernising age-old farmer practice. *Pestic. Outlook* **2001**, *6*, 233–238.
8. Nyirenda, S.P.; Sileshi, G.W.; Belmain, S.R.; Kamanula, J.F.; Mvumi, M.; Sola, P.; Nyirenda, G.K.C.; Stevenson, P.C. Farmers' ethno-ecological knowledge of vegetable pests and pesticidal plant use in northern Malawi and eastern Zambia. *Afr. J. Agric. Res.* **2011**, *6*, 1525–1537.
9. Kamanula, J.; Sileshi, G.W.; Belmain, S.R.; Sola, P.; Mvumi, B.M.; Nyirenda, G.K.C.; Nyirenda, S.P.; Stevenson, P.C. Farmers' insect pest management practices and pesticidal plant use in the protection of stored maize and beans in Southern Africa. *Int. J. Pest Manag.* **2011**, *57*, 41–49. [CrossRef]
10. Isman, M.B. Botanical insecticides, deterrents, and repellents in modern agriculture and an increasingly regulated world. *Annu. Rev. Entomol.* **2006**, *51*, 45–66. [CrossRef]

11. Dougoud, J.; Toepfer, S.; Bateman, M.; Jenner, W.H. Efficacy of homemade botanical insecticides based on traditional knowledge. A review. *Agron. Sustain. Dev.* **2019**, *39*, 37. [CrossRef]
12. Isman, M.B. Botanical insecticides: For richer, for poorer. *Pest Manag. Sci.* **2008**, *64*, 8–11. [CrossRef] [PubMed]
13. Mkindi, A.; Mpumi, N.; Tembo, Y.; Stevenson, P.C.; Ndakidemi, P.A.; Mtei, K.; Machunda, R.; Belmain, S.R. Invasive weeds with pesticidal properties as potential new crops. *Ind. Crops Prod.* **2017**, *110*, 113–122. [CrossRef]
14. Tembo, Y.; Mkindi, A.G.; Mkenda, P.A.; Mpumi, N.; Mwanauta, R.; Stevenson, P.C.; Ndakidemi, P.A.; Belmain, S.R. Pesticidal Plant Extracts Improve Yield and Reduce Insect Pests on Legume Crops without Harming Beneficial Arthropods. *Front. Plant Sci.* **2018**, *9*, 1425. [CrossRef] [PubMed]
15. Mkenda, P.; Mwanauta, R.; Stevenson, P.C.; Ndakidemi, P.; Mtei, K.; Belmain, S.R. Extracts from field margin weeds provide economically viable and environmentally benign pest control compared to synthetic pesticides. *PLoS ONE* **2015**, *10*, e0143530. [CrossRef]
16. Belmain, S.R.; Amoah, B.A.; Nyirenda, S.P.; Kamanula, J.F.; Stevenson, P.C. Highly Variable Insect Control Efficacy of *Tephrosia vogelii* Chemotypes. *J. Agric. Food Chem.* **2012**, *60*, 10055–10063. [CrossRef]
17. Stevenson, P.C.; Kite, G.C.; Lewis, G.P.; Forest, F.; Nyirenda, S.P.; Belmain, S.R.; Sileshi, G.W.; Veitch, N.C. Distinct chemotypes of *Tephrosia vogelii* and implications for their use in pest control and soil enrichment. *Phytochemistry* **2012**, *78*, 135–146. [CrossRef]
18. Green, P.W.C.; Belmain, S.R.; Ndakidemi, P.A.; Farrell, I.W.; Stevenson, P.C. Insecticidal activity of *Tithonia diversifolia* and *Vernonia amygdalina*. *Ind. Crops Prod.* **2017**, *110*, 15–21. [CrossRef]
19. Mkenda, P.P.A.; Stevenson, P.C.P.; Ndakidemi, P.; Farman, D.I.; Belmain, S.R. Contact and fumigant toxicity of five pesticidal plants against *Callosobruchus maculatus* (Coleoptera: Chrysomelidae) in stored cowpea (*Vigna unguiculata*). *Int. J. Trop. Insect Sci.* **2015**, *35*, 1–13. [CrossRef]
20. Munyemana, F.; Alberto, A.L. Evaluation of larvicidal activity of selected plant extracts against *Plutella xylostella* (Lepidoptera: Plutellidae) larvae on cabbage. *Adv. Med. Plant Res.* **2017**, *5*, 11–20. [CrossRef]
21. Pavela, R.; Benelli, G. Ethnobotanical knowledge on botanical repellents employed in the African region against mosquito vectors—A review. *Exp. Parasitol.* **2016**, *167*, 103–108. [CrossRef]
22. Trdan, S.; Cirar, A.; Bergant, K.; Andjus, L.; Kač, M.; Vidrih, M.; Rozman, L. Effect of temperature on efficacy of three natural substances to Colorado potato beetle, *Leptinotarsa decemlineata* (Coleoptera: Chrysomelidae). *Acta Agric. Scand. Sect. B Soil Plant Sci.* **2007**, *57*, 293–296. [CrossRef]
23. Stevenson, P.C.; Belmain, S.R. Pesticidal plants in African agriculture: Local uses and global perspectives. *Outlooks Pest Manag.* **2016**, *27*, 226–230. [CrossRef]
24. Hernández-Lambraño, R.; Caballero-Gallardo, K.; Olivero-Verbel, J. Toxicity and antifeedant activity of essential oils from three aromatic plants grown in Colombia against *Euprosterna elaeasa* and *Acharia fusca* (Lepidoptera: Limacodidae). *Asian Pac. J. Trop. Biomed.* **2014**, *4*, 695–700. [CrossRef]
25. Martinez, S.S.; van Emden, H.F. Sublethal concentrations of azadirachtin affect food intake, conversion efficiency and feeding behaviour of *Spodoptera littoralis* (Lepidoptera: Noctuidae). *Bull. Entomol. Res.* **1999**, *89*, 65–71. [CrossRef]
26. Liang, G.-M.; Chen, W.; Liu, T.-X. Effects of three neem-based insecticides on diamondback moth (Lepidoptera: Plutellidae). *Crop Prot.* **2003**, *22*, 333–340. [CrossRef]
27. Roel, A.R.; Dourado, D.M.; Matias, R.; Porto, K.R.A.; Bednaski, A.V.; Costa, R.B. da The effect of sub-lethal doses of *Azadirachta indica* (Meliaceae) oil on the midgut of *Spodoptera frugiperda* (Lepidoptera, Noctuidae). *Rev. Bras. Entomol.* **2010**, *54*, 505–510. [CrossRef]
28. Amoabeng, B.W.; Gurr, G.M.; Gitau, C.W.; Nicol, H.I.; Munyakazi, L.; Stevenson, P.C. Tri-trophic insecticidal effects of African plants against cabbage pests. *PLoS ONE* **2013**, *8*, e78651. [CrossRef]
29. Amoabeng, B.W.; Gurr, G.M.; Gitau, C.W.; Stevenson, P.C. Cost:benefit analysis of botanical insecticide use in cabbage: Implications for smallholder farmers in developing countries. *Crop Prot.* **2014**, *57*, 71–76. [CrossRef]
30. Okrikata, E.; Bukar, M.; Ali, B. Economic Viability of Chilli Pepper and Neem Seed Kernel Powdered Formulations Vis-à-vis Sevin Dust (85%) in the Management of Lepidopterous Stemborers on Sorghum in North Eastern Nigeria. *J. Biol. Agric. Healthc.* **2016**, *6*, 99–103.
31. Prasanna, B.M.; Huesing, J.E.; Eddy, R.; Peschke, V.M. *Fall Armyworm in Africa: A guide for Integrated Pest Management*; International Maize and Wheat Improvement Center: Mexico City, Mexico, 2018.

32. Miranda, M.A.F.M.; Varela, R.M.; Torres, A.; Molinillo, J.M.G.; Gualtieri, S.C.J.; Macías, F.A. Phytotoxins from *Tithonia diversifolia*. *J. Nat. Prod.* **2015**, *78*, 1083–1092. [CrossRef]
33. Rabe, T.; Mullholland, D.; van Staden, J. Isolation and identification of antibacterial compounds from *Vernonia colorata* leaves. *J. Ethnopharmacol.* **2002**, *80*, 91–94. [CrossRef]
34. Koul, O.; Wahab, S. (Eds.) *Neem: Today and in the New Millennium*; Springer: Dordrecht, The Netherlands, 2004; ISBN 978-1-4020-1229-7.
35. Singh, P.; Jayaramaiah, R.H.; Sarate, P.; Thulasiram, H.V.; Kulkarni, M.J.; Giri, A.P. Insecticidal potential of defense metabolites from *Ocimum kilimandscharicum* against *Helicoverpa armigera*. *PLoS ONE* **2014**, *9*, e104377. [CrossRef] [PubMed]
36. Adeniyi, S.A.; Orjiekwe, C.L.; Ehiagbonare, J.E.; Arimah, B.D. Preliminary phytochemical analysis and insecticidal activity of ethanolic extracts of four tropical plants (*Vernonia amygdalina*, *Sida acuta*, *Ocimum gratissimum* and *Telfaria occidentalis*) against beans weevil (*Acanthscelides obtectus*). *Int. J. Phys. Sci.* **2010**, *5*, 753–762.
37. Kamanula, J.F.; Belmain, S.R.; Hall, D.R.; Farman, D.I.; Goyder, D.J.; Mvumi, B.M.; Masumbu, F.F.; Stevenson, P.C. Chemical variation and insecticidal activity of *Lippia javanica* (Burm. f.) Spreng essential oil against *Sitophilus zeamais* Motschulsky. *Ind. Crops Prod.* **2017**, *110*, 75–82. [CrossRef]
38. Mkindi, A.G.; Tembo, Y.; Ndakidemi, P.A.; Belmain, S.R.; Stevenson, P.C. Phytochemical Analysis of *Tephrosia vogelii* across East Africa Reveals Three Chemotypes that Influence Its Use as a Pesticidal Plant. *Plants* **2019**, *8*, 597. [CrossRef]
39. Dos Santos, A.C.V.; de Almeida, W.A.; Fernandes, C.C.; de Sousa, A.H. Extractos hidroalcohólicos de plantas propias de la Amazonía suroccidental como alternativa de control de la oruga militar tardía. *Idesia* **2016**, *34*, 63–67.
40. Abbott, W.S. A Method of Computing the Effectiveness of an Insecticide. *J. Econ. Entomol.* **1925**, *18*, 265–267. [CrossRef]
41. Green, P.W.C.; Veitch, N.C.; Stevenson, P.C.; Simmonds, M.S.J. Cardenolides from *Gomphocarpus sinaicus* and *Pergularia tomentosa* (Apocynaceae: Asclepiadoideae) deter the feeding of *Spodoptera littoralis*. *Arthropod. Plant. Interact.* **2011**, *5*, 219–225. [CrossRef]
42. Hummelbrunner, L.A.; Isman, M.B. Acute, sublethal, antifeedant, and synergistic effects of monoterpenoid essential oil compounds on the tobacco cutworm, *Spodoptera litura* (Lep., Noctuidae). *J. Agric. Food Chem.* **2001**, *49*, 715–720. [CrossRef]
43. Silva, M.S.; Broglio, S.M.F.; Trindade, R.C.P.; Ferrreira, E.S.; Gomes, I.B.; Micheletti, L.B. Toxicity and application of neem in fall armyworm. *Comun. Sci.* **2015**, *6*, 359–364. [CrossRef]
44. Williams, W.P.; Buckley, P.M.; Daves, C.A. Identifying resistance in corn to southwestern corn borer (lepidoptera: Crambidae), fall armyworm (Lepidoptera: Noctuidae), and corn earworm (lepidoptera; Noctuidae). *J. Agric. Urban Entomol.* **2006**, *23*, 87–95.

© 2020 by the authors. Licensee MDPI, Basel, Switzerland. This article is an open access article distributed under the terms and conditions of the Creative Commons Attribution (CC BY) license (http://creativecommons.org/licenses/by/4.0/).

Article

United Forces of Botanical Oils: Efficacy of Neem and Karanja Oil against Colorado Potato Beetle under Laboratory Conditions

Kateřina Kovaříková * and Roman Pavela

Group Secondary Metabolites in Crop Protection, Crop Research Institute, Dmovská 507, 161 06 Praha 6-Ruzyně, Czech Republic; pavela@vurv.cz
* Correspondence: kovarikova@vurv.cz

Received: 14 November 2019; Accepted: 12 December 2019; Published: 14 December 2019

Abstract: Neem and karanja oil are the most promising botanical insecticides in crop protection nowadays. Given that information about the insecticidal abilities of these oils is lacking, the aim was to explore the effects of neem and karanja oil binary mixtures. The insecticidal activity of NeemAzal T/S (Trifolio-M GmbH, Lahnau, Germany) (neem oil), Rock Effect (Agro CS a.s., Česká Skalice, Czech Republic) (karanja oil), and their binary mixes (at 1:1, 1:2, and 2:1 volume ratios) against the larvae of the Colorado potato beetle (CPB; *Leptinotarsa decemlineata*) was studied. In our bioassays, a synergistic effect of the mixtures, which was dose-dependent, was observed for the first time against this pest. The most effective blend was the 1:1 ratio. Its efficacy was more or less the same as, or even greater than, the neem oil alone. The LC_{50} of neem oil two days after application was (0.075 g·L^{-1}) and the LC_{50} of the mixture was (0.065 g·L^{-1}). The LC_{50} of karanja oil was (0.582 g·L^{-1}), which was much higher than the LC_{50} of neem oil. The LC_{90} of neem oil five days after application was (0.105 g·L^{-1}) and the LC_{90} of the mixture was (0.037 g·L^{-1}). The LC_{90} of karanja oil was (1.032 g·L^{-1}). The results demonstrate that it is possible to lower the doses of both oils and get improved efficacy against CPB larvae; nevertheless, further verification of the results in field conditions is necessary.

Keywords: synergism; neem; karanja; Colorado potato beetle; botanical insecticides

---

1. Introduction

The Colorado potato beetle (CPB), (*Leptinotarsa decemlineata* (Say, 1824), Coleoptera: Chrysomelidae) is one of the most important potato pests. This species is native to North America, from where it has gradually spread across Europe and Asia [1] alongside the expansion of potato cultivation. *L. decemlineata* is now considered to be the most important insect defoliator of potatoes. Through defoliation, the yield of tubers can be reduced by more than 50% [2]. Moreover, if the pest appears early and in intense numbers, it can destroy the entire production of a potato growing operation. The values of "intense numbers" vary from author to author and correspond with economic (action) thresholds, which were established for optimizing the use of insecticide applications and are unique to certain conditions. For example, Senanayake and Holliday [3] suggest 0.14 to 0.82 larvae per plant. Mailloux and Bostanian [4] stated that a measure based on the level of defoliation is better than one based on pest abundance. These levels were estimated, for example, by Zehnder et al. [5] at 20% leaf loss for young plants, 30% for plants from early bloom to late bloom, and 60% for plants from late bloom to harvest for fields in eastern Virginia. Nevertheless, the CPB larval stage is considered to be the most harmful stage to potato production [4], so we focused our research on the CPB larvae.

Various non-chemical control methods have been introduced since the 19th century to reduce the impact of CPB. Crop rotation, trap crops (eggplant), and other agrotechnical practices were recommended to farmers at that time [6]. In a study by Reed [6], it was concluded that the only two reliable methods were hand picking and the use of Paris Green (a toxic pigment). Hand picking, especially before mating, was considered very effective, but impractical on a larger scale. As such, the majority of alternative methods failed, and the most common method became the use of pesticides. Although pests have developed resistance to many active ingredients of insecticides, the use of chemicals remains the most widely used method against CPB to date (although biological controls are also often applied [7–9]).

According to the Arthropod Pesticide Resistance Database [10], CPB has shown resistance to 56 active ingredients of insecticides to date, among them spinosad [11] and *Bacillus thuringiensis* [12]. Mota-Sanchez et al. [11] also observed cross resistance to nine other neonicotinoids in an imidacloprid-resistant adult population of CPB. As of now, resistance has been recorded to most of the commonly used insecticides, such as pyrethroids, neonicotinoids, diamides, and organophosphates. As Yamamoto et al. [13] stated, it is possible to delay or prevent resistance development of pests via rotating insecticides with different modes of action and using certain combinations of insecticides. Barnes et al. [14] proved that a strategy using mixtures of insecticides is even more effective than the rotation of insecticides, and this is the very essence of botanical insecticides (BIs), which are complex mixtures of many functional secondary metabolites [15,16], in contrast to chemical substances, which are often characterized by one active ingredient, complemented by a number of inactive ingredients to facilitate the application. Moreover, the excessive use of chemical pesticides harms the environment, non-target organisms, and humans. By contrast, botanical pesticides are biodegradable and leave no harmful residues. Unfortunately, the scale of products applicable in organic production is insufficient as well. It is necessary to search for additional environmentally acceptable substances for effective control of CPB, and the use of botanical pesticides is a promising possibility. Chaudhary et al. [17] stated that the use of botanical pesticides is the most efficient means to replace synthetic pesticides, and among those, extracts and oils are the best choices. Neem and karanja oil, in particular, have great potential for use in sustainable integrated pest management.

Neem oil is one of the most promising substances in the current approach to pest control. Neem oil is a product of the Indian neem tree *Azadirachta indica* (A.) Juss. It possesses a variety of insecticidal properties, such as repellency [18], antifeedancy, toxicity, and growth disruption, against numerous pest species [19]. For example, the biochemical effect (growth inhibition, feeding deterrence, oviposition inhibition) of neem oil against more than 30 Lepidopteran pests [20] has been well documented. The main active ingredient is azadirachtin, a tetranortriterpenoid, which was isolated from the seeds of *Azadirachta indica* by Butterworth and Morgan [21] and is known to disrupt insects' metamorphosis [22]. Neem oil is a contact insecticide, but even systemic activity has been documented [23]. Pavela et al. [24] also proved that azadirachtin can be taken up by plant roots and thus affect the population of immature aphids feeding on the treated plant.

Karanja oil is a product of the seeds of a widespread tropical and subtropical tree called *Pongamia pinnata* (L.) Pierre [25]. Karanja oil is rich in furano-flavonoids [26]. Al Muqarrabun et al. [27] summarized the known attributes of up to 70 flavones and their derivatives that had been isolated from *Pongamia pinnata* by various authors. Of these, karanjin, which was first discovered by Limaye [28], is particularly effective against a large number of insects [29]. The oil and extract of karanja act as insecticides [30], repellents, antifeedants, and growth regulators [31], and even oviposition deterrents [32].

Some insecticides, in combinations, may exhibit greater-than-additive toxicity, but the prediction of mixture toxicity using a response addition model is not always accurate for active ingredients with different modes of action [33]. In the case of neem and karanja oils, a synergistic effect was found by Kumar et al. [34] in bioassays on aphids and mites. Later, this phenomenon was confirmed by Packiam and Ignacimuthu [35] during experiments on *Spodoptera litura*. However, although the synergistic effect of insecticides is of great importance in practice, there is still much to be explored in this research area; for example, application doses may be reduced (economic and environmental benefits) and, moreover, multi-substance mixtures prevent the development of resistant populations of pests (anti-resistant strategy).

Although the efficacy of neem extracts on the Colorado potato beetle has already been studied, the efficacy of karanja oil against this pest has not yet been reported. Similarly, a possible synergistic effect of binary mixtures of these oils has also not been studied sufficiently to date. Thus the aim of our research was to evaluate the insecticidal properties of two commercial products (NeemAzal T/S - Trifolio-M GmbH, Lahnau, Germany, Rock Effect - Agro CS a.s., Česká Skalice, Czech Republic)), based on *Azadirachta indica* seed kernel oil and *Pongamia pinnata* oil, respectively, against the most destructive stage of CPB and to verify any possible synergistic effects of their binary mixtures in three different ratios with regard to practical use.

2. Results

The insecticidal activity of both tested BIs against CPB larvae was very good; however, significant differences in efficacies were found. From the LC_{50} (LC_{90}) of Rock Effect (Agro CS a.s., Česká Skalice, Czech Republic) (0.582 g·L^{-1}) and NeemAzal T/S (Trifolio-M GmbH, Lahnau, Germany) (0.075 g·L^{-1}), it is obvious that karanja oil by itself is significantly less effective against larvae of *Leptinotarsa decemlineata* in comparison with neem oil. Complete results are presented in Table 1.

A comparison of the LC values revealed that both BIs showed significant chronic toxicity, which resulted in an apparent decrease in lethal concentrations over time (Table 1). The LC_{50} (LC_{90}) values for NeemAzal T/S (Trifolio-M GmbH, Lahnau, Germany) were estimated at 0.075 g·L^{-1} (0.618 g·L^{-1}) after 48 h and 0.005 g·L^{-1} (0.029 g·L^{-1}) after 8 days, which is an obviously lower range of concentrations than the estimate for Rock Effect (Agro CS a.s., Česká Skalice, Czech Republic), where the LC_{50} (LC_{90}) was 0.582 g·L^{-1} (1.692 g·L^{-1}) after 48 h and 0.259 g·L^{-1} (0.774 g·L^{-1}) after 8 days.

The insecticidal activity of the mixtures was assessed by the activity of the pure BIs. The LC range of the mixtures was usually within the LC range of NeemAzal T/S (Trifolio-M GmbH, Lahnau, Germany) and Rock Effect (Agro CS a.s., Česká Skalice, Czech Republic). As such, all the mixtures were more efficient than Rock Effect (Agro CS a.s., Česká Skalice, Czech Republic) alone. An exception was found in the case of mixtures after 48 h (acute toxicity), when the LC_{90} was higher for the mixtures than for the original substances. The LC values of the mixtures decreased over time, as did the LC values of the pure BIs. The LC_{50} (LC_{90}) of the mixtures (1:1, 1:2, and 2:1 respectively) were estimated as follows: 0.065–0.001 g·L^{-1} (1.651–0.011 g·L^{-1}), 0.543–0.012 g·L^{-1} (6.553–0.028 g·L^{-1}), and 0.387–0.015 g·L^{-1} (6.361–0.059 g·L^{-1}) 48 h to 8 days after treatment.

The order of the tested BIs according to efficacy (LC) is summarized as follows: mixture (1:1), NeemAzal T/S (Trifolio-M GmbH, Lahnau, Germany), mixture (1:2), mixture (2:1), and Rock Effect (Agro CS a.s., Česká Skalice, Czech Republic). The improved efficacy of the mixtures indicates an obvious synergistic effect.

Table 1. Toxicity of botanical insecticides (Bis) and their mixtures against *Leptinotarsa decemlineata* larvae. LC_{50} and LC_{90}: concentration causing 50% and 90% mortality of insects, respectively. CI_{95}: 95% confidence intervals, insecticide activity is considered significantly different when the 95% CI fails to overlap. Chi = Chi-square value, significant at $p < 0.05$ level.

Insecticides	Days after Treatment	LC_{50}	CI^{95}	LC^{90}	CI^{95}	Chi-Square	p-Value
Neem Azal T/S	2	0.075 ± 0.011	0.066–0.101	0.618 ± 0.215	0.372–1.323	0.988	0.804
	5	0.021 ± 0.002	0.011–0.021	0.105 ± 0.011	0.082–0.156	2.031	0.565
	8	0.005 ± 0.001	0.003–0.008	0.029 ± 0.004	0.022–0.041	0.007	0.999
Rock Effect	2	0.582 ± 0.093	0.344–0.858	1.692 ± 0.428	1.112–4.319	2.316	0.677
	5	0.371 ± 0.071	0.190–0.553	1.032 ± 0.263	0.672–2.813	5.065	0.281
	8	0.259 ± 0.054	0.119–0.379	0.774 ± 0.176	0.523–1.165	1.821	0.768
Rock Effect and NeemAzal T/S 1:2	2	0.543 ± 0.151	0.367–0.611	6.553 ± 2.441	5.369–7.921	6.591	0.252
	5	0.021 ± 0.005	0.015–0.031	0.151 ± 0.027	0.094–0.211	5.398	0.369
	8	0.012 ± 0.001	0.012–0.017	0.028 ± 0.002	0.024–0.034	0.677	0.954
Rock Effect and NeemAzal T/S 1:1	2	0.065 ± 0.016	0.042–0.113	1.651 ± 0.325	1.354–1.923	2.088	0.552
	5	0.008 ± 0.015	0.001–0.060	0.037 ± 0.023	0.016–0.065	1.507	0.681
	8	0.001 ± 0.001	0.002–0.026	0.011 ± 0.001	0.007–0.016	0.052	0.996
Rock Effect and NeemAzal T/S 2:1	2	0.387 ± 0.052	0.302–0.512	6.361 ± 2.585	3.391–7.182	5.126	0.401
	5	0.053 ± 0.002	0.012–0.091	0.267 ± 0.085	0.168–0.378	3.771	0.582
	8	0.015 ± 0.002	0.011–0.019	0.059 ± 0.007	0.047–0.077	0.336	0.996

3. Discussion

This experiment involved testing the insecticidal activity of NeemAzal T/S (Trifolio-M GmbH, Lahnau, Germany) (neem oil), Rock Effect (Agro CS a.s., Česká Skalice, Czech Republic) (karanja oil), and three binary mixtures. In the course of the experiment, it was observed that the larvae treated with the oils also showed a reduction in food intake, which resulted in little leaf damage compared to leaves in the control variant. This effect was not a primary target of the testing; nevertheless, antifeedancy has previously been reported for karanja oil [36] as well as neem oil [37].

The efficacy of neem oil was very good even at low concentrations such as 0.1–2 g·L^{-1} in our bioassays. The efficacy of neem products may vary with respect to concentrations of azadirachtin [38], and thus some authors have stated lower insecticidal activity against CPB [39,40]. However, that was not the case in our study, because the commercial product NeemAzal T/S (Trifolio-M GmbH, Lahnau, Germany) used in the bioassay contains 1% of the purified active ingredient Azadirachtin A, which is a very potent insect growth inhibitor [41] (reviewed in [19,42]). In the case of karanja oil, higher doses were needed to achieve the same mortality of CPB larvae as with neem oil. According to the literature, karanja oil is generally effective at higher doses, from 10 g·L^{-1} upwards [43–45]. Deshmukh and Borle [46] also mentioned that karanja oil has some limits for use at the farmer's level, because its aqueous suspension is not as effective as that of neem. The higher efficacy of neem oil over karanja oil was also reported by Biswas et al. [47] in the case of *Helicoverpa armigera*.

Zehnder and Warthen [37] found that azadirachtin caused mortality, and they also observed an antifeedant effect in the case of CPB. Feuerhake and Schmutterer [48] tested neem seed extract formulations and were able to achieve 100% mortality of CPB. Other observed effects of neem oil on CPB are disruption of egg hatching and larvae molting [49]. The repellent effects of neem oil were even observed in laboratory as well as in field conditions [50]. Kaethner [51] observed the effect of fitness reduction in the treated beetles and a lower fecundity of females. Murray et al. [52] confirmed the oviposition effects of citrus limonoids, which are structurally and functionally similar to azadirachtin. The effect of karanja oil on CPB has not been studied to date.

With respect to other insect pests, synergistic effects were reported for both neem and karanja oils. Synergism refers to when the combined effect of two factors is greater than the sum of individual factors. It can occur between various insecticides [14], insecticides and fungicides [53], insecticides and poor nutritional conditions [54], and so on. For example, a synergism was found for the combination of neem oil and *Beauveria bassiana* on *Spodoptera litura* Fabricius [55], *Tribolium castaneum* [56], and aphids [57]. The combination of both oils with some pyrethrins shows synergism as well [36,58]. Synergism was also observed in the combination of neem and karanja oil by Kumar et al. [34], where both oils were found to be highly effective individually and also in combination against mites and aphids. In our case, the most significant synergistic effect was found for the mixture of neem and karanja oil in a 1:1 ratio. In addition, this effect appeared chronically and not acutely, and thus it is possible to use lower doses of insecticides. A notable observation in this experiment was that the individual neem and karanja oil treatments were less effective compared to the mixtures. For example, the toxicity (LC$_{90}$) five days after treatment with the 1:1 mixture was almost 3-fold higher than for neem oil alone, and up to 28-fold higher than for karanja. The other mixtures showed synergism as well, although the effect was significantly weaker. Packiam and Ignacimuthu [35] came to the same conclusions when testing various combinations of neem and karanja oils against larvae of *Spodoptera litura*. Thus, it is clear that the strength of the synergistic effect is ratio dependent, and this is the first research paper to describe such a phenomenon in the case of CPB.

Neem and karanja oil seem to provide the perfect solution to all the problems of contemporary crop protection. Nevertheless, the possible cons should be discussed as well. Because it possesses a variety of insecticidal properties, even against some beneficial insects in the juvenile stage [59], the use of neem oil carries a potential risk. Any decision to use neem oil should therefore be made carefully. The same situation occurs in the case of karanja oil, which is considered a broad-spectrum insecticide [36]. On the other hand, Koss et al. [60] and Radkova et al. [61] found that beneficial arthropods in potato fields were more abundant when botanical and selective pesticides were sprayed in contrast to when chemical pesticides were used. Neem oil is considered environmentally friendly because it is free of chloramine, phosphorous, and nitrogen atoms, which are commonly found in synthetic pesticides. Moreover, using neem oil in the field can prevent some other pests from ovipositing [62], thus providing improved crop protection, and due to the synergistic effect of the compounds, a reduced dose is used, significantly reducing toxicity to non-target organisms while also reducing the cost of applications. Generally, neem products can be recommended for many integrated pest management (IPM) programs [63]. The effect of an insecticide based on a combination of neem and karanja oils (PONNEEM#) has already been tested against the hymenopteran parasitoid wasp *Trichogramma chilonis*, which was not affected from applications with up to a 0.5% concentration [35]. The recommended dose for NeemAzal T/S (Trifolio-M GmbH, Lahnau, Germany) in potatoes is 2.5 L for 300–700 L of water/ha (0.8–0.35% concentration). Because the 1:1 mixture was almost 3 times more effective, we should theoretically recommend the use of a 0.3–0.1% concentration for field application, and thus parasitoids should not be affected.

Field efficacy, however, will still need to be verified by a series of tests before its implementation, because some authors [64] have demonstrated the low stability of neem oil under field conditions due to photodegradation. This issue can be resolved theoretically via nano and micro encapsulation, which improves the stability and efficacy of oils exposed to UV light [65]; on the other hand, this solution is more suitable for soil applications of neem seed oil [66]. Moreover, the cost of such a product will be affected as well. The greater variability of karanja oil efficacy against CPB larvae reveals that this compound is less reliable than neem oil, especially when acute toxicity is concerned. However, Kumar and Singh [36] stated that the persistence of karanja oil is greater than for other botanical insecticides. In addition, the karanja extract is a highly effective UV absorbent [67] and is now being used in cosmetics as a component of modern sunscreen preparations, even for humans. As Wanyika et al. [68] have indicated, stabilization can be developed by adding solar radiation protectants, and therefore the combination of neem and karanja oils seems to be a perfect match, because what is missing in one is replaced by the other.

4. Materials and Methods

4.1. Insecticides

Two botanical insecticides that are suitable for organic farming were used in the bioassays. The first one (NeemAzal T/S) (Trifolio-M GmbH, Lahnau, Germany) was tested at five different doses and the second one (Rock Effect) (Agro CS a.s., Česká Skalice, Czech Republic) was tested at six different doses. Both substances were tested as contact insecticides.

NeemAzal® T/S (Trifolio-M GmbH, Lahnau, Germany) is a commercial formulation of the seed kernel extract of the tree *Azadirachta indica* (A.) Juss., which contains 1% of the purified active ingredient Azadirachtin A. It is generally used as a 0.3% to 0.5% aqueous solution. In the following text, the term "neem oil" is used.

Rock Effect (Agro CS a.s., Česká Skalice, Czech Republic) is a commercial formulation of *Pongamia pinnata* (L.) Pierre oil. The oil content is declared as 868.5 g·L^{-1}. It is generally used as a 1% to 3% aqueous solution. In the following text, the term "karanja oil" is used.

4.2. Insects and Plant Material

Newly enclosed second instar larvae of *Leptinotarsa decemlineata* (Say) (Coleoptera: Chrysomelidae) were obtained from a colony reared on potato, *Solanum tuberosum*, cv. Agria, in a climate chamber at 22 ± 1 °C, a relative humidity of 40–60%, and a 16:8 (L:D) photoperiod. The colony was established from eggs and adults collected from a potato culture at a local field of the Crop Research Institute in Prague, Czech Republic (50°05′N, 14°18′E, 340 m a.s.l.) in June and August 2019.

4.3. Laboratory Bioassay

A potato leaf with five to seven leaflets was inserted into floral foam supporting moisture (Oasis, Belgium) and placed in a plastic box (16 × 11 × 7 cm). Twenty to twenty-five individuals (L2) of CPB were transferred carefully using a fine brush on each potato leaf. The tested BIs were dissolved in water right before use. NeemAzal T/S (Trifolio-M GmbH, Lahnau, Germany), Rock Effect (Agro CS a.s., Česká Skalice, Czech Republic), and their binary mixtures in three ratios (1:1, 2:1, and 1:2, by volume) were tested. For each BI, at least five concentrations that resulted in more than 0% and less than 100% mortality, based on preliminary assays, were used. The concentration series were 0.1–2 $g \cdot L^{-1}$ for NeemAzal T/S (Trifolio-M GmbH, Lahnau, Germany), 2–25 $g \cdot L^{-1}$ for Rock Effect (Agro CS a.s., Česká Skalice, Czech Republic), 0.05–2 $g \cdot L^{-1}$ for the 1:1 mixture, 0.1–15 $g \cdot L^{-1}$ for the 2:1 Rock Effect:Neem Azal (RE:NA) mixture, and 0.05–5 $g \cdot L^{-1}$ for the 1:2 (RE:NA) mixture. To obtain the LC value, the insecticidal activity of the concentration series of each insecticide was tested under laboratory conditions.

Once the larvae began to feed on the leaves, the prepared solutions were sprayed on both sides of the leaves using a hand sprayer. The control variant was sprayed with water only. Each variant was performed in three repetitions. The experiment was run in a climate chamber set at 22 ± 1 °C and 69% ± 6% relative humidity. The mortality rate was evaluated at 48 h, 5 days, and 8 days later. Individuals unable to right themselves when disturbed were counted as dead.

4.4. Statistical Analysis

Abbott's formula was used to correct the data for control mortality [69]. Probit analysis [70] was used to estimate the doses needed to cause 50% mortality (LC_{50}) and 90% mortality (LC_{90}), as well as the associated 95% confidence limits for each BI, including binary mixtures, using BioStat (version 5.9.8.5).

5. Conclusions

Because of their natural origin and environmental friendliness, botanical pesticides currently have great potential. Furthermore, botanical pesticides are also very potent insecticides and, due to their composition, they can help to fight the global problem of insects developing resistance to insecticides.

Both of the tested BIs (insecticides based on neem oil and karanja oil) were efficient against *L. decemlineata* larvae at different concentrations. However, the efficacy of karanja oil against CPB larvae showed higher variability than for neem oil, on account of the high azadirachtin content of the commercial neem product and also the slightly different mode of action of both oils. Our experiment demonstrated a synergistic effect of neem and karanja oils against CPB larvae under laboratory conditions—one that was ratio-dependent. The most potent mixture (1:1 ratio) was equally, or even more effective, than neem oil itself. This effect intensified with respect to exposure time and appeared chronically. Five days after application of the mixture, LC_{90} values were 3-fold higher for neem oil and up to 28-fold higher for karanja oil. Due to these results, the recommended field dose of the mixture was estimated as high as 1.4 L in 500 L of water/ha (0.3%). Nevertheless, further verification in field conditions of the results achieved by this experiment is necessary in order to reach reliable conclusions and implement them into practice, and the effect on beneficial and non-target organisms should be verified.

Author Contributions: Conceptualization: R.P. and K.K.; methodology: R.P. and K.K.; software: R.P.; investigation: K.K.; data curation: R.P.; writing—original draft preparation: K.K.; writing—review and editing: K.K.; visualization: K.K.; supervision: R.P.; project administration: K.K.; funding acquisition: R.P.

Funding: This research was funded by the Ministerstvo Zemědělství - Národní Agentura pro Zemědělský Výzkum, grant number QK1920214.

Acknowledgments: The authors would like to thank the team members for providing the necessary background for running the experiments.

Conflicts of Interest: The authors declare no conflict of interest.

References

1. Weber, D. Colorado beetle: Pest on the move. *Pestic. Outlook* **2003**, *14*, 256–259. [CrossRef]
2. Ozturk, G.; Yildrim, Z. Effect of bio activators on the tuber yield and tuber size of potatoes. *Turk. J. Field Crops* **2013**, *18*, 82–86.
3. Senanayake, D.G.; Holliday, N.J. Economic injury levels for Colorado potato beetle (Coleoptera: Chrysomelidae) on 'Norland' potatoes in Manitoba. *J. Econ. Entomol.* **1989**, *83*, 2058–2064. [CrossRef]
4. Mailloux, G.; Bostanian, N.J. Effect of manual defoliation on potato yield at maximum abundance of different stages of Colorado potato beetle, *Leptinotarsa decemlineata* (Say), in the field. *J. Agric. Entomol.* **1989**, *6*, 217–226.
5. Zehnder, G.; Encill, A.M.; Speese, J. Action thresholds based on plant defoliation for management of Colorado potato beetle (Coleoptera: Chrysomelidae) in potato. *J. Econ. Entomol.* **1995**, *88*, 155–161. [CrossRef]
6. Reed, E.B. Insects injurious to the potato. In *Annual Report—Entomological Society of Ontario*; Bethune, C.J.S., Saunders, W., Reed, E.B., Eds.; The Society: Toronto, ON, Canada, 1872; Volumes 2–8, pp. 65–79.
7. Trdan, S.; Vidrih, M.; Laznik, Ž. Activity of four entomopathogenic nematode species against different developmental stages of Colorado potato beetle, *Leptinotarsa decemlineata* (Coleoptera, Chrysomelidae). *Helminthologia* **2009**, *46*, 14–20. [CrossRef]
8. Laznik, Ž.; Tóth, T.; Lakatos, T.; Vidrih, M.; Trdan, S. Control of the Colorado potato beetle (*Leptinotarsa decemlineata* [Say]) on potato under field conditions: A comparison of the efficacy of foliar application of two strains of *Steinernema feltiae* (Filipjev) and spraying with thiamethoxam. *J. Plant Dis. Protect.* **2010**, *117*, 129–135. [CrossRef]
9. Weber, D.C. Biological control of potato insect pests. In *Insect Pests of Potato*; Giordanengo, P., Vincent, C., Alyokhin, A., Eds.; Elsevier: Waltham, MA, USA, 2013; Volume 1, pp. 399–405.
10. Arthropod Pesticide Resistance Database. Available online: http://www.pesticideresistance.org (accessed on 18 September 2019).
11. Mota-Sanchez, D.; Hollingworth, R.M.; Grafius, E.J.; Moyer, D.D. Resistance and cross-resistance to neonicotinoid insecticides and spinosad in the Colorado potato beetle, *Leptinotarsa decemlineata* (Say) (*Coleoptera*: Chrysomelidae). *Pest Manag. Sci.* **2006**, *62*, 30–37. [CrossRef]
12. Whalon, M.E.; Miller, D.L.; Hollingworth, R.M.; Grafius, E.J.; Miller, J.R. Selection of a Colorado potato beetle (*Coleoptera*: Chrysomelidae) strain resistant to *Bacillus thuringiensis*. *J. Econ. Entomol.* **1993**, *86*, 226–233. [CrossRef]
13. Yamamoto, I.; Kyomura, N.; Takahashi, Y. Negatively correlated cross resistance: Combinations of N-methylcarbamate with N-propylcarbamate or oxadiazolone for green rice leafhopper. *Arch. Ins. Biochem. Phys.* **1993**, *22*, 277–288. [CrossRef]
14. Barnes, E.H.; Dobson, R.J.; Barger, I.A. Worm control and anthelmintic resistance: Adventures with a model. *Parasitol. Today* **1995**, *11*, 56–63. [CrossRef]
15. Lahlou, M. Methods to study the photochemistry and bioactivity of essential oils. *Phytother. Res.* **2004**, *18*, 435–448. [CrossRef] [PubMed]
16. Miresmailli, S.; Bradbury, R.; Isman, M.B. Comparative toxicity of *Rosmarinus officinalis* L. essential oil and blends of its major constituents against *Tetranychus urticae* Koch (*Acari*: Tetranychidae) on two different host plants. *Pest Manag. Sci.* **2006**, *62*, 366–371. [CrossRef] [PubMed]
17. Chaudhary, S.; Kanwar, R.K.; Sehgal, A.; Cahill, D.M.; Barrow, C.J.; Sehgal, R.; Kanwar, J.R. Progress on *Azadirachta indica* based biopesticides in replacing synthetic toxic pesticides. *Front. Plant. Sci.* **2017**, *8*, 610. [CrossRef]
18. Bina, S.; Javadi, I.; Iravani, O. Evaluation of the repellency effect of neem (*Melia azedarach*) plant extracts based on the Mittler&Dadd method. *J. Agric. Chem. Environ.* **2017**, *6*, 165–174. [CrossRef]
19. Saxena, R.C. Insecticides from neem. In *Insecticides of Plant Origin*; Arnason, J.T., Philogène, B.J.R., Morand, P., Eds.; American Chemical Society: Washington, DC, USA, 1989; Volume 387, pp. 110–135. [CrossRef]
20. Senthil-Nathan, S. Physiological and biochemical effect of neem and other Meliaceae plants secondary metabolites against Lepidopteran insects. *Front. Physiol.* **2013**, *4*, 1–17. [CrossRef]
21. Butterworth, J.H.; Morgan, E.D. Isolation of a substance that suppresses feeding in locusts. *Chem. Commun.* **1968**, *1*, 23–24. [CrossRef]

22. Tomlin, C. *The Pesticide Manual: A World Compendium: Incorporating the Agrochemicals Handbook*, 11th ed.; British Crop Protection Council: Farnham, Surrey; The Royal Society of Chemistry: Cambridge, UK, 1997.
23. Osman, M.Z.; Port, G.R. Systemic action of neem seed substances against *Pieris brassicae*. *Entomol. Exp. Appl.* **1990**, *54*, 297–300. [CrossRef]
24. Pavela, R.; Barnet, M.; Kocourek, F. Effect of azadirachtin applied systemically through roots of plants on the mortality, development and fecundity of the cabbage aphid (*Brevicoryne brassicae*). *Phytoparasitica* **2004**, *32*, 286–294. [CrossRef]
25. Belide, S.; Sajjalaguddam, R.R.; Paladugu, A. Cytokinin preconditioning enhances multiple shoot regeneration in *Pongamia pinnata* (L.) Pierre—A potential, non-edible tree seed oil source for biodiesel. *Electron. J. Biotechn.* **2010**, *13*, 1–8. [CrossRef]
26. Bringi, N.V.; Mukerjee, S.K. Karanja seed (*Pongamia glabra*) oil. In *Non-Traditional Oil Seeds and Oils in India*; Bringi, N.V., Ed.; Oxford IBH Publishing Co.: New Delhi, India, 1987; pp. 143–166.
27. Al Muqarrabun, L.M.R.; Ahmat, N.; Ruzaina, S.A.S.; Ismail, N.H.; Sahidin, I. Medicinal uses, phytochemistry and pharmacology of *Pongamia pinnata* (L.) Pierre: A review. *J. Ethnopharmacol.* **2013**, *150*, 395–420. [CrossRef] [PubMed]
28. Limaye, D.B. Karanjin part I: A crystalline constituent of the oil from *Pongamia glabra*. *Proc. 12th Indian Acad. Sci. Congr.* **1925**, 118–125.
29. Mathur, Y.K.; Srivastava, J.P.; Nigam, S.K.; Banerji, R. Juvenomimetic effects of karanjin on the larval development of flesh fly *Sarcophaga ruficornis* (Cyclorrhapha: Diptera). *J. Ent. Res.* **1990**, *14*, 44–51.
30. Parmar, B.S.; Gulati, K.C. Synergists for pyrethrins (II)-karanjin. *Indian J. Entomol.* **1969**, *31*, 239–243.
31. Kumar, V.; Chandrashekar, K.; Sidhu, O.P. Efficacy of karanjin and different extracts of *Pongamia pinnata* against selected insect pests. *J. Ent. Res.* **2006**, *31*, 121–124.
32. Pavela, R.; Herda, G. Effect of pongam oil on adults of the greenhouse whitefly *Trialeurodes vaporariorum* (Homoptera: Trialeurodidae). *Entomol. Gener.* **2007**, *30*, 193–201. [CrossRef]
33. Pape-Lindstrom, P.A.; Lydy, M.J. Synergistic toxicity of atrazine and organophosphate insecticides contravenes the response addition mixture model. *Environ. Toxicol. Chem.* **1997**, *16*, 2415–2420. [CrossRef]
34. Kumar, V.; Chandrashekar, K.; Sidhu, O.P. Synergistic action of neem and karanj to aphids and mites. *J. Ent. Res.* **2007**, *31*, 121–124.
35. Packiam, S.M.; Ignacimuthu, S. Effect of PONNEEM[#] on *Spodoptera litura* (Fab.) and its compatibility with *Trichogramma chilonis* Ishii. *Braz. Arch. Biol. Technol.* **2012**, *55*, 291–298.
36. Kumar, M.; Singh, R. Potential of *Pongamia glabra* Vent as an insecticide of plant origin. *Biol. Agric. Hortic.* **2002**, *20*, 29–50. [CrossRef]
37. Zehnder, G.; Warthen, J.D. Feeding inhibition and mortality effects of neem-seed extract on the Colorado potato beetle (Coleoptera, Chrysomelidae). *J. Econ. Entomol.* **1988**, *81*, 1040–1044. [CrossRef]
38. Isman, M.B.; Koul, O.; Luczynski, A.; Kaminski, J. Insecticidal and antifeedant bioactivities of neem oils and their relationship to azadirachtin content. *J. Agr. Food Chem.* **1990**, *38*, 1406–1411. [CrossRef]
39. Trdan, S.; Cirar, A.; Bergant, K.; Andjus, L.; Kač, M.; Vidrih, M.; Rozman, L. Effect of temperature on efficacy of three natural substances to Colorado potato beetle, *Leptinotarsa decemlineata* (Coleoptera: Chrysomelidae). *Acta Agr. Scand.* **2007**, *57*, 293–296. [CrossRef]
40. Ropek, D.; Kołodziejczyk, M. Efficacy of selected insecticides and natural preparations against *Leptinotarsa decemlineata*. *Potato Res.* **2019**, *62*, 85–95. [CrossRef]
41. Ruscoe, C.N.E. Growth disruption effects of an insect antifeedant. *Nat. N. Biol.* **1972**, *236*, 159–160. [CrossRef]
42. Rembold, H.; Sharma, G.K.; Czoppelt, C.; Schmutterer, H. Azadirachtin: A potent insect growth regulator of plant origin. *Z. Ang. Ent.* **2009**, *93*, 12–17. [CrossRef]
43. Ketkar, C.M. Use of tree-derived non edible oils as surface protectants for stored legumes against Callosobruchus maculatus and C. chinensis. In *Natural Pesticides from the Neem Tree and Other Tropical Plants, In Proceedings of the 3rd International Neem Conference, Nairobi, Kenya, 10–15 July 1968*; Deutsche Gesellschaft für Technische Zusammenarbeit: Eschborn, Germany, 1986; pp. 535–542.
44. Negi, R.S.; Srivastava, M.; Saxena, M.M. Egg laying and adult emergence of *Callospbrichus chinensis* on green gram (*Vigna radiata*) treated with pongam oil. *Indian J. Entomol.* **1997**, *59*, 362–365.
45. Satpathi, C.R.; Ghatak, S.S.; Bhusan, T.K. Efficacy of some plant extracts against the larvae of Indian meal moth *Corcyra cephalonica* Staint (Gelechiidae: Lepidoptera). *Environ. Ecol.* **1991**, *9*, 687–689.

46. Deshmukh, S.D.; Borle, M.N. Studies on the insecticidal properties of indigenous plant products. *Indian J. Entomol.* **1976**, *37*, 11–18.
47. Biswas, D.; Uddin, M.M.; Ahmad, M. Biorational management of tomato fruit borer, *Helicoverpa armigera* (Hübner) in winter under field condition of Bangladesh. *Fundam. Appl. Agric.* **2019**, *4*, 792–797. [CrossRef]
48. Feuerhake, K.; Schmutterer, H. Use of simple methods for extraction of neem seed, formulation of extracts and their effects on various insect pests. *Z. Pflanzenkr. Pflanzenschutz* **1982**, *89*, 737–747.
49. National Research Council (US) Panel on Neem. Effects on Insects. In *Neem: A Tree for Solving Global Problems*; National Academies Press: Washington, DC, USA, 1992.
50. Zabel, A.; Manojlovic, B.; Rajkovic, S.; Stanković, S.; Kostic, A.M. Effect of Neem extract on *Lymantria dispar* L. (*Lepidoptera: Lymantriidae*) and *Leptinotarsa decemlineata* Say. (*Coleoptera: Chrysomelidae*). *Anz. Schädlingskd. J. Pest Sci.* **2002**, *75*, 19–25. [CrossRef]
51. Kaethner, M. Fitness reduction and mortality effects of neem-based pesticides on the Colorado potato beetle *Leptinotarsa decemlineata* Say (Col., Chrysomelidae). *J. Appl. Ent.* **1992**, *113*, 456–465. [CrossRef]
52. Murray, K.D.; Groden, E.; Drummond, F.A.; Alford, A.R.; Conley, S.; Storch, R.H.; Bentley, M.D. Citrus limonoid effects on Colorado potato beetle (*Coleoptera: Chrysomelidae*) colonization and oviposition. *Environ. Entomol.* **1995**, *24*, 1275–1283. [CrossRef]
53. Thompson, H.M.; Fryday, S.L.; Harkin, S.; Milner, S. Potential impacts of synergism in honeybees (*Apis mellifera*) of exposure to neonicotinoids and sprayed fungicides in crops. *Apidologie* **2014**, *45*, 545–553. [CrossRef]
54. Tosi, S.; Nieh, J.C.; Sgolastra, F.; Cabbri, R.; Medrzycki, P. Neonicotinoid pesticides and nutritional stress synergistically reduce survival in honey bees. *Proc. R. Soc. B* **2017**, *284*, 20171711. [CrossRef]
55. Mohan, M.C.; Reddy, N.P.; Devi, U.K.; Kongara, R.; Sharma, H.C. Growth and insect assays of *Beauveria bassiana* with neem to test their compatibility and synergism. *Biocontrol. Sci. Techn.* **2007**, *17*, 1059–1069. [CrossRef]
56. Akbar, W.; Lord, J.C.; Nechols, J.R.; Loughin, T.M. Efficacy of *Beauveria bassiana* for red flour beetle when applied with plant essential oils or in mineral oil and organosilicone carriers. *J. Econ. Entomol.* **2005**, *98*, 683–688. [CrossRef]
57. Filotas, M.; Sanderson, J.; Wraight, S.P. Compatibility and potential synergism between the entomopathogenic fungus *Beauveria bassiana* and the insect growth regulator azadirachtin for control of the greenhouse pests *Myzus persicae* and *Aphis gossypii*. In Proceedings of the Society for Invertebrate Pathology Annual Meeting Proceedings, Anchorage, Alaska, 7–11 August 2005; p. 81.
58. Rao, G.R.; Dhingra, S. Synergistic activity of some vegetable oils in mixed formulations with cypermethrin against different instars of *Spodoptera litura* Fabricius. *J. Ent. Res.* **1997**, *21*, 153–160.
59. Zanuncio, J.C.; Mourão, S.A.; Martínez, L.C.; Wilcken, C.F.; Ramalho, F.S.; Plata-Rueda, A.; Soares, M.A.; Serrão, J.E. Toxic effects of the neem oil (*Azadirachta indica*) formulation on the stink bug predator, *Podisus nigrispinus* (*Heteroptera: Pentatomidae*). *Sci. Rep.* **2016**, *6*, 30261. [CrossRef]
60. Koss, A.M.; Jensen, A.S.; Schreiber, A.; Pike, K.S.; Snyder, W.E. Comparison of predator and pest communities in Washington potato fields treated with broad-spectrum, selective, or organic insecticides. *Environ. Entomol.* **2005**, *34*, 87–95. [CrossRef]
61. Radkova, M.; Kalushkov, P.; Chehlarov, E.; Gueorguiev, B.; Naumova, M.; Ljubomirov, T.; Stoichev, S.; Slavov, S.; Djilianov, D. Beneficial arthropod communities in commercial potato fields. *Compt. Rend. Acad. Bulg. Sci.* **2017**, *70*, 309–316.
62. Shah, F.M.; Razaq, M.; Ali, Q.; Shad, S.A.; Aslam, M.; Hardy, I.C.W. Field evaluation of synthetic and neem-derived alternative insecticides in developing action thresholds against cauliflower pests. *Sci. Rep.* **2019**, *9*, 1–13. [CrossRef] [PubMed]
63. Schmutterer, H. Properties and potential of natural pesticides from the neem tree, *Azadirachta indica*. *Annu. Rev. Entomol.* **1990**, *35*, 271–297. [CrossRef]
64. Barek, S.; Paisse, O.; Grenier-Loustalot, M.-F. Analysis of neem oils by LC-MS and degradation kinetics of azadirachtin-A in a controlled environment. Characterization of degradation products by HPLC-MS-MS. *Anal. Bioanal. Chem.* **2004**, *378*, 753–763. [CrossRef]
65. Riyajan, S.-A.; Sakdapipanich, J.T. Encapsulated neem extract containing Azadirachtin-A within hydrolysed poly(vinylacetate) for controlling its release and photodegradation stability. *Chem. Eng. J.* **2009**, *152*, 591–597. [CrossRef]

66. Devi, N.; Maji, T.K. A novel microencapsulation of neem (*Azadirachta indica* A. Juss.) seed oil (NSO) in polyelectrolyte complex of j-carrageenan and chitosan. *J. Appl. Polym. Sci.* **2009**, *113*, 1576–1583. [CrossRef]
67. Buddepu, M.; Sabithadevi, K.; Ashok, V.; Ramprasad, M.V.N.S. Determination of in vitro sunscreen activity of *Pongamia pinnata* (L.) essential oil. *Drug Invent. Today* **2011**, *3*, 197–199.
68. Wanyika, H.N.; Kareru, P.G.; Keriko, J.M.; Gachanja, A.; Kenji, G.M.; Mukiira, N.J. Contact toxicity of some fixed plant oils and stabilized natural pyrethrum extracts against adult maize weevils (*Sitophilus zeamais* Motschulsky). *Afr. J. Pharm. Pharmaco.* **2009**, *3*, 66–69.
69. Abbott, W.S. A method of computing the effectiveness of an insecticide. *J. Econ. Entomol.* **1925**, *18*, 265–267. [CrossRef]
70. Finney, D.J. *Probit Analysis*, 3rd ed.; University Press: Cambridge, UK, 1971; p. 333.

© 2019 by the authors. Licensee MDPI, Basel, Switzerland. This article is an open access article distributed under the terms and conditions of the Creative Commons Attribution (CC BY) license (http://creativecommons.org/licenses/by/4.0/).

Article

Phytochemical Analysis of *Tephrosia vogelii* across East Africa Reveals Three Chemotypes that Influence Its Use as a Pesticidal Plant

Angela G. Mkindi [1], Yolice Tembo [2], Ernest R. Mbega [1], Beth Medvecky [3], Amy Kendal-Smith [4,5], Iain W. Farrell [4], Patrick A. Ndakidemi [1], Steven R. Belmain [4] and Philip C. Stevenson [4,6,*]

1. Department of Sustainable Agriculture, Biodiversity and Ecosystems Management, Centre for Research, Agricultural Advancement, Teaching Excellence and Sustainability (CREATES), The Nelson Mandela African Institution of Science and Technology, P.O. Box 447 Arusha, Tanzania; angela.mkindi@nm-aist.ac.tz (A.G.M.); ernest.mbega@nm-aist.ac.tz (E.R.M.); patrick.ndakidemi@nm-aist.ac.tz (P.A.N.)
2. Bunda College, Lilongwe University of Agriculture and Natural Resources-Malawi, P.O. Box 219 Lilongwe, Malawi; ytembo@bunda.luanar.mw
3. Innovations in Development, Education and The Mathematical Sciences (IDEMS) International, 15 Warwick Road, Reading, RG2 7AX, UK; bethmedvecky@gmail.com
4. Royal Botanic Gardens, Kew, Richmond, Surrey TW9 3DS, UK; bs16a3s@leeds.ac.uk (A.K.-S.); i.farrell@kew.org (I.W.F.); s.r.belmain@greenwich.ac.uk (S.R.B.)
5. Faculty of Biological Sciences, University of Leeds, Leeds LS2 9JT, UK
6. Natural Resources Institute, University of Greenwich, Central Avenue, Chatham Maritime, Kent ME4 4TB, UK
* Correspondence: P.Stevenson@kew.org

Received: 27 November 2019; Accepted: 10 December 2019; Published: 12 December 2019

Abstract: *Tephrosia vogelii* is a plant species chemically characterized by the presence of entomotoxic rotenoids and used widely across Africa as a botanical pesticide. Phytochemical analysis was conducted to establish the presence and abundance of the bioactive principles in this species across three countries in East Africa: Tanzania, Kenya, and Malawi. Analysis of methanolic extracts of foliar parts of *T. vogelii* revealed the occurrence of two distinct chemotypes that were separated by the presence of rotenoids in one, and flavanones and flavones that are not bioactive against insects on the other. Specifically, chemotype 1 contained deguelin as the major rotenoid along with tephrosin, and rotenone as a minor component, while these compounds were absent from chemotype 2, which contained previously reported flavanones and flavones including obovatin-3-*O*-methylether. Chemotype 3 contained a combination of the chemical profiles of both chemotype 1 and 2 suggesting a chemical hybrid. Plant samples identified as chemotype 1 showed chemical consistency across seasons and altitudes, except in the wet season where a significant difference was observed for samples in Tanzania. Since farmers are unable to determine the chemical content of material available care must be taken in promoting this species for pest management without first establishing efficacy. While phytochemical analysis serves as an important tool for quality control of pesticidal plants, where analytical facilities are not available simple bioassays could be developed to enable extension staff and farmers to determine the efficacy of their plants and ensure only effective materials are adopted.

Keywords: spatial-temporal variation; chemotype 3; deguelin; rotenoids; botanical insecticides

1. Introduction

Tephrosia vogelii Hook. f. (*Leguminosae*) is a plant species reported to be used widely for its medicinal, insecticidal, and soil enrichment potential in tropical Africa [1–6]. Specifically, research on *T. vogelii* reported medicinal properties such as anti-cancer activity [7–9] and efficacy as an ectoparasite treatment for domestic animals including poultry [10–14]. A number of studies have sought to validate the reported use of *T. vogelii* as a botanical insecticide under laboratory and field conditions and have reported its effectiveness for crop protection and reduced impacts on beneficial ecosystem services [15–17]. Likewise, *Tephrosia* is reported to have high biomass and is therefore important as a soil amendment and is compatible with food crops when intercropped in addition to its nitrogen fixing property [18,19]. Hence, using *T. vogelii* for small scale farmers may support reduced industrial fertilizer and synthetic pesticides application all of which bear cost and safety implications.

Deguelin was reported to be the major active compound in *T. vogelii* occurring in all plant parts along with the minor components of tephrosin and rotenone [20–23]. However, a previous study reported that some *T. vogelii* did not contain rotenoids, and was less effective as an insecticide [15,22]. This highlighted the need to ensure that effective chemotypes of pesticidal plants were available when promoting their use to farmers to ensure effective control of pests. However, this is challenging in the absence of suitable local facilities to undertake such quality control and establish variation when pesticidal plants are harvested, processed, and used locally [24]. Natural variation in the chemistry of bioactive components in pesticidal plants is reported [25] and can have consequences for use and ultimately trust in pesticidal plants as an alternative to synthetic inputs by farmers.

Variability in the chemistry of *T. vogelii* could lead to farmers unknowingly using ineffective material and influence negatively the wider adoption and commercialization of botanical insecticides. Small scale farming communities, who are the main beneficiaries of *T. vogelii*, identify the plant using locally acquired knowledge through morphological features. However, *T. vogelii* chemotypes may not be identified and distinguished morphologically [22]. Likewise, traditional use does not have the capacity to determine effectiveness prior to use. Phytochemical analysis is an essential tool for the selection of elite provenances of plant materials [26] and should be used for the identification of effective *T. vogelii* provenances prior to propagation. Here we collected samples from local farmers who had *Tephrosia* growing in their fields and were using it for some non-food purpose. We sought to understand the key applications of *Tephrosia* through a survey of farmers who used the plants. From the collected samples, we evaluated the presence and concentration of deguelin in *T. vogelii* leaf materials across 91 locations in three East African countries to establish the extent of chemotype variation and identify elite materials for propagation of improved seed material of *T. vogelii*. Variation of *Tephrosia* chemotypes with flower colors, seasons, rainfall, and altitudes were also assessed to establish whether these traits could be used as markers of effectiveness or for the presence of active chemicals in the plant. Further recommendations are also presented to help farmers more easily identify bioactive plants for improved efficacy.

2. Results and Discussion

2.1. Status of Use of Tephrosia vogelii by Small Scale Farmers

Eight questions were used to assess the extent of *T. vogelii* use among farmers in locations where samples were collected in Tanzania. All interviewed farmers were aware of *T. vogelii*, commonly known as "Utupa" (Figure 1.). Farmers reported using *T. vogelii* mostly for field pest control (59%), fishing (45%), and storage pest control (36%). Other uses such as, human medication, soil fertility and as mole repellants and for ectoparasites control were also reported. A few farmers (5%) reported awareness of the plant from witnessing institutional researchers collecting the plant for use in research and also planting *Tephrosia* in research institutions such as the Uyole Agricultural Research—Mbeya and Tanzania Coffee Research Institute—Kilimanjaro. The specific research activity was not clearly communicated.

Our results indicate that *T. vogelii* is most widely used for pest control among other uses as shown in the survey results. Farmers' responses in this study, align with reports about the use of *T. vogelii* in controlling pests in vegetables and in stored products [15] and ectoparasite control in domestic animals [12,24,27]. The wider use of *T. vogelii* for small-scale farmers could be associated with previous projects that promoted integrated pest management using *T. vogelii* and research on soil improvement [28,29].

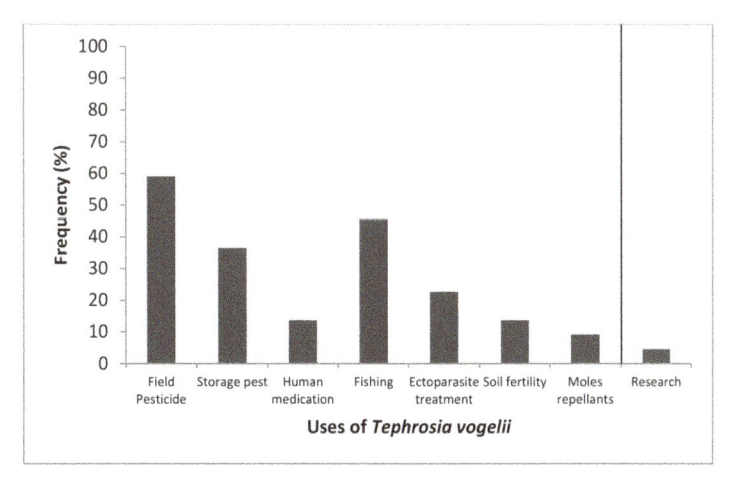

Figure 1. Ethno botanical uses of *T. vogelii* among local small-scale farmers from the six Tanzanian regions. Data are frequencies of responses on eight key uses of *T. vogelii* from the sample size ($n = 22$).

2.2. Phytochemical Analysis of T. vogelii Leaf Samples

Analysis of methanolic extracts of the leaf samples identified 2 chemotypes. Figure S1, shows the LC-MS chromatograms of chemotype 1 characterised by the presence of rotenoids with corresponding peaks between 19 and 20 min. These were determined from a comparison of their spectral data to in house standards [22]. Rotenone was identified from UV (LC-PDA) λmax nm, 301; (MS) m/z, 395.4 [M + H]$^+$, while tephrosin was identified from UV (LC-PDA) λmax nm, 272, 300, 314 sh; (MS) m/z, 433.4 [M + Na]$^+$ and deguelin from UV (LC-PDA) λmax nm, 270, 301, 319; (MS) m/z, 395.4 [M + H]$^+$. Chemotype 2 with peaks between 18 and 21 min were determined to contain obovatin 3-*O*-methylether as the major component from an in house standard and had UV (LC-PDA) λmax nm, 270, 295, 348; (MS) m/z, 337.4 [M + H]$^+$. Other similar components having ions with $m/z = 337$ and 367 corresponding to flavones and flavanones are reported earlier including Z-tephrostachin [22].

We also identified a third chemotype (Figure S1). Chemotype 3 was a chemical hybrid of chemotypes 1 and 2 showing the presence of both the rotenoids and the flavanones and flavones reported in 1 and 2 respectively in equivalent quantities. A further finding recorded plants as chemotype 1 but indicated trace quantities of flavones and flavanones from chemotype 2, suggesting that a variety of potential chemical variants may exist in natural and propagated materials. The analyses were undertaken on a single leaflet so were not a consequence of sample mixing of chemotypes 1 and 2. Furthermore individual leaves from the same plant were chemically very similar.

The presence of chemotypes 1 and 2 corresponded with findings [22] in which chemotypes 1 and 2 were first reported. However, here we provide a comprehensive regional assessment of their occurrence to determine what potential impacts the occurrence of chemotype 2 might have on the application and uptake of this species for pest control and other uses. Chemotype 3 is a previously unreported chemotype in this species which we report here. Studies have shown that the production of new flavonoids such as those found in *T. vogelii* could be influenced by environmental factors whereby compounds changed after exposure to conditions such as carbohydrates and light [20]. However, chemical variation in these varieties is most likely genetic since different chemotypes were first reported from the same location and in the same soil in adjacent fields [22]. There would therefore be value in analysing for further chemotypes and determining if the hybrids produce lower quantities of both compound groups.

2.3. Frequency of T. vogelii Chemotypes.

Plant material was collected from specific locations in three countries (Tanzania, Kenya, and Malawi) where *T. vogelii* is used and their geographical references are presented in Table S1. Approximately, 7% of samples were identified as chemotype 3, while 20% were chemotype 2. Most samples (74%) were chemotype 1. A higher proportion of chemotype 1 was also reported by [15] in Malawi from the analysis of 12 samples. Table 1, illustrates the proportions of chemotypes by countries. The abundance in plant materials with chemotype 1 coincided with efficacy studies of *T. vogelii* on medicinal [10,13,21,30] and insecticidal properties of rotenoids [15–17,31] which revealed that rotenoids were the compounds most frequently found in *T. vogelii* sampled and are responsible for the plants' biological activity.

2.4. Spatial Distribution of Plants Samples Chemotypes

In this study, deguelin, the most abundant pesticidal rotenoid in *T. vogelii*, was used as an indicator compound and its concentration in the plant was assessed across the study zone. The chemical composition of *T. vogelii* was presented with reference to location across the three countries (Figure 2). Samples collected from Malawi and Kenya contained chemotypes 1, 2, and 3 and were located in Lilongwe and in 12 Kenyan counties respectively, while only chemotype 1 was recorded from 14 locations across five regions in Tanzania. The results, from this study present the potential for understanding the diversity of pesticidal plant chemotypes across a whole region. Local efficacy testing of pesticidal activity in *T. vogelii* using a simple assay would potentially be conducted across various locations where the plants grow to ensure reliable efficacy results for local farmers using the plant.

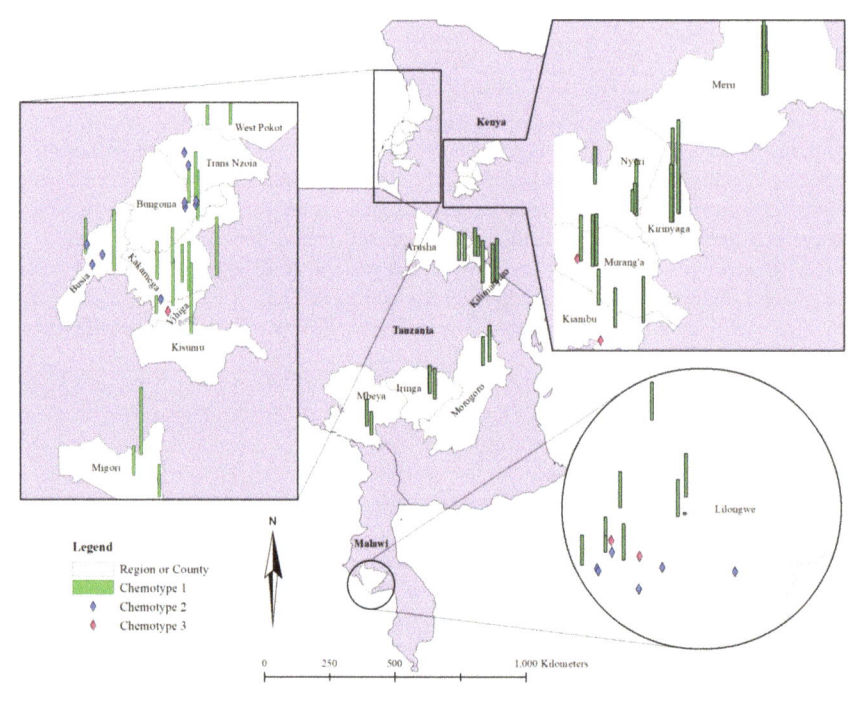

Figure 2. Spatial variation of *T. vogelii* chemotypes in Tanzania, Kenya, and Malawi, indicating presence of Chemotypes 1, 2, and 3. Green bars depict the presence of deguelin while blue marks indicate the presence of chemotype 2. The purple marks indicates the presence of chemotype 3, a chemical hybrid of chemotype 1 and 2.

Table 1. Summary distribution of chemotype within the study area.

Variables	No. of Observations	No. of Missing Values	No. of Categories	Mode	Mode Frequency	Categories	Frequency Per Category	Rel. Frequency Per Category (%)	Proportion Per Category
Overall	91	0	3	Chemotype 1	67	Chemotype 1	67	74	1
						Chemotype 2	18	20	0
						Chemotype 3	6	7	0
Kenya	57	0	3	Chemotype 1	44	Chemotype 1	44	77	1
						Chemotype 2	10	18	0
						Chemotype 3	3	5	0
Malawi	20	0	3	Chemotype 1	9	Chemotype 1	9	45	0
						Chemotype 2	8	40	0
						Chemotype 3	3	15	0
Tanzania	14	0	1	Chemotype 1	14	Chemotype 1	14	100	1
						Chemotype 2	0	0	0
						Chemotype 3	0	0	0

2.5. Spatial Temporal Variation of Chemotype 1 in T. vogelii

Linear regression analysis was performed to test the variation of deguelin content in *T. vogelii* based on altitude. The linear regression for Malawi ($r^2 = 0.178$, F = 3.9, df = 18, $p = 0.064$), Kenya ($r^2 = 0.03$, F = 1.74, df = 56, $p = 0.193$), dry season in Tanzania ($r^2 = 0.008$, F = 0,096, df = 12, $p = 0.762$), and wet season in Tanzania ($r^2 = 0.122$, F = 1.665, df = 12, $p = 0.221$) showed no significant relationship between changing altitude and the concentrations of deguelin. Further analysis of data from Tanzania revealed no significant correlation between rainfall recorded in the wet ($r^2 = 0.005$, F = 0.016, df = 3, $p = 0.9$), and dry seasons ($r^2 = 0.72$, F = 7.725, df = 56, $p = 0.069$). The results concur with findings reported by [15] although these earlier data were of just a few samples.

Analysis of variance on the samples collected over two seasons in Tanzania showed that there was no significant variation in the deguelin concentration with locations in the dry season (ANOVA F = 0.272, df = 8, $p = 0.916$). In the wet season, however, a significant variation (ANOVA F = 7.092, df = 8, $p = 0.008$) was observed (Table 2) where the highest and lowest levels of deguelin were observed in samples collected from Same and Mbeya districts respectively. Seasonal variation of deguelin was also reported by [15,32] where higher concentrations occurred in the wet season compared with the dry season.

Table 2. Spatial and temporal variation of deguelin in *T. vogelii* from locations in Tanzania. The values presented are means ± SE. **, = significant at $P \leq 0.01$, ns = not significant. Means followed by the same letter in a column are not significantly different.

Location	Dry Season Deguelin (ppm)	Wet Season Deguelin (ppm)
Same	6841 ± 523 a	8756 ± 197 a
Iringa	5644 ± 1202 a	4879 ± 132 bc
Morogoro	5423 ± 1621 a	6229 ± 207 b
Kilimanjaro	5144 ± 682 a	6377 ± 791 b
Mbeya	5699 ± 314 a	3385 ± 196 c
Arusha	5339 ± 139 a	4803 ± 4 bc
One way ANOVA F statistics	0.27 ns	7.09 **

2.6. Association between T.vogelii Flower Color and Chemotypes

One simple morphological feature for identification of *T. vogelii* and potentially distinguishing chemotypes is flower color, with colors typically white or purple. In this study, plants that had flowers at the time of sample collection were recorded along with leaf samples for chemical analysis. Generally, a high percent of plants with white colored flowers were recorded compared to those with purple flower. A higher percent of occurrence of white flowers was associated with the presence of chemotype 1. Chemotype 3 was associated with only purple color while a lower percent of white color was associated with chemotype 2 (Figure 3). Regression analysis showed a strong correlation ($r^2 = 0.43$, F = 22.02, df = 29, $p = 0.0001$) between chemotype and flower color where chemotype 1 was related to white and chemotype 2 to purple color. In contrast to earlier findings [22] who found no correlation. Flower colors could be used as initial tool for identification of chemotypes. However, the fact that some purple flowers (with smaller percent) were also associated with chemotype 1, the decision to use a plant for pest control purposes should be guided with simple assays for evidence of effective chemotypes. Small scale farmers could adopt simpler tests of plant materials against storage pests such as cowpea weevils (*Callosobruchus maculatus*) as already done by Belmain et al. [15] although this was under laboratory setting.

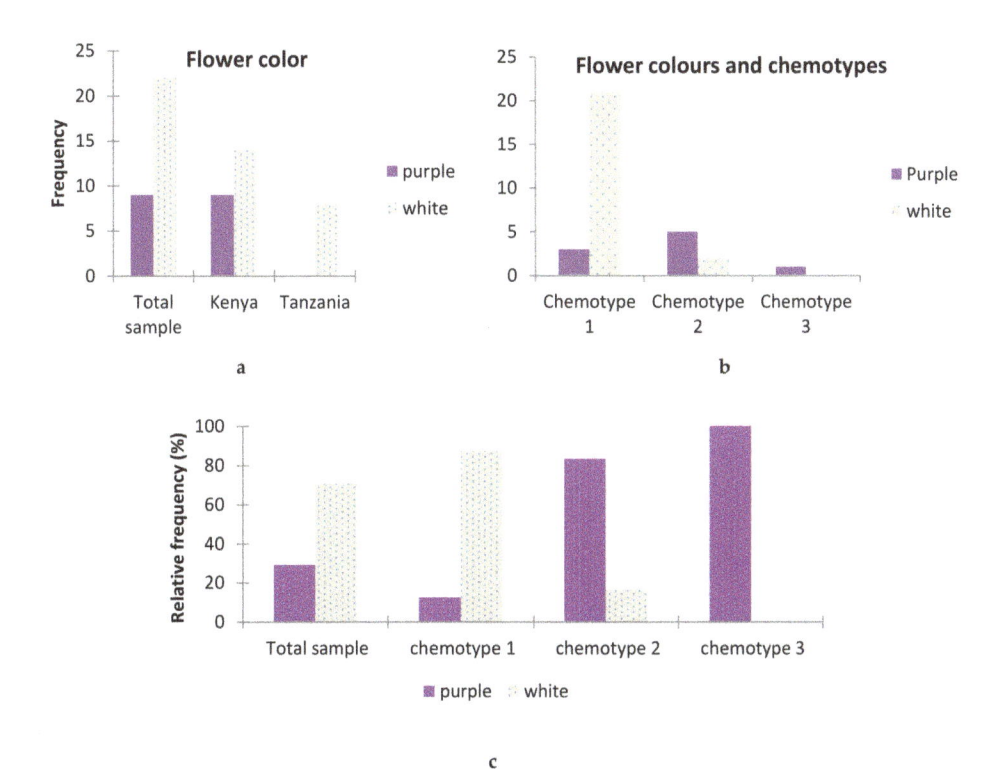

Figure 3. The figure above shows (**a**) a descriptive statistics results showing the presence of purple and white flowers in Tanzania and Kenya; (**b**) numerical distribution of *T. vogelii* flower color with the chemotype of the plant; and (**c**) the frequency in percentage (%) of the occurrence of flower color with chemotype.

2.7. Summary of Indicators for Chemotype Identification

Small-scale farmers require hands-on information to enable them to decide on suitable *T. vogelii* materials for pest control. From this study, proposed and tested indicators would be used. Table 3 below highlights what farmers would need to consider while harvesting and using the materials.

Table 3. Tested and proposed options that farmers would need consider to select effective *T. vogelii* plant material.

Option	Results	Reliability for Chemotypes Identification
Tested Options		
Elevation	No correlation	Not reliable
Season	No correlation	Not reliable: Although wet season enhances higher content of bioactive compounds in chemotype 1
Flower Color	Positive correlation	Somewhat reliable: Could be used to decide on the chemotype where white flowers are known to be related with chemotype 1. N.B., a few plants with chemotype 1 had purple flowers.
Proposed Options		
Simple assays	Report from Belmain et al., 2012	Reliable: Test assessment of plant (10% leaf powder in small test container with bruchids), could be a rapid, simple and affordable tool. Pesticidal properties of *Tephrosia* are fast acting and chemotype could be determined in 48 h.

3. Materials and Methods

3.1. Analysis of Tephrosia Vogelii Leaf Samples

3.1.1. Samples Collection

Plant leaf samples were collected from farmers' fields. Farmers identified the specific *T. vogelii* plants that were used for controlling crop pests and diseases and for medical uses. In Tanzania, leaf samples of *T. vogelii* were collected over two seasons, the wet season and dry season in 14 sites located across five regions: Arusha, Kilimanjaro, Morogoro, Mbeya, and Iringa. The five regions were identified after revising *T. vogelii* collections preserved in the National Herbarium in Tanzania to identify possible areas where the plants could be growing. Samples were collected in March and September 2018 during the wet and dry season respectively. In each region, two sites were identified where samples were collected depending on the availability of the plant at that time.

Herbarium samples were collected and assigned voucher numbers, processed and stored in the National herbarium. Identified plant in each point of sample collection was used for the two seasons to justify analysis of variation with seasons. Rainfall data for the particular months of sample collection were obtained from the Tanzania Meteorological Agency (TMA)—Tanzania. In Malawi, samples were collected in the Lilongwe area on farmers' fields. Likewise in Kenya collections were made on farmers' fields in Kisumu, Homa Bay, Migori, Siaya counties, the western region Kakamega, Busia, Bungoma, Mumias, and Central Kenyan counties.

A total of 28 samples were collected in Tanzania that included 14 samples for each of the dry and wet seasons. In Malawi, 20 samples were collected from Lilongwe area between May and November 2018, while in Kenya, a total of 57 samples were collected between February and April 2019. Collected samples were dried under the shed, packed into plastic zip bags and stored in dark and dry conditions under ambient temperature before being processed and analyzed.

3.1.2. Survey of Farmers Awareness on the Use of *T. vogelii*

Twenty two farmers from six Tanzanian regions were interviewed in the survey to determine the uses of *T. vogelii* in the household. In order to identify *T. vogelii* uses with reference to the type of sample collected, only farmers who owned the plant or neighbors to the famer owning the plants were interviewed. The selection of farmers therefore did not follow specific social survey protocols for sample sizes selection.

3.1.3. Sample Analysis

Dried *T. vogelii* samples were powdered using an electric grinder (SALTER, Model No EK2311ROFB distributed by UP Global Sourcing, Victoria Street, Manchester, OL9 0DD, UK Made in China). *Tephrosia* powder (50 mg/mL) was extracted in methanol. Each extract was left to stand for 24 h at room temperature before chemical analysis. Plant leaf extracts were transferred to Eppendorf tubes and centrifuged for 20 min at 5000 rpm. Supernatant (300 uL) was transferred into HPLC vials for analysis. Extracts were analyzed by liquid chromatography (LC)-Electrospray Ionization Mass Spectroscopy (ESIMS) and UV spectroscopy using Thermo Fisher Velos Pro LC-MS. Samples (5 µL) were injected directly on to a Phenomenex Luna C18 (2) column (150 Å~3 mm i.d., 3 µm particle size) at 400 µL min^{-1} and eluted using a linear gradient of 90:0:10 (t = 0 min) to 0:90:10 (t = 20–25 min), returning to 90:0:10 (t = 27–30 min). Solvents were water: methanol: 1% formic acid in acetonitrile, respectively. The column was maintained at 30 °C. Compounds were detected on a Thermo Fisher Velos Pro Dual-Pressure Linear Ion Trap Mass Spectrometer. Samples were scanned, using FTMS, from *m/z* 200–600 corresponding to the range of molecular ions expected in samples of *T. vogelii*. UV peak area were quantified against a calibration curve of an authentic in-house standard [22]. The resulting peak areas of deguelin were measured at a wavelength of 300 nm and arranged in an excel file for statistical analysis.

3.2. Presentation of Data and Sampling Points

Graphical presentation of chemotype and variation in amounts of chemotype 1 in *T. vogelii* was performed using ARC GIS, ARCMAP version 10.3.

3.3. Statistical Analysis

Analysis of Variance and descriptive statistics, proportion analysis and regression analysis were performed using XLSTAT version 2019.2.2.59614 (Addinsoft (2019). XLSTAT statistical and data analysis solution. Boston, MA, USA. https://www.xlstat.com).

4. Conclusions

This study has demonstrated the chemical variation in *T. vogelii*, across a variety of location types in three countries of East Africa and revealed considerable variation in chemistry influencing the bioactivity of plants materials. The study has also highlighted key uses of the plant, hence indicating its importance to farmers' livelihoods. Correlation of key factors with the effectiveness of the plant materials are also discussed along with the identification of options that farmers would consider when selecting elite materials. From this study we realize the potential of a region wide study that provide an expanded perspective of plant chemistry for a wider community use and uptake. To mitigate the variations under local conditions, simple and locally tailored assays, where farmers could test plant materials against storage pests would provide a rapid assessment tool of plants efficacy. However, further research on possible propagation strategies that ensure the availability and use of elite materials as well as investigation of more indicators for chemotype identification is required.

Supplementary Materials: The following are available online at http://www.mdpi.com/2223-7747/8/12/597/s1, Table S1: Location and chemotype data for samples of *T. vogelii* in Tanzania, Kenya, and Malawi, Figure S1: Chromatograms showing: (**a**) Chemotype 1, presence of deguelin; (**b**) Chemotype 2 absence of rotenoids; (**c**) chemotype 3 signifying presence of chemotypes 1 and 2 suggesting a hybrid, and (**d**) presence of Chemotype 1 with minimal quantitative presence of chemotype 2.

Author Contributions: Conceptualization, A.G.M., P.C.S., E.R.M., S.R.B., Y.T., and P.A.N.; methodology, A.G.M, P.C.S., I.W.F., and S.R.B.; software, P.C.S.; validation, P.C.S.; formal analysis, A.G.M., P.C.S., A.K.-S., and I.W.F.; investigation, A.G.M., B.M.; resources, P.C.S. and S.R.B.; data curation, P.C.S., I.W.F., and A.G.M.; writing—original draft preparation, A.G.M.; writing—review and editing, P.A.N., E.R.M., P.C.S., S.R.B., and B.M.; visualization, P.C.S.; supervision, P.A.N. and P.C.S.; project administration, S.R.B.; funding acquisition, S.R.B., P.A.N., and P.C.S.

Funding: This research was funded by grants from the McKnight foundation to SRB and PCS Grant No: 17-070, GCRF-BBSRC grant BB/R020361/1 grant to PCS and the World Bank to PAN Grant No.5799-TZ.

Conflicts of Interest: The authors declare no conflict of interest.

References

1. Burkill, H.M. *The Useful Plants of West Tropical Africa. Volume 2: Families EI.*; Royal Botanic Gardens: KEW, UK, 1994.
2. Kamanula, J.; Sileshi, G.W.; Belmain, S.R.; Sola, P.; Mvumi, B.M.; Nyirenda, G.K.; Nyirenda, S.P.; Stevenson, P.C. Farmers' insect pest management practices and pesticidal plant use in the protection of stored maize and beans in Southern Africa. *Int. J. Pest Manag.* **2010**, *57*, 41–49. [CrossRef]
3. Mafongoya, P.L.; Kuntashula, E. Participatory evaluation of *Tephrosia* species and provenances for soil fertility improvement and other uses using farmer criteria in eastern Zambia. *Exp. Agric.* **2005**, *41*, 69–80. [CrossRef]
4. Neuwinger, H.D. Plants used for poison fishing in tropical Africa. *Toxicon* **2004**, *44*, 417–430. [CrossRef] [PubMed]
5. Nyirenda, S.P.; Sileshi, G.W.; Belmain, S.R.; Kamanula, J.F.; Mvumi, B.M.; Sola, P.; Nyirenda, G.K.; Stevenson, P.C. Farmers' ethno-ecological knowledge of vegetable pests and pesticidal plant use in Malawi and Zambia. *Afr. J. Agric. Res.* **2011**, *6*, 1525–1537.
6. Sileshi, G.; Mafongoya, P.L.; Kwesiga, F.; Nkunika, P. Termite damage to maize grown in agroforestry systems, traditional fallows and monoculture on nitrogen-limited soils in eastern Zambia. *Agric. For. Entomol.* **2005**, *7*, 61–69. [CrossRef]

7. Gbadamosi, I.T.; Erinoso, S.M. A review of twenty ethnobotanicals used in the management of breast cancer in Abeokuta, Ogun State, Nigeria. *Afr. J. Pharm. Pharmacol.* **2016**, *10*, 546–564. [CrossRef]
8. Touqeer, S.; Saeed, M.A.; Ajaib, M. A review on the phytochemistry and pharmacology of genus *Tephrosia*. *Phytopharmacology* **2013**, *4*, 598–637.
9. Varughese, R.S.; Lam, W.S.-T.; Marican, A.A.B.H.; Viganeshwari, S.H.; Bhave, A.S.; Syn, N.L.; Wang, J.; Wong, A.L.-A.; Kumar, A.P.; Lobie, P.E.; et al. Biopharmacological considerations for accelerating drug development of deguelin, a rotenoid with potent chemotherapeutic and chemopreventive potential. *Cancer* **2019**, *125*, 1789–1798.
10. Antonio, A.P.; Perera, L.M.S.; García, J.A.; Rehmana, M.U.; Choudhary, M.I. Anthelminthic Activity Against Sheep Gastrointestinal Nematodes in Chemical Compounds from *Tephrosia Vogelii* Leaves. *J. Anim. Vet. Sci.* **2019**, *6*, 8–17.
11. Jacques, D.T.; Safiou, A.; Jédirfort, H.; Souaïbou, F. In Vitro effect of the ethanolic extract of *Tephrosia vogelii* on Rhipicephalus Sanguineus in Abomey-Calavi. *Avicenna J. Phytomed.* **2015**, *5*, 247–259.
12. Kalume, M.K.; Losson, B.; Angenot, L.; Tits, M.; Wauters, J.N.; Frédérich, M.; Saegerman, C. Rotenoid content and in vitro acaricidal activity of *Tephrosia vogelii* leaf extract on the tick Rhipicephalus appendiculatus. *Vet. Parasitol.* **2012**, *190*, 204–209. [CrossRef] [PubMed]
13. Marango, S.N.; Khayeka-Wandabwa, C.; Makwali, J.A.; Jumba, B.N. Experimental therapeutic assays of *Tephrosia vogelii* against Leishmania major infection in murine model: In vitro and in vivo. *BMC Res. Notes* **2017**, *10*, 698. [CrossRef] [PubMed]
14. Okitoi, L.O.; Ondwasy, H.O.; Siamba, D.N.; Nkurumah, D. Traditional herbal preparations for indigenous poultry health management in Western Kenya. *Livest. Res. Rural Dev.* **2007**, *19*, 72.
15. Belmain, S.R.; Amoah, B.A.; Nyirenda, S.P.; Kamanula, J.F. Highly variable insect control efficacy of *Tephrosia vogelii* chemotypes. *J. Agric. Food Chem.* **2012**, *60*, 10055. [CrossRef]
16. Mkenda, P.A.; Stevenson, P.C.; Ndakidemi, P.; Farman, D.I.; Belmain, S.R. Contact and fumigant toxicity of five pesticidal plants against *Callosobruchus maculatus* (Coleoptera: Chrysomelidae) in stored cowpea (*Vigna unguiculata*). *Int. J. Trop. Insect Sci.* **2015**, *35*, 172–184. [CrossRef]
17. Tembo, Y.; Mkindi, A.G.; Mkenda, P.A.; Mpumi, N. Pesticidal Plant Extracts Improve Yield and Reduce Insect Pests on Legume Crops Without Harming Beneficial Arthropods. *Front. Plant Sci.* **2018**, *9*, 1425. [CrossRef]
18. Mhango, W.G.; Snapp, S.S.; Phiri, G.Y.K. Opportunities and constraints to legume diversification for sustainable maize production on smallholder farms in Malawi. *Renew. Agric. Food Syst.* **2013**, *28*, 234–244. [CrossRef]
19. Snapp, S.; Kanyama-Phiri, G.; Kamanga, B.; Gilbert, R.; Wellard, K. Farmer and Researcher Partnerships in Malawi: Developing Soil Fertility Technologies for the Near-Term and Far-Term. Available online: /core/journals/experimental-agriculture/article/farmer-and-researcher-partnerships-in-malawi-developing-soil-fertility-technologies-for-the-nearterm-and-farterm/0E20BDD9E4EFDFC4A74F915536410653 (accessed on 6 August 2019).
20. Lambert, N.; Trouslot, M.-F.; Nef-Campa, C.; Chrestin, H. Production of rotenoids by heterotrophic and photomixotrophic cell cultures of *Tephrosia vogelii*. *Phytochemistry* **1993**, *34*, 1515–1520. [CrossRef]
21. Lokhande, K.B.; Nagar, S.; Swamy, K.V. Molecular interaction studies of deguelin and its derivatives with Cyclin D1 and Cyclin E in cancer cell signaling pathway: The computational approach. *Sci. Rep.* **2019**, *9*, 1–13. [CrossRef]
22. Stevenson, P.C.; Kite, G.C.; Lewis, G.P.; Forest, F.; Nyirenda, S.P.; Belmain, S.R.; Sileshi, G.W.; Veitch, N.C. Distinct chemotypes of *Tephrosia vogelii* and implications for their use in pest control and soil enrichment. *Phytochemistry* **2012**, *78*, 135–146. [CrossRef]
23. Wang, A.; Wang, W.; Chen, Y.; Ma, F.; Wei, X.; Bi, Y. Deguelin induces PUMA-mediated apoptosis and promotes sensitivity of lung cancer cells (LCCs) to doxorubicin (Dox). *Mol. Cell. Biochem.* **2018**, *442*, 177–186. [CrossRef] [PubMed]
24. Dougoud, J.; Toepfer, S.; Bateman, M.; Jenner, W.H. Efficacy of homemade botanical insecticides based on traditional knowledge. A review. *Agron. Sustain. Dev.* **2019**, *39*, 37. [CrossRef]
25. Kamanula, J.F.; Belmain, S.R.; Hall, D.R.; Farman, D.I.; Goyder, D.J.; Mvumi, B.M.; Masumbu, F.F.; Stevenson, P.C. Chemical variation and insecticidal activity of *Lippia javanica* (Burm. f.) Spreng essential oil against *Sitophilus zeamais* Motschulsky. *Ind. Crops Prod.* **2017**, *110*, 75–82. [CrossRef]

26. Sarasan, V.; Kite, G.C.; Sileshi, G.W.; Stevenson, P.C. Applications of phytochemical and in vitro techniques for reducing over-harvesting of medicinal and pesticidal plants and generating income for the rural poor. *Plant Cell Rep.* **2011**, *30*, 1163–1172. [CrossRef] [PubMed]
27. Gadzirayi, C.; Mutandwa, E.; Mwale, M.; Chindundu, T. Utilization of *Tephrosia vogelii* in controlling ticks in dairy cows by small-scale commercial farmers in Zimbabwe. *Afr. J. Biotechnol.* **2009**, *8*, 17.
28. Mihale, M.J.; Deng, A.L.; Selemani, H.O.; Kamatenesi, M.M.; Kidukuli, A.W.; Ogendo, J.O. Use of indigenous knowledge in the management of field and storage pests around Lake Victoria basin in Tanzania. *Afr. J. Environ. Sci. Technol.* **2009**, *3*, 9.
29. Snapp, S.S.; Rohrbach, D.D.; Simtowe, F.; Freeman, H.A. Sustainable soil management options for Malawi: Can smallholder farmers grow more legumes? *Agric. Ecosyst. Environ.* **2002**, *91*, 159–174. [CrossRef]
30. Negi, P.S. Plant extracts for the control of bacterial growth: Efficacy, stability and safety issues for food application. *Int. J. Food Microbiol.* **2012**, *156*, 7–17. [CrossRef]
31. Mkindi, A.; Mpumi, N.; Tembo, Y.; Stevenson, P.C.; Ndakidemi, P.A.; Mtei, K.; Machunda, R.; Belmain, S.R. Invasive weeds with pesticidal properties as potential new crops. *Ind. Crops Prod.* **2017**, *110*, 113–122. [CrossRef]
32. Irvine, J.E.; Freyre, R.H. Effect of Planting Time and Photoperiod on *Tephrosia vogelii*. *Agron. J.* **1966**, *58*, 49–51. [CrossRef]

© 2019 by the authors. Licensee MDPI, Basel, Switzerland. This article is an open access article distributed under the terms and conditions of the Creative Commons Attribution (CC BY) license (http://creativecommons.org/licenses/by/4.0/).

Article
Insect Antifeedant Components of *Senecio fistulosus* var. *fistulosus*—Hualtata

Liliana Ruiz-Vásquez [1,2], Matías Reina [2], Víctor Fajardo [3], Matías López [4] and Azucena González-Coloma [5,*]

[1] Natural Resources Research Center (CIRNA), National University of the Peruvian Amazon (UNAP), Iquitos, Peru; lilyruizv@gmail.com
[2] Institute of Natural Products and Agrobiology (IPNA), Spanish Research Council (CSIC), 38206 Tenerife, Spain; mreina@ipna.csic.es
[3] Faculty of Sciences, University of Magallanes (UMAG), Punta Arenas 01855, Chile; victor.fajardo@umag.cl
[4] University Institute of Bio-Organic Antonio González (IUBO), University of La Laguna, 38206 Tenerife, Spain; mlopez@ull.es
[5] Institute of Agricultural Sciences (ICA), Spanish Research Council (CSIC), 28006 Madrid, Spain
* Correspondence: azu@ica.csic.es; Tel.: +34-917-452-500

Received: 22 April 2019; Accepted: 5 June 2019; Published: 15 June 2019

Abstract: From a bioactive methanolic extract of *Senecio fistulosus*, the antifeedant effects of the alkaloidal and non-alkaloidal fractions were tested against the insects *Spodoptera littoralis*, *Myzus persicae* and *Rhopalosiphum padi*, with the non-alkaloidal fraction being antifeedant. The phytochemical study of the non-alkaloidal fraction of *S. fistulosus*, resulted in the isolation of four compounds, two 9-oxo-furanoeremophilanes (**1**, **2**), an eremophilanolide, 1β,10β-epoxy-6-acetoxy-8α-hydroxy-eremofil-7(11)-en-8β,12-olide (**3**) and a maaliol derivative (**4**). The alkaloidal fraction yielded two known pyrrolizidine alkaloids (**5**, **6**). Compounds **1**, **3** and **4** are new natural products. Furanoeremophilane **2** was a strong antifeedant against *S. littoralis* and maaliane **4** inhibited the settling of *M. persicae*.

Keywords: *Senecio fistulosus*; antifeedant; sesquiterpene; pyrrolizidine alkaloid; structure-activity relationships

---

1. Introduction

The genus, *Senecio* (Asteraceae), is distributed worldwide and contains pyrrolizidine alkaloids (PAs). PAs are toxic to mammals and feeding deterrents for insect herbivores [1]. Compounds present in the non-alkaloidal fraction of *Senecio* spp have been described as part of their defense [1–3]. The most frequent chemical groups found in the non-alkaloidal fraction of *Senecio* are eremophilane-type sesquiterpenes of the furanoeremophilane and eremophilanolide type [1]. Some of these compounds have insect antifeedant, acaricidal, fungicidal, cytotoxic, phytotoxic, antioxidant, anti-inflammatory, and antimicrobial effects [1–6] and have been proposed as being an important part of *Senecio* defense [1–5].

In Chile, the genus *Senecio* is abundant (~210 species) [7]. There are several reports on eremophilane sesquiterpenes from Chilean *Senecio* species with defensive properties [1–3]. The species, *Senecio fistulosus*, grows from the western area of Patagonia to central Chile and it is used in folk medicine for its effects on the heart [8,9]. A previous study on the phytochemistry of *S. fistulosus*, from the central region of Chile, reported the presence of furanoeremophilanes 4α-hydroxy-6β-angeloxy-10βacetoxy-9-oxo-furanoeremophilane and 4α-hydroxy-6β-angeloxy-9-oxo-furanoeremophilane [10], but there are no reports on the defensive chemistry of this species.

In this work, the authors studied the chemical defenses of *S. fistulosus* var. *fistulosus* from the Magallanic region, containing a large number of the Chilean *Senecio* species and subspecies distributed in the Patagonic Cordillera and the coastal areas [7].

From a bioactive methanolic extract of *S. fistulosus*, the alkaloidal and non-alkaloidal fractions were tested against the insects *Spodoptera littoralis* (Boisd), *Myzus persicae* (Sulzer) and *Rhopalosiphum padi*, with the non-alkaloidal being antifeedant. Two furanoeremophilanes (**1, 2**), one eremophilanolide (**3**) and maaliol derivative (**4**) have been isolated from the non-alkaloidal fraction, with compounds **1, 3** and **4** being reported as natural products for the first time. Additionally, two pyrrolizidine alkaloids (**5, 6**) were isolated from the alkaloidal fraction.

2. Results and Discussion

Extracts of *S. fistulosus* (methanolic, MeOH, non-alkaloidal and alkaloidal) were tested against the phytophagous insects *Spodoptera littoralis*, *Myzus persicae* and *Rhopalosiphum padi*. The MeOH extract showed a significant effect on *M. persicae* (SI = 78 ± 4%; EC_{50} = 1.69 µg/cm^2, 0.69–4.15, 95% confidence limits—CL), the non-alkaloidal extract was active on *S. littoralis* (FI = 76 ± 6%) and *M. persicae* (SI = 75 ± 6%; EC_{50} = 2.25 µg/cm^2, 1.19–4.24, 95% CL). The alkaloidal extract showed moderate activity against *S. littoralis* (SI = 62 ± 6%) and *R. padi* (SI = 69 ± 6%), indicating that *S. fistulosus* defense chemistry is mainly due to compounds present in the non-alkaloidal fraction, as previously suggested for PA producing plants [2–4,11].

The phytochemical study of the non-alkaloidal fraction of *S. fistulosus* resulted in the isolation of four compounds, two 9-oxo-furanoeremophilanes (**1, 2**) [10,12], an eremophilanolide, 1β,10β-epoxy-6-acetoxy-8α-hydroxy-eremofil-7(11)-en-8β,12-olide (**3**) and a maaliol derivative (**4**). The alkaloidal fraction yielded two known pyrrolizidine alkaloids (**5, 6**) [13,14] (Figure 1).

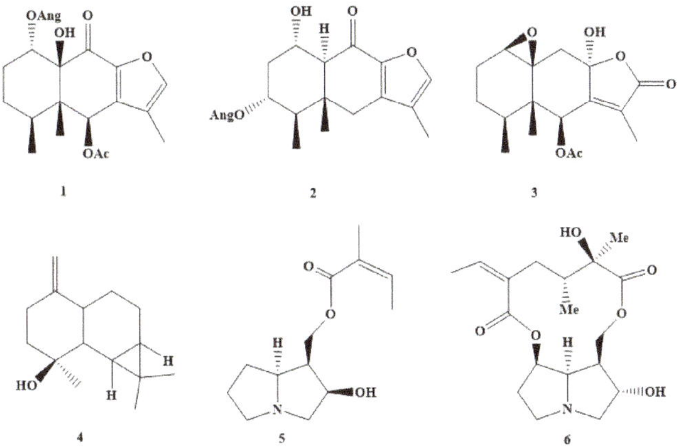

Figure 1. Chemical structures of compounds **1–6**.

Compounds **1, 3** and **4** are described here for the first time as natural products. A previous study on *S. fistulosus* reported the presence of furanoeremophilanes 4α-hydroxy-6β-angeloxy-10β-acetoxy-9-oxo-furanoeremophilane and 4α-hydroxy-6β-angeloxy-9-oxo-furanoeremophilane [10]. The difference in furanoeremophilane composition could be related to the different origin of the plant populations studied (Magallanes versus the central region of Chile).

The structural elucidation was carried out based on their ^{1}H and ^{13}C NMR spectra including (1D) and (2D) (COSY, HSQC, HMBC and NOESY) experiments, X-ray diffraction, as well as its physical, spectrometric (EIMS and HREIMS) and comparison with the chemical bibliography reported for similar compounds.

Compound **1** was isolated in crystalline form. Its infrared (IR) spectrum showed absorption bands at 3452, 1748, 1720 and 1679 cm^{-1} attributable to a hydroxyl, ester, and carbonyl groups. Its HR-EI-MS showed a molecular-ion peak at *m/z* 404.1838 (M$^+$, calculated for $C_{22}H_{28}O_7$, 404.1835) and a major fragment in the upper part of the spectrum at 345 (3%) [M-OCOCH$_3$]$^+$. The ^{1}H and ^{13}C NMR spectra of compound **1** (Table 1) showed signals of an olefinic proton at δ(H) 7.41 (br s, *J* = 1.3, H-C(12)) and one methyl group on a double bond at δ(H) 1.91 (d, *J* = 1.1, Me(13)), indicating the presence of a furan ring with a methyl group at C(11). The chemical shifts of signals δ(H) 0.98 (s), 1.17 (d, *J* = 7.5) assigned to Me(14) and Me(15), indicated a cis-decalin system [10]. Signals at δ(H) 3.98 (br s) were assigned to a proton on a hydroxyl group, and signals at δ(H) 5.92 (qq, *J* = 1.5, 7.2, H-C(3′)), 1.88 (dq, *J* = 1.6, 7.4, Me(4′)), 1.55 (quint., *J* = 1.5 Hz, Me(5′)); δ(C) 126.6 (s, C(2′)), 140.5 (d, C(3′)), 15.8 (q, C(4′)), 19.9 (q, C(5′)) corresponded to an angelate group. The chemical shift at δ(H) 7.03 (s) was attributed to H$_\alpha$-(6), a geminal proton of an acetate group [δ(H) 2.19 (s, OCOCH$_3$); δ(C) 20.9 (q) and 170.9 (s)]. The HMBC experiment showed correlations between H-C(1) with C(3), C(5), C(10) and C(1′), which allowed for the location of the angelate group at C(1). Correlations of the OH proton with C(5), C(9) and C(10) confirmed the location of the hydroxyl group at C(10). Correlation of H$_\alpha$-C(6) with C(4), C(8), C(11), C(14) and OCOCH$_3$ located the acetate group at C(6) (Figure 2). The remaining correlations were in agreement with the proposed structure. The relative stereochemistry of **1** was established by a NOESY experiment (Figure 3). H$_\beta$-C(1) gave a positive NOE effect with the H$_\beta$-C(2) and H$_\beta$-C(3) signals, confirming the α-configuration of the angelate group. In the same way, H$_\alpha$-C(6) presented a NOE effect with protons H$_\alpha$-C(4) and Me(13), establishing the configuration of the acetate group as C-6β. The NOE effect of H-C(3′) with Me(4)′/Me(5′) and the chemical shift of Me(5′) (δ(C) = 20.6 ppm) suggested the Z-geometry for the double bond of the H-C(2′)/H-C(3′) of the angelate group. The molecular structure of **1** was confirmed by X-ray diffraction (Figure 3), resolved by direct methods with SIR97, and was established as 1α-angeloyloxy-6β-acetoxy-10β-hydroxy-9-oxo-furanoeremophilane.

The HR-EI-MS of compound **2**, showed a molecular-ion peak at *m/z* 346.1785 (M$^+$, calculated for $C_{20}H_{26}O_5$, 346.1780), and its IR spectrum showed the presence of absorption bands at 3446, 1733, 1716 and 1699 cm^{-1} attributable to hydroxyl, ester, and carbonyl groups. The analysis of ^{1}H and ^{13}C NMR spectroscopic data of **2** (Table 1) indicated the presence of a trans-decalin based on the chemical shift at δ(H) 0.88 (s, Me(14)), and by comparison with published data. Therefore, the structure of **2** was confirmed as 1α-hydroxy-3α-angeloyloxy-10α H-9-oxo-furanoeremophilane, previously isolated from *Senecio smithii*, [12].

Table 1. ^{1}H (500 MHz) and ^{13}C (125 MHz) NMR data of compounds 1 and 2 in CDCl$_3$.

Position	1		2	
	δ$_H$ in ppm, Multiplicity, *J* (in Hz)	δ$_C$ in ppm	δ$_H$ in ppm, Multiplicity, *J* (in Hz)	δ$_C$ in ppm
1β	4.84 br s	74.5 d	4.25 ddd (4.8, 9.7, 11.7)	65.5 d
2α	2.30 m	20.7 t	1.50 q (11.8)	39.1 t
2β	1.64 m		2.47 m	
3α	1.40 m	23.9 t	-	71.4 d
3β	2.32 m		4.90 dt (4.6, 11.5)	
4α	1.65 m	32.3 d	1.84 m	46.9 d
5	-	50.2 s	-	43.9 s
6α	7.03 s	68.6 d	2.52 d (16.6)	35.9 t
6β	-		2.71 d (16.6)	
7	-	139.4 s	-	138.1 s
8	-	145.9 s	-	146.5 s
9	-	186.9 s	-	188.7 s
10β	-	79.8 s	2.42 d (9.8)	60.6 d
11	-	121.8 s	-	121.7 s
12	7.41 br s	146.9 d	7.41 s	145.8 d
13	1.91 d (1.1)	8.3 q	1.99 d (1.0)	7.8 q

Table 1. Cont.

Position	1		2	
	δ_H in ppm, Multiplicity, J (in Hz)	δ_C in ppm	δ_H in ppm, Multiplicity, J (in Hz)	δ_C in ppm
14	0.98 s	15.5 q	0.88 s	14.3 q
15	1.17 d (7.5)	16.0 q	0.98 d (6.7)	10.5 q
1'	-	165.7 s	-	167.6 s
2'	-	126.6 s	-	128.1 s
3'	5.92 qq (1.5, 7.2)	140.5 d	6.05 qq (1.4, 7.2)	137.8 d
4'	1.88 dq (1.6, 7.4)	15.8 q	1.97 dq (1.5, 7.2)	15.8 q
5'	1.55 quint. (1.5)	19.9 q	1.88 quint. (1.5)	20.6 q
OCOCH$_3$	2.19 s	20.9 q	-	-
OCOCH$_3$	-	170.9 s	-	-
OH-10	3.98 br s	-	-	-

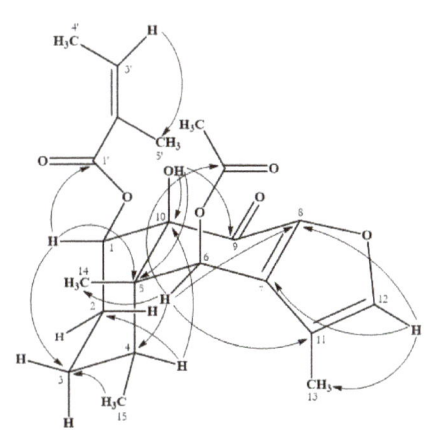

Figure 2. HMBC correlations observed for compound 1.

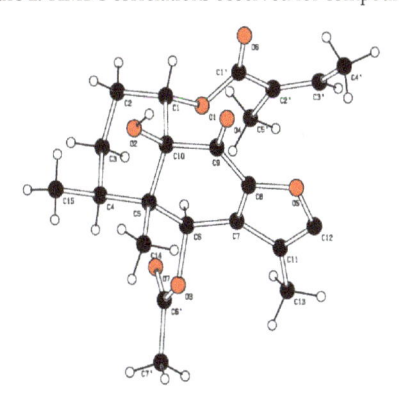

Figure 3. NOESY and ORTEP of compound 1.

The HR-EI-MS of compound 3 showed a molecular ion peak at m/z 322.1425 (M$^+$, calculated for C$_{17}$H$_{22}$O$_6$, 322.1416) and fragments in the upper part of the spectrum at m/z 262 (100%) [M-CH$_3$COOH]$^+$. Signals for seventeen carbon atoms were observed in its ^{13}C NMR spectrum. Their multiplicities were analyzed by a DEPT experiment which determined four methyl, three methylenes, three methines and six quaternary carbons. The ^{1}H and ^{13}C NMR spectra of 3 showed signals at δ(H) 3.18 (d, J = 4.6, H-C(1)); δ(C) 62.7 (d, C-(1)) and 60.9 (s, C-(10)), attributed to chemical shifts characteristic of an epoxide group at C-(1)-C-(10); signals at δ(H) 2.20 (s, OCOCH$_3$); δ(C) 20.9 (q, OCOCH$_3$) and 170.6 (s, OCOCH$_3$), corresponding to an acetate group at C(6), a signal for a geminal proton of an acetate group at δ(H) 5.92 (q, J = 1.8); δ(C) 73.8 (d, C(6)); and a signal at δ(H) 1.87 (d, J = 1.8); δ(C) 8.2 (q) corresponding to a Me(13). The HSQC and HMBC experiments (Table 2) confirmed the presence of an eremophilanolide skeleton and the localization of the epoxide and acetate groups, respectively. The relative stereochemistry of 3 was determined by a NOESY experiment (Figure 4). The positive NOE effect of H$_\alpha$-(6) with the methyl Me(13) signal was consistent with a β configuration of the γ-lactone, which agrees with the observed homoalilic coupling constant J$_{6-13}$ = 1.8 [15,16] with an angle between the two bonds of about 90°. Therefore, an α-configuration for the hydroxyl group at C(8) was determined. The observed NOE effect of H$_\alpha$-(1) with H$_\beta$-(9) confirmed the β-configuration of the epoxide. Compound 3 was identified based on its spectroscopic data as 1β,10β-epoxy-6β-acetoxy-8α-methoxy-eremofil-1(10),7(11)-diene-12,8β-olide, previously obtained by epoxidation and subsequent acetylation of the compound 6β-hydroxy-8α-methoxy-eremophil-1(10),7(11)-dien-12,8β-olide, isolated from S. magellanicus [2].

Table 2. ^{1}H (500 MHz) and ^{13}C (125 MHz) NMR data of compounds 3 and 7 * in CDCl$_3$.

Position	3		7 *	
	δ$_H$ in ppm, Multiplicity, J (in Hz)	δ$_C$ in ppm	δ$_H$ in ppm, Multiplicity, J (in Hz)	δ$_C$ in ppm
1β	3.18 d (4.6)	62.7 d	3.14 d (4.3)	63.0 d
2α	1.96 dd (6.8, 10.8)	20.3 t	2.04 m	21.0 t
2β	2.04 dd (5.9, 10.5)		2.21 m	
3α	1.36 dc (3.5, 9.4)	23.9 t	1.38 m	24.2 t
3β	1.63 m		1.61 m	
4α	1.62 m	32.5 d	1.62 m	33.0 d
5	-	43.4 s	-	43.5 s
6α	5.92 c (1.8)	73.8 d	5.69 t (1.7)	74.3 d
7	-	155.0 s	-	153.9 s
8	-	101.3 s	-	104.4 s
9α	1.79 d (13.6)	43.4 t	1.80 d (13.6)	43.6 t
9β	2.31 d (13.6)		2.27 d (13.6)	
10	-	60.9 s	-	61.0 s
11	-	124.6 s	-	126.8 s
12	-	170.8 s	-	170.9 s
13	1.87 d (1.8)	8.2 q	1.92 d (1.2)	8.6 q
14	1.09 s	14.5 q	1.09 s	14.5 q
15	1.04 d (7.2)	16.1 q	1.03 d (7.0)	16.5 q
OMe-8	-	-	3.23 s	50.9 q
OCOCH$_3$	2.20 s	20.9 q	2.21 s	21.0 q
OCOCH$_3$	-	170.6 s	-	170.3 s

* Source: Reina et al. [2].

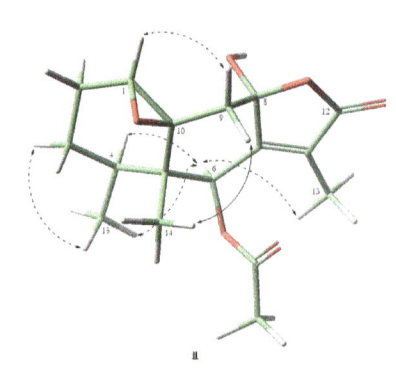

Figure 4. NOESY of compound 3.

Compound **4** was isolated as a colorless oil. Its HR-EI-MS mass spectrum showed a molecular-ion peak at *m/z* 220.1831 (M$^+$, calculated for C$_{15}$H$_{24}$O, 220.1827) and fragments at 205 (55%) [M-CH$_3$]$^+$ and 187 (19%) [M-CH$_3$ + H$_2$O]$^+$. The presence of three tertiary methyl groups at δ(H) 1.04 (s, Me(13)), 1.06 (s, Me(12)) and 1.28 (s, Me(15)) were observed in the ^{1}H and ^{13}C NMR spectra (Table 3). An HSQC experiment showed their correlations with carbons at δ(C) 16.5 (q, Me(13)), 28.8 (q, C(12)), and 26.2 (q, C(15)), and signals from two hydrogens of a cyclopropane at δ(H) 0.47 (dd, *J* = 9.6, 11.4, H-C(6)); δ(C) 30.1 (d, C(6)) and 0.71 (ddd, *J* = 6.1, 9.5, 11.4, H-C(7)); δ(C) 27.7 (d, C(7)) [16]. Additional signals attributable to two protons at δ(H) 4.69 (t, *J* = 1.6, H$_a$–C(14)) and 4.63 (q, *J* = 1.7 H$_b$-C(14)); a δ(C) 106.4 (t) signal assigned to an exocyclic methylene and a proton signal at δ(H) 2.20 (m, H$_β$-C(10)); δ(C) 53.6 (d, C(10)) were observed. The latter signal presented HMBC correlations at δ$_C$ 153.6 (t, C-(14)), 39.1 (t, C(2)) and 54.6 (d, C(5)), suggesting the location of the exocyclic methylene at C-1. The remaining signals were assigned by analysis of the 1D and 2D NMR spectra and by comparing these data with similar compounds. The structure of **4** was identified as the maaliol derivative (+)-1(14)-en-maaliol [17] which has not been previously isolated as a natural product.

Table 3. ^{1}H (500 MHz), ^{13}C (125 MHz) and HMBC NMR data of compound **4** in CDCl$_3$.

Position	δ$_H$ in ppm, Multiplicity, *J* (in Hz)	δ$_C$ in ppm	HMBC
1	-	153.6 s	-
2a	2.42 ddd (1.3, 6.3, 13.3)	39.1 t	C-1, C-4, C-10, C-14
2b	2.05 m		
3a	1.77 m	41.9 t	C-2, C-4
3b	1.56 d		
4	-	81.1 s	-
5α	1.32 m	54.6 d	C-1, C-4, C-6, C-10, C-11
6β	0.47 dd (9.6, 11.4)	30.1 d	C-4, C-8, C-11, C-13
7β	0.71 ddd (6.1, 9.5, 11.4)	27.7 d	C-5, C-11, C-13
8a	1.98 m	24.9 t	-
8b	1.01 m		
9a	1.90 m	26.9 t	C-5, C-10
9b	1.63 m		
10β	2.20 m	53.6 d	C-1, C-2, C-5, C-6, C-14
11	-	20.4 s	-
12	1.06 s	28.8 q	C-7, C-6, C-11, C-13
13	1.04 s	16.5 q	C-7, C-6, C-11, C-12
14a	4.69 t (1.6)	106.4 t	C-2, C-10
14b	4.63 q (1.7)		
15	1.28 s	26.2 q	C-3, C-4, C-5

Two unsaturated pyrrolizidine alkaloids (PAs) were isolated from the alkaloidal fraction, 9-O-angeloylpetasinecine (hectorine **5**), and rosmarinine (**6**). These alkaloids were identified by comparison of their spectral data (^{1}H a ^{13}C NMR and EIMS) with previous reports [13,14].

The antifeedant effects of compounds **1, 2** and **4** are shown in Table 4. Furanoeremophilane **2** was a strong antifeedant to *S. littoralis* (EC$_{50}$ = 0.64 μg/cm^2) while the maaliane **4** affected *M. persicae* (EC$_{50}$ = 0.97 μg/cm^2). Antifeedant furanoeremophilanes have been described in *Senecio* species such as *S. magellanicus* (against *M. persicae* and *S. littoralis*) [2] and *S. otites* (against *M. persicae* and *R. padi*) [11,18].

Table 4. Antifeedant activity of *S. fistulosus* compounds **1, 2, 4**.

Compound	S. littoralis %FI (50 µg/cm^2) [a]	EC$_{50}$ (µg/cm^2) [b]	M. persicae %SI [b] (50 µg/cm^2) [a]	EC$_{50}$ (µg/cm^2) [b]
1	65 ± 6 *		52 ± 7	
2	83 ± 6 *	0.64 (0.36–1.16)	52 ± 7	
4	64 ± 7 *		90 ± 3 *	0.97 (0.71–1.32)

[a] %FI/%SI = [1 − (T/C)] × 100, where T and C are the consumption/settling of treated and control leaf disks, respectively. [b] Effective antifeedant dose (EC$_{50}$) and 95% confidence (lower, upper). * $p < 0.05$, Wilcoxon paired test.

Furanoeremophilanes are less abundant in *Senecio* than eremophilanolides. Therefore, the studies on their structure-activity relationships (SAR) are limited. Table 5 shows a compilation of the available information on the SAR of these structures, including the results presented in this work. The active compounds against the aphid *M. persicae* are characterized by the absence of substituents in C-1, C-3 and C-10, regardless of the substituent in C-6 (**8–10, 11, 12**). The presence of β-OH/C-1 and the α-OAng/C-3 group (compound **2**) resulted in an important antifeedant activity against *S. littoralis*. In addition, the C-6 substitution pattern together with the C-1/C-10 unsaturation determined post-ingestion effects on *S. littoralis* [11].

Table 5. Antifeedant structure-activity relationships of *Senecio* furanoeremophilanes against *S. littoralis* and *M. persicae*.

Compound	Substituent				S. littoralis	M. persicae
	C-1	C-3	C-6	C-10	%FI [c] (EC$_{50}$) [d]	%SI [c] (EC$_{50}$) [d]
1	α-OAng	H$_2$	β-OAc	β-OH	65.0	52.0
2	α-OH	α-OAng	H$_2$	α-H	83.0 (0.64)	52.0
8 [a]	Δ^1	H$_2$	β-OAng	Δ^{10}	32.0	67.0
9 [a]	Δ^1	H$_2$	β-OH	Δ^{10}	62.0	71.0
10 [a]	Δ^1	H$_2$	β-OCOCH$_2$CH$_3$	Δ^{10}	45.0	75.0
11 [b]	H$_2$	H$_2$	β-OAc	α-H	51.0	74.0 (21.9)
12 [b]	H$_2$	H$_2$	β-OTigl	α-H	65.0	74.0 (12.2)

[a] Compounds **8–10** from Domínguez et al. [11]. [b] Compounds **11, 12** from Reina et al. [2]. [c] %FI / %SI values at 50µg/cm^2. [d] Effective antifeedant dose (µg/cm^2).

Maalianes have been isolated from a range of organisms, such as liverworts, marine sponges, soft corals and bacteria, however, they are not abundant in nature. A small amount of biological activity has been reported and includes fish toxicity, in vitro antimalarial activity, cytotoxicity and antimicrobial [19]. This is the first report on the insect antifeedant effects of a maaliane sesquiterpene.

PAs **5** and **6**, with necines of the rosmarinecine and petasinecine type (1,2-saturated base), were isolated from the alkaloidal extract of *S. fistulosus*. The role of PAs as plant defenses against phytophagous insects has been widely documented [20], however, this alkaloidal extract showed moderate-low antifeedant activity (62 ± 6%FR against *S. littoralis* and 69 ± 6%SI against *R. padi*).

PAs with unsaturated retronecines are potentially more toxic than rosmarinecine and petasinecine type (1,2-saturated base) PAs [21]. For example, rosmarinine with a petasinecine, did not form hepatotoxic reactive pyrrole intermediates [22,23] and cytotoxic assays have demonstrated a higher toxicity of retronecine and otonecine PAs compared with platynecine PAs [24]. Therefore, PAs **5** and **6** have a low risk of associated toxicity.

3. Materials and Methods

3.1. General

For column chromatography (CC), Si-gel (107734, 107741, and 107749, Merck) and Sephadex LH-20 (Sigma–Aldrich) were used. For TLC chromatography, Si-gel (105554 and 105715; Merck) plates were used and visualized with óleum solution (sesquiterpenes) and Dragendorff's reagent (alkaloids). The prep. HPLC chromatography was carried out on a Beckman 125P system equipped with an Ultrasphere semiprep column (10 × 250 mm) and a UV/visible diode array detector 168. Optical rotations were determined at 20 °C on a Perkin-Elmer 343 Plus polarimeter. IR Spectra were recorded in CHCl$_3$ on a Perkin Elmer 1600 spectrophotometer. NMR spectra were recorded on a pulsed-field gradient Bruker Advance II-500 MHz spectrometer (solvent as internal standard CDCl$_3$, at δ_H 7.26 and δ_C 77.0) and the Bruker software was used for DEPT, ^{1}H, ^{1}H-COSY (Homonuclear correlation spectroscopy), NOESY (Nuclear Overhauser Effect Spectroscopy), HSQC (Heteronuclear single quantum coherence spectroscopy) and HMBC (Heteronuclear Multiple Bond Correlation). EI and HR-EI-MS spectra were recorded in m/z on a Micromass Autospec spectrometer.

3.2. Extraction and Isolation

Aerial parts of *S. fistulosus* (Asteraceae), identified by Orlando Dollenz, were collected in Sierra Baguales (March 2009, Punta Arenas, Magallanes, Chile,) during the flowering period. A voucher specimen (# 7569) has been deposited in the Herbarium of the Patagonian Institute, Magallanes University (UMAG), Punta Arenas, Chile.

Grounded dried aerial plant parts (2.50 kg) were extracted with MeOH (20 L) at room temperature for a week to give a crude MeOH extract (190.5 g, 7.62% yield of plant dry weight). The MeOH extract (157.6 g) was treated with a f H$_2$SO$_4$ 0.5 M and CH$_2$Cl$_2$ (1:1) solution. Zinc dust was used to reduce the aqueous phase under continuous stirring (4–6 hours) and then filtered, basified (30% NH$_4$OH, pH = 8–9) and extracted with CH$_2$Cl$_2$ (236.0 mg of alkaloids, 9.4 × 10^{-3}%). The organic phase, dried over anhydrous Na2SO$_4$ (non-alkaloidal fraction, 9.0 g, 0.36%), was chromatographed on a SiO$_2$ vacuum-liquid chromatography column (VLC) and eluted with a hexane/EtOAc/MeOH gradient to give seven fractions. Fr-0 (hexane 100%, 360.7 mg), Fr-1, (hexane/EtOAc 95:5%, 2.2 g), Fr-2 (hexane/EtOAc 90:10%, 1.7 g), Fr-3 (hexane/EtOAc 75:25%, 1.1 g), Fr-4 (hexane/EtOAc 50:50%, 966.6 mg), Fr-5 (EtOAc 100%, 478.2 mg), Fr-6 (MeOH 100%, 2.6 g). Fr-1 (2.2 g, 8.8 × 10^{-2}%) was further chromatographed on a CC Sephadex LH-20 column, CC silica gel and semi-preparative normal-phase HPLC eluted with an isocratic mixture of hexane/EtOAc at 3 ml/min flow rate to give compound **4** (16.1 mg, 6.4 × 10^{-4}%). Fr-2 (1.7 g) was chromatographed on CC Sephadex LH-20, CC silica gel, circular chromatography and semi-preparative normal phase HPLC eluted with an isocratic mixture of hexane/EtOAc at a flow rate of 3 ml/min to give compounds **1** (41.0 mg, 1.6 × 10^{-3}%), **2** (29.9 mg, 1.2 × 10^{-3}%) and **3** (3.5 mg, 1.4 × 10^{-4}%).

The alkaloidal fraction was submitted to neutral alumina CC, eluted with an EtOAc/MeOH gradient and PTLC (20 × 20 cm, 0.25 mm) to give compounds **5** (1.3 mg; 5.2 × 10^{-5}%) and **6** (0.9 mg; 3.6 × 10^{-5}%).

3.2.1. α-Angeloyloxy-6β-acetoxy-10β-hydroxy-9-oxo-furanoeremophilane (1)

Colorless crystal, mp 127–130 °C (hexane/EtOAc); $[\alpha]_D^{20}$ −28.2 (*c*, 0.82, CHCl$_3$). IR (CHCl$_3$) $\nu_{máx.}$: 3452, 1748, 1720, 1679, 1232 cm^{-1}. EI-MS: 404 (1, M$^+$), 345 (3), 260 (20), 262 (12), 178 (55), 136 (4), 91 (4), 83 (100), 57 (8), 55 (39). HR-EI-MS: 404.1838 (M$^+$, C$_{22}$H$_{28}$O$_7$; calculated for 404.1835). For ^{1}H and ^{13}C NMR data see Table 1.

3.2.2. α-hydroxy-3α-angeloyloxy-10αH-9-oxo-furanoeremophilane (2)

White amorphous solid; $[\alpha]_D^{20}$ −37.5 (*c*, 0.59, CHCl$_3$). IR (CHCl$_3$) $\nu_{máx.}$: 3446, 1733, 1716, 1699, 1456 cm^{-1}. EI-MS: 346 (3, M$^+$), 246 (9), 228 (10), 213 (18), 191 (8), 163 (100), 135 (9), 105 (5), 91 (13), 83 (19), 77 (7), 55 (24). HR-EI-MS: 346.1785 (M$^+$, C$_{20}$H$_{26}$O$_5$; calculated for 346.1780). For ^{1}H and ^{13}C NMR data see Table 1 [12].

3.2.3. β,10β-epoxy-6β-acetoxy-8α-hydroxy-eremophil-7(11)-en-8β,12-olide (3)

Colorless oil; $[\alpha]_D^{20}$ −72.8 (c, 0.25, CHCl$_3$). IR (CHCl$_3$) ν$_{máx}$: 3392, 1771, 1749, 1717 cm^{-1}. EI-MS: 322 (0.4, M$^+$), 298 (2), 280 (3), 262 (100), 244 (7), 216 (6), 142 (100), 124 (50), 95 (63). HR-EI-MS: 322.1425 (M$^+$, C$_{17}$H$_{22}$O$_6$; calculated for 322.1416). For ^{1}H and ^{13}C NMR data see Table 2.

3.2.4. (+)-1(14)-en-maaliol (4)

Colorless oil; $[\alpha]_D^{20}$ +4.0 (c, 0.78, CHCl$_3$). IR (CHCl$_3$) ν$_{máx}$: 3421, 2930, 2868, 1653, 1457, 1375, 1152, 913, 889, 668 cm^{-1}. EI-MS: 220 (5, M$^+$), 205 (55), 187 (19), 177 (12), 162 (24), 159 (34), 147 (36), 133 (29), 121 (38), 119 (60), 105 (75), 93 (78), 91 (100), 79 (93), 69 (85). HR-EI-MS: 220.1831 (M$^+$, C$_{15}$H$_{24}$O; calculated for 220.1827). For ^{1}H and ^{13}C NMR spectral data see Table 3.

3.2.5. 9-O-angelylpetasinecine (hectorine) (5)

Colorless oil; $[\alpha]_D^{20}$ −62.86 (c, 0.07, CHCl$_3$). EI-MS: 239 (9, M$^+$), 222 (9), 190 (3), 188 (11), 140 (61), 122 (9), 111 (9), 83 (100), 70 (11), 68 (6), 55 (27). HR-EI-MS: 239.1512 (M$^+$, C$_{13}$H$_{21}$NO$_3$; calculated for 239.1521). ^{1}H NMR (CDCl$_3$, 500 MHz): δ$_H$ 2.51 (1H, m, H$_\alpha$-C(1)), 4.23 (1H, t, J = 3.9 Hz, H$_\alpha$-C(2)), 3.42 (1H, dd, J = 4.0, 13.0 Hz, H$_\alpha$-C(3)), 3.03 (1H, d, J = 13.0, Hz, H$_\beta$-C(3)), 3.50 (1H, t, J = 8.5 Hz, H$_\alpha$-C(5)), 2.98 (1H, m, H$_\beta$-C(5)), 1.84 (1H, m, H$_\alpha$-(6)), 2.06 (1H, m, H$_\beta$-C(6)), 1.80 (1H, m, H$_\alpha$-C(7)), 1.99 (1H, m, H$_\beta$-C(7)), 3.90 (1H, m, H$_\alpha$-C(8)), 4.73 (1H, dd, J = 10.0, 11.5 Hz, H$_d$-C(9)), 4.15 (1H, dd, J = 4.9, 11.5 Hz, H$_u$-C(9)), 6.15 (1H, cc, J = 1.5, 7.2 Hz, H-(3′)), 1.99 (3H, d, J = 7.0 Hz, H-C(4′)), 1.90 (3H, quint., J = 1.6 Hz, H-C(5′)). ^{13}C-NMR: δ(C) 46.3 (d, C-1), 73.2 (d, C-2), 61.6 (t, C-3), 56.9 (t, C-5), 27.2 (t, C-6), 27.9 (t, C-7), 66.7 (d, C-8), 60.3 (t, C-9), 169.2 (s, C-1′), 127.4 (s, C-2′), 139.8 (d, C-3′), 16.1 (q, C-4′), 20.7 (q, C-5′) [13].

3.2.6. Rosmarinine (6)

As a white resin; $[\alpha]_D^{20}$ −51.1 (c, 0.09, CHCl$_3$). EI-MS: 353 (5, M$^+$), 282 (2), 227 (4), 180 (6), 156 (43), 154 (87), 138 (100), 122 (32), 98 (27), 82 (86), 81 (21), 55 (41). HR-EI-MS: 353.1835 (M$^+$, C$_{18}$H$_{27}$NO$_6$; calculated for 353.1838). ^{1}H NMR (CDCl$_3$, 500 MHz): δ(H) 2.54 (1H, m, H$_\alpha$-C(1)), 4.26 (1H, m, H$_\beta$-C(2)), 3.15 (1H, dd, J = 7.3, 11.2 Hz, H$_\alpha$-C(3)), 2.94 (1H, dd, J = 7.9, 11.3 Hz, H$_\beta$-C(3)), 3.33 (1H, t, J = 8.9 Hz, H$_\alpha$-C(5)), 2.60 (1H, m, H$_\beta$-C(5)), 2.08 (1H, m, H$_\alpha$-C(6)), 2.27 (1H, m, H$_\beta$-C(6)), 5.08 (1H, t, J = 2.9 Hz, H$_\alpha$-C(7)), 3.71 (1H, dd, J = 3.4, 7.8 Hz, H$_\alpha$-C(8)), 4.89 (1H, dd, J = 5.4, 12.6 Hz, H$_d$-C(9)), 4.11 (1H, dd, J = 1.1, 12.6 Hz, H$_u$-C(9)), 1.80 (1H, m, H$_\beta$-C(13)), 2.27 (1H, m, H$_a$-C(14)), 1.96 (1H, m, H$_b$-C(14)), 1.34 (3H, s, H$_\alpha$-C(18)), 0.97 (3H, d, J = 6.7 Hz, H$_\alpha$-C(19)), 5.80 (1H, c, J = 7.1 Hz, H-(20)), 1.85 (3H, dd, J = 1.5, 7.2 Hz, H-C(21)). ^{13}C-NMR: δ(C) 49.1 (d, C-1), 69.3 (d, C-2), 60.9 (t, C-3), 53.5 (t, C-5), 34.6 (t, C-6), 75.2 (d, C-7), 69.6 (d, C-8), 62.3 (t, C-9), 180.7 (s, C-11), 77.6 (s, C-12), 38.0 (d, C-13), 39.7 (t, C-14), 132.7 (s, C-15), 167.6 (s, C-16), 25.8 (q, C-18), 11.9 (q, C-19), 134.9 (d, C-20), 15.3 (q, C-21) [14].

3.2.7. Crystal Structure Analysis

Intensity data, for both compounds, were collected at 293 K on an Oxford Diffraction Supernova dual Atlas CCD diffractometer, using Cu Kα (λ = 1.5418 Å) radiation. Data collection, cell refinement and data reduction were performed with the CrysAlisPRO [25] set of programs. The structure was solved by direct methods using SIR97 [26]. Refinements were performed with SHELXL-97 [27] using full-matrix least squares, with anisotropic displacement parameters for all the non-hydrogen atoms. The H-atoms were placed at calculated positions with C-H distances 0.95-1.00 Å and refined using a riding model. Calculations were mainly performed with WinGX [28] and molecular graphics were computed with PLATON [29].

X-ray crystal data: C$_{22}$H$_{28}$O$_7$, Mw = 404.44, orthorhombic, space group, P2$_1$2$_1$2$_1$, Z = 8, a = 8.7493(2), b = 13.4234, c = 37.0309(11) Å; V = 4349.1(2) Å^3, μ(Cu Kα) = 0.76 mm^{-1}, ρ$_{calc}$ = 1.23 g.cm^{-3}; S = 1.06, final R indices: R$_1$ = 0.0678 and Rw = 0.1860 for 7134 observed from 8339 independent and 15715 measured reflections (θ$_{max}$ = 70.99, I > 2σ(I) criterion and 536 parameters); maximum and minimum residues are and 0.30 and −0.24 e.Å^{-3} respectively. There are two independent molecules in the asymmetric unit with minor conformational differences between them. The absolute structure is based on the refinement of the Flack [30] (Flack 1983), x = 0.0 (3), parameter against 3610 CuKα Bijvoet pairs. The Hooft [31] analysis yielded y = 0.06(8) and P2 (true) = 1.000.

Crystallographic data (excluding structure factor tables) has been deposited with de Cambridge Crystallographic Data Center as supplementary publications no. CCDC1455588. Copies of the data can be obtained free of charge on application to The Director, CCDC, 12 Union Road, Cambridge CB1EZ, UK ((Fax: Int. + (1223) 336 033); e-mail: deposit@ccdc.cam.ac.uk)).

3.3. Insect Bioassays

S. littoralis, *M. persicae* and *R. padi* colonies were reared on an artificial diet [32], bell pepper (*Capsicum annuum*) and barley (*Hordeum vulgare*) plants, respectively. The plants are grown from seeds in pots with commercial substrate. The plants for rearing aphids are infected regularly (bell pepper plants with 4 leaves, barley plants of 10 cm length). The insect colonies and host plants were maintained at 22 ± 1 °C, > 70% relative humidity with a photoperiod of 16:8 h (L:D) in a growth chamber.

Antifeedant bioassays: The upper surface of *C. anuum* and *H. vulgare* leaf disks or fragments (1.0 cm^2) were treated with 10 μl of the test substance. The crude extracts and products were tested at an initial dose of 100 or 50 μg/cm^2 respectively. Five Petri dishes (9 cm diam.) or twenty ventilated plastic boxes (2 × 2 cm) with two newly molted *S. littoralis* L6 larvae (≤24 h) or ten apterous aphid adults (24–48 h old) each were allowed to feed at room temperature for *S. littoralis* (<2 h) or in a growth chamber for the aphids (24 h, environmental conditions as above). Each experiment was repeated 2-3 times (SE < 10%) and terminated when the consumption of the control disks reached 65–75% for *S. littoralis* or after 24 h for aphids. The leaf disk area consumed was measured on their digitalized images (Image J, http://imagej.nih.gov/ij). Settling was measured by counting the number of aphids settled on each leaf fragment. Feeding or settling inhibition (%FI or %SI) was calculated as % FI/%SI = [1 − (T/C) × 100], where T and C are the consumption/settling of treated and control leaf disks, respectively. The antifeedant effects (% FI/SI) were analyzed for significance by the nonparametric Wilcoxon signed-rank test. Extracts and compounds with an FI/SI ≤ 75% were further tested in a dose-response experiment (3–4 serial dilutions) to calculate their relative potency (EC$_{50}$, the effective dose to give a 50% feeding/settling reduction) from a linear regression analysis (% FI/SI on Log-dose) [33].

4. Conclusions

Senecio fistulosus is characterized by their content in sesquiterpenes (furanoeremophilanes, eremophilanolides and maaliane type) and pyrrolizidine alkaloids. The antifeedant properties of ethanolic, non-alkaloidal, alkaloidal extracts and compounds have been studied. Most of the insect antifeedant effects were found in the ethanolic and non-alkaloidal extracts, containing mainly sesquiterpenes with low amounts of PAs. The isolated furanoeremophilanes sesquiterpenes type had structure-dependent antifeedant effects. In addition to their antifeedant action, these sesquiterpenes could play a role in insect-plant interactions.

Author Contributions: Conceptualization, A.G.-C., M.R. and V.F.; funding acquisition, A.G.-C.; investigation, A.G.-C., L.R.-V., M.L. and M.R.; methodology, A.G.-C., L.R.-V and M.R.; resources, V.F.; writing—original draft, A.G.-C., M.L. and M.R.; writing—review & editing, A.G.-C. and L.R.-V.

Funding: This work has been supported by grants CTQ2015-64049-C3-1-R, (MINECO/FEDER), UMAG 027103-026703 (Dirección de Investigación, Chile) and a JAEPRE-DOC-CSIC predoctoral fellowship to L.R.V.

Conflicts of Interest: The authors declare no conflicts of interest.

References

1. Portero, A.G.; González-Coloma, A.; Reina, M.; Díaz, C.E. Plant-defensive sesquiterpenoids from *Senecio* species with biopesticide potential. *Phytochem. Rev.* **2012**, *11*, 391–403. [CrossRef]
2. Reina, M.; Santana, O.; Domínguez, D.M.; Villarroel, L.; Fajardo, V.; López-Rodríguez, M.; González-Coloma, A. Defensive sesquiterpenes from *Senecio candidans* and *S. magellanicus*, and their structure-activity relationships. *Chem. Biodivers.* **2012**, *9*, 625–643. [CrossRef] [PubMed]
3. Ruiz-Vásquez, L.; Reina, M.; López-Rodríguez, M.; Giménez, C.; Cabrera, R.; Cuadra, P.; Fajardo, V.; González-Coloma, A. Sesquiterpenes, flavonoids, shikimic acid derivatives and pyrrolizidine alkaloids from *Senecio kingii* Hook. *Phytochemistry* **2015**, *117*, 245–253. [CrossRef] [PubMed]
4. Ruiz-Vásquez, L.; Olmeda, A.S.; Zuñiga, G.; Villarroel, L.; Echeverri, L.F.; González-Coloma, A.; Reina, M. Insect Antifeedant and Ixodicidal Compounds from *Senecio adenotrichius*. *Chem. Biodivers.* **2017**, *14*, e1600155. [CrossRef] [PubMed]
5. Ruiz-Vásquez, L.; Ruiz Mesia, L.; Reina-Artiles, M.; López-Rodríguez, M.; González-Platas, J.; Giménez, C.; Cabrera, R.; González-Coloma, A. Benzofurans, benzoic acid derivatives, diterpenes and pyrrolizidine alkaloids from Peruvian *Senecio*. *Phytochem. Lett.* **2018**, *28*, 47–54. [CrossRef]
6. Agullo-Ortuño, M.T.; Diaz, C.E.; Gonzalez-Coloma, A.; Reina, M. Structure-Dependent Cytotoxic Effects of Eremophilanolide Sesquiterpenes. *Nat. Prod. Commun.* **2017**, *12*, 663–665. [CrossRef] [PubMed]
7. Henríquez, J.M.; Pisano, E.; Marticorena, C. *Catálogo de la Flora Vascular de Magallanes (XII Región)*; Anales Instituto de la Patagonia, Sección Ciencias Naturales, Universidad de Magallanes: Punta Arenas, Chile, 1995; pp. 5–30.
8. Arias Cassará, M.L.; Borkosky, S.A.; González Sierra, M.; Bardón, A.; Ybarra, M.I. Two new furanoeremophilanes from *Senecio santelisis*. *Chem. Biodivers.* **2010**, *7*, 1745–1753. [CrossRef] [PubMed]
9. Bolzan, A.A.; Silva, C.M.; Francescato, L.N.; Murari, A.L.; Silva, G.N.; Heldwein, C.G.; Heinzmann, B. Espécies de *Senecio* na Medicina Popular da América Latina e Toxicidade Relacionada a sua Utilização. *Lat. Am. J. Pharm.* **2007**, *26*, 619–625.
10. Villarroel, L.; Torres, R.J. A new furanoeremophilane from *Senecio fistulosus*. *J. Nat. Prod.* **1985**, *48*, 841–842. [CrossRef]
11. Domínguez, D.M.; Reina, M.; Villarroel, L.; Fajardo, V.; González-Coloma, A. Bioactive Furanoeremophilanes from *Senecio otites* Kunze ex DC.Z. *Naturforsch* **2008**, *63c*, 837–842. [CrossRef]
12. Bohlmann, F.; Zdero, C.; King, R.M.; Robinson, H. Furanoeremophilanes from *Senecio smithii*. *Phytochemistry* **1981**, *20*, 2389–2391. [CrossRef]
13. Bai, Y.; Benn, M.; Duke, N.; Gul, W.; Huangand, Y.Y.; Rüeger, H. The alkaloids of *Brachyglottis hectori*. *Arkivoc* **2006**, *3*, 34–42.
14. Were, O.; Benn, M.; Munavu, R.M. The pyrrolizidine alkaloids of *Senecio syringifolius* and *S. hadiensis* from Kenya. *Phytochemistry* **1993**, *32*, 1595–1602. [CrossRef]
15. Naya, K.; Nogi, N.; Makiyama, Y.; Takashima, H.; Imagawa, T. The Photosensitized Oxygenation of Furanoeremophilanes. II. The Preparation and Stereochemistry of the Isomeric Hydroperoxides and the Corresponding Lactones from Furanofukinin and Furanoeremophilane. *Bull. Chem. Soc. Jpn.* **1977**, *50*, 3002–3006. [CrossRef]
16. Harrie, J.M.; Joannes, B.P.A.; Wijnberg, C.R.; Groom, A. Rearrangement reactions of aromadendrane derivatives. The synthesis of (+)-maaliol, starting from natural (+)-aromadendrene-IV. *Tetrahedron* **1994**, *50*, 4733–4744.
17. De Pascual, T.; Urones, J.G.; Fernández, A. An aristolochic acid derivative from *Aristolochia longa*. *Phytochemistry* **1983**, *22*, 2753–2754. [CrossRef]
18. Gutierrez, C.; Fereres, A.; Reina, M.; Cabrera, R.; González-Coloma, A. Behavioral and sublethal effects of structurally related lower terpenes on *Myzus persicae*. *J. Chem. Ecol.* **1997**, *23*, 1641–1650. [CrossRef]
19. Duran-Peña, M.J.; Ares, J.M.B.; Hanson, J.R.; Collado, I.G.; Hernandez-Galan, R. Biological activity of natural sesquiterpenoids containg a gem-dimethylcyclopropane unit. *Nat. Prod. Rep.* **2015**, *32*, 1236–1248. [CrossRef]
20. Hartmann, T. Chemical ecology of pyrrolizidine alkaloids. *Planta* **1999**, *207*, 483–495. [CrossRef]
21. Castells, E.; Mulder, P.P.J.; Pérez-Trujillo, M. Diversity of pyrrolizidine alkaloids in native and invasive *Senecio pterophorus* (Asteraceae): Implications for toxicity. *Phytochemistry* **2014**, *108*, 137–146. [CrossRef]

22. Culvenor, C.C.J.; Edgar, J.A.; Smith, L.W.; Jago, M.V.; Peterson, J.E. Active metabolites in the chronic hepatotoxicity of pyrrolizidine alkaloids, including otonecine esters. *Nat. New Biol.* **1971**, *229*, 255–256. [CrossRef]
23. Styles, J.; Asbey, J.; Mattocks, A.R. Evaluation in vitro of several pyrrolizidine alkaloid carcinogens: Observations on the essential pyrrolic nucleus. *Carcinogenesis* **1980**, *1*, 161–164. [CrossRef] [PubMed]
24. Li, Y.H.; Kan, W.L.T.; Li, N.; Lin, G. Assessment of pyrrolizidine alkaloid induced toxicity in an in vitro screening model. *J. Ethnopharmacol.* **2013**, *150*, 560–567. [CrossRef] [PubMed]
25. *CrysAlis RED Version 1.171.32.5*; Oxford Diffraction Ltd.: Abingdon, UK, 2007.
26. Altomare, A.; Cascarano, G.; Giacovazzo, C.; Guagliardi, A.; Moliterni, A.G.G.; Burla, M.C.; Polidor, G.; Camalli, M. *Spagna R. SIR97*; University of Bari: Bari, Italy, 1997.
27. Sheldrick, G.M. A short history of SHELX. *Acta Cryst.* **2008**, *A64*, 112–122. [CrossRef] [PubMed]
28. Farrugia, L.J. WinGX suite for small-molecule single-crystal crystallography. *J. Appl. Cryst.* **1999**, *32*, 837–838. [CrossRef]
29. Spek, A.L. Single-crystal structure validation with the program PLATON. *J. Appl. Cryst.* **2003**, *36*, 7–13. [CrossRef]
30. Flack, H.D. On enantiomorph-polarity estimation. *Acta Cryst.* **1983**, *A39*, 876–881. [CrossRef]
31. Hooft, R.W.W.; Straver, L.H.; Spek, A.L. Determination of absolute structure using Bayesian statistics on Bijvoet differences. *J. Appl. Cryst.* **2008**, *41*, 96–103. [CrossRef]
32. Poitout, S.; Bues, S. Elevage de plusieursespeces de Lepidopteres Noctuidae sur milleu artificiel simplifié. *Ann. Zool. Ecol. Anim.* **1970**, *2*, 79–91.
33. Burgueño-Tapia, E.; Castillo, L.; González-Coloma, A.; Joseph-Nathan, P. Antifeedant and phytotoxic activity of the sesquiterpene p-benzoquinone perezone and some of its derivatives. *J. Chem. Ecol.* **2008**, *34*, 766–771. [CrossRef]

© 2019 by the authors. Licensee MDPI, Basel, Switzerland. This article is an open access article distributed under the terms and conditions of the Creative Commons Attribution (CC BY) license (http://creativecommons.org/licenses/by/4.0/).

Review

Opportunities and Scope for Botanical Extracts and Products for the Management of Fall Armyworm (*Spodoptera frugiperda*) for Smallholders in Africa

Naomi B. Rioba [1,*] and Philip C. Stevenson [2,3]

1. School of Agriculture and Biotechnology, University of Kabianga, Kericho P.O. Box 2030-20200, Kenya; naomirioba@kabianga.ac.ke
2. Natural Resources Institute, University of Greenwich, Chatham Maritime, Kent ME4 4TB, UK; p.c.stevenson@gre.ac.uk
3. Jodrell Laboratory, Royal Botanic Gardens, Kew, Richmond, Surrey TW9 3AB, UK
* Correspondence: naomirioba@kabianga.ac.ke; Tel.: +254-720-361-466

Received: 16 January 2020; Accepted: 2 February 2020; Published: 6 February 2020

Abstract: Fall Armyworm (FAW) (*Spodoptera frugiperda*) is a polyphagous and highly destructive pest of many crops. It was recently introduced into Africa and now represents a serious threat to food security, particularly because of yield losses in maize, which is the staple food for the majority of small-scale farmers in Africa. The pest has also led to increased production costs, and threatens trade because of quarantines imposed on produce from the affected countries. There is limited specific knowledge on its management among smallholders since it is such a new pest in Africa. Some synthetic insecticides have been shown to be effective in controlling FAW, but in addition to the economic, health and environmental challenges of pesticide use insecticide resistance is highly prevalent owing to years of FAW management in the Americas. Therefore, there is a need for the development and use of alternatives for the management of FAW. These include plant-derived pesticides. Here we review the efficacy and potential of 69 plant species, which have been evaluated against FAW, and identify opportunities for use among small-scale maize farmers with a focus on how pesticidal plants might be adopted in Africa for management of FAW. The biological activities were diverse and included insecticidal, insectistatic (causing increased larval duration), larvicidal, reduced growth and acute toxicity (resulting in adverse effects within a short time after exposure). While most of these studies have been conducted on American plant taxa many South American plants are now cosmopolitan weeds so these studies are relevant to the African context.

Keywords: biopesticides; botanicals; corn; insects; pests; prospects

---

1. Introduction

Fall Armyworm (FAW) (*Spodoptera frugiperda* Hurst) (Lepidoptera: Noctuidae) is a highly polyphagous pest having been reported on more than 80 species in 23 families [1] including cotton (*Gossypium hirsutum* L.) (Malvales: Malvaceae), corn (*Zea mays* L.) (Cyperales: Poaceae) and many other grass crops [2]. Originally native and restricted to the Americas, FAW was recorded for the first time in Africa in 2016 [3] and now it has spread to over 30 countries in Africa.

These invasive populations are now well established and causing severe destruction to important crops that underpin the livelihoods of many farmers across Africa [4], due to the variety of host plants and the favorable environment and climate. The pest has characteristics that means it presents a wider-reaching threat to Africa [3]. For example, in comparison with the African armyworm *(Spodoptera exempta)*, FAW larvae have unique mouthparts with notched cutting edges, enabling it to feed on flora that are rich in silica content. More so, the older larvae feed on the younger ones and can dominate the competitors of the same species and others of different species within the same genus hence ensuring its survival [5]. FAW has raised greater concern among farmers than related African *Spodoptera* species because it causes especially severe damage to maize, feeding on virtually all parts of the plant leading to considerable damage, and sometimes results in total crop failure [6].

Sustainable approaches to managing this new African pest should ideally be integrated, tailored and appropriate for smallholders with mixed cropping farming systems and reduced input costs. While the use of chemical pesticides dominates existing approaches [7], several alternative control options exist and are being considered including resistant varieties [8–10], biological control [11,12], crop management practices [13,14], plant diversity [14], and mechanical methods [15]. However, none of these methods has yet delivered a viable option for effective control of FAW, hence the search for alternative approaches including those from plant extracts and their products. Some pesticidal plants and botanical insecticides are effective and their use could reduce reliance on synthetic insecticides since they have lower non-target impacts and could even boost growth [16–19]. Here we review existing research on plant extracts that have been evaluated for the management of FAW with the aim of identifying those with potential for use by small-scale farmers in Africa, or informing approaches to identifying and evaluating untested native African plant taxa since pesticidal plants are already used as crudely produced products among smallholder farming communities in Africa with notable success [20–22]. While one recent study has specifically sought to evaluate African plant taxa for activity against FAW [23], most of the studies reviewed here have been conducted on South American taxa but many of these species are now cosmopolitan weeds so are relevant to the African context. For example, *Ageratum conyzoides* L. is a widely used plant for a multitude of uses in Africa including pest control but originates from South America where it has been evaluated for efficacy against FAW [24,25]. Similarly, *Dysphania* (syn. *Chenopodium*) *ambrosioides* (L.) Mosyakin & Clemants, has been shown to be biologically active vs FAW [26,27] but is also considered for use in Africa [28], while species such as *Corymbia* (syn. *Eucalyptus*) *citriodora* (Hook.) K.D. Hill & L.A.S.Johnson are widespread in both Africa and America but non-native and have been evaluated for activity against FAW [29].

2. Opportunities and Potential of Botanical Extracts and Products

Interest in using plant extracts for pest control is increasing since these can: 1) reduce the cost of production of the crop, 2) reduce the environmental damage and non-target effects, and 3) reduce dependence on synthetic insecticides [30,31]. There are many researchers studying insecticidal plants for the control of FAW with several reporting promising results, although many do not since they do not establish the chemical basis of activity or store any reference specimens [32]. Some of the pesticidal plant species that have been shown to be effective in the management of FAW are presented in Table 1.

FAW larvae ingesting maize leaves treated with the essential oil of *Ageratum conyzoides* were killed with 70% mortality caused at the concentration of 0.5%. The essential oil contained precocene as the major active component (87%) [24]. This finding is highly relevant to the African context where this plant grows widely on farmland. This means that it is easily available to farmers. It has already been used to control lepidopteran and other pests by some small holders in Africa. It has been shown to have reduced non-target effects on natural enemies of pests [85].

Ruta graveolens, Cymbopogon citratus, Zingiber officinale, Malva sylvestris, Petiveria alliaceae, Bacharis genisterlloides and Artemisia verlotorum were also shown to cause mortality for caterpillars of FAW [36], but active components in these plants were not identified, meaning that the work has limited value in the efforts to develop new approaches for FAW control unless more research is conducted to identify the active compounds that are responsible for the biological activity.

Table 1. Plant species that have been evaluated for their activity against Fall Armyworm (FAW) and potential for use in its management.

Family	Plant Species	Action	Refs
Amaranthaceae	*Dysphania* (syn. *Chenopodium*) *ambrosioides* L. Mosyakin & Clemants	Mortality, decreased pupal weight	[26,27]
Anacardiaceae	*Schinus molle* L.	High mortality	[26]
Annonaceae	*Annona squamosa* L.	Decreased pupa weight, increased larval mortality	[33]
Apiaceae	*Foeniculum vulgare* Mill.	Sublethal effects	[34]
Apocynaceae	*Calotropis procera* (Aiton) W.T. Aiton	Decreased pupa weight, increased larval mortality	[33]
Asparagaceae	*Yucca periculosa* Baker	Growth regulating activity, increased developmental period, insecticidal activity, reduced pupation survival, reduced insect growth	[35]
Asteraceae	*Ageratum conyzoides* L.	Insecticidal (70% mortality)	[24]
Asteraceae	*Baccharis genistelloides* (Lam.) Pers.	Mortality	[36]
Asteraceae	*Artemisia verlotiorum* Lamotte	Mortality	[36]
Asteraceae	*Roldana barba-johannis* (DC.) H. Rob. & Brettell	Insecticidal	[37]
Asteraceae	*Gutierrezia microcephala* DC. A. Gray	Longer time for pupation and emergence of adults, severe toxicity against adults, insect growth inhibitory activity	[38]
Asteraceae	*Lychnophora ericoides* Mart.	Egg mortality	[39]
Asteraceae	*Trichogonia villosa* (Spreng.) Sch. Bip. Ex Baker	Egg mortality	[39]
Asteraceae	*Lychnophora ramosissima* Gardner	Larvicidal	[39]
Asteraceae	*Vernonia holosenicea* Mart. Ex DC.) L.	87% mortality	[39]
Asteraceae	*Senecio salignus* DC.	Antifeedant, insecticidal, juvenomimetric activity	[40]
Asteraceae	*Tagetes erecta* L.	Antifeedant effect causing 50% reduction of larval weight, 40–80% pupal mortality, 48–72% larval mortality	[41]
Cactaceae	*Myrtillocactus geometrizans* Mart. Ex Pfeiff	Insect growth regulating, larvicidal, delayed pupation	[41]
Cardiopteridaceae	*Cymbopogon winterrianus* Jowitt.	Alters biochemical profile of larvae, diminished reproduction, reproductive failure	[42,43]
Caricaceae	*Carica papaya* L.	90% mortality	[44–46]
Celastraceae	*Maytenus disticha* (Hook. F) Urb.	Insecticidal activity	[47]

Table 1. Cont.

Family	Plant Species	Action	Refs
Celastraceae	M. boaria (Molina)	Insecticidal activity	[47]
Convovulaceae	Ipomoea murucoides Roem. And Schult	46.16% mortality, reduced larval weight increased pupation time	[48,49]
Euphorbiaceae	Ricinus communis L.	Insecticidal and insectistatic, larvicidal, growth inhibition	[50,51]
Euphorbiaceae	Jatropha curcas L.	High mortality	[26]
Euphorbiaceae	Jatropha gossypiifolia L.	Antifeedant to larva, synergistic with pesticide	[52]
Euphorbiaceae	Euphorbia pulcherrima Willd. Ex Klotzsch	58.5% mortality, reduced larva and pupae weight, increased larva period, reduced egg viability	[53,54]
Leguminosae	Copaifera langsdorffii Desf.	Low fertility and fecundity, low viability of eggs, larval growth reduction, inhibited trypsin activity, egg abnormalities	[55]
Leguminosae	Militia ferruginea Hochst.	High mortality	[26]
Lamiaceae	Ocimum basilicum L.	Toxicity, non-preference, knockdown	[23,56,57]
Lamiaceae	Ocimum gratissimum L.	Sublethal effects	[58]
Lamiaceae	Salvia keerlii Benth.	Insecticidal	[59]
Lamiaceae	Salvia ballotiflora Benth.	Insecticidal, insectistatic, increased larval and pupal duration, reduced pupa weight	[59,60]
Lamiaceae	Salvia connivens Epling	Insecticidal, insectistatic	[59]
Lamiaceae	Salvia microphylla Kunth	Antifeedant, insecticidal, juvenomimetric activity	[40]
Malvaceae	Malva sylvestris L.	Mortality	[36]
Meliaceae	Melia azedarach L.	Reduced larval feeding, reduced larval growth, synergistic with pesticide	[52,61]
Meliaceae	Trichilia pallens C. de Candolle	Mortality	[62,63]
Meliaceae	Trichilia pallida Sw.	Mortality	[63]
Meliaceae	Cedrela salvadorensis Standl.	Larval mortality, growth reduction, inhibited larval growth, reduced pupal weights and adult emergence	[64]
Meliaceae	Cedrela dugessi S. Watson	Larval mortality, growth reduction, inhibited larval growth, reduced pupal weights and adult emergence	[64]
Meliaceae	Melia abyssinica	High mortality	[26]
Meliaceae	Trichilia pallida Sw.	No egg deformities	[65]
Meliaceae	Azadirachta indica Juss.	Reduced insect growth, increased development period, mortality, low egg laying, antifeedant activity, growth regulating activity, mortality, larvicidal	[24,26,65–75]

Table 1. Cont.

Family	Plant Species	Action	Refs
Monimiaceae	*Peumus boldus* Molina	75% mortality	[76]
Moringaceae	*Moringa oleifera* Lam.	Low feeding ratio, (antifeedant activity) mortality	[77]
Myrtaceae	*Eucalyptus citriodora* Hook	Growth regulating activity	[29]
Myrtaceae	*Eucalyptus staigeriana* F. Muell. Ex Bailey	Sublethal effects	[34]
Myrtaceae	*Eucalyptus globulus* Labill.	High mortality	[26]
Myrtaceae	*Siphoneugena densiflora* Berg	100% larval mortality	[78]
Petiveriaceae	*Petiveria alliacea* L.	Mortality	[36]
Phytolaccaceae	*Phytolacca dodecandra* L'Herit.	High mortality	[26]
Piperaceae	*Piper tuberculatum* Jacq.	Insecticidal	[79]
Piperaceae	*Piper hispidinervum* C. DC.	Affects spermatogenis and egg laying	[79]
Poaceae	*Cymbopogon citratus* (DC.) Stapf	Mortality	[36]
Poaceae	*Cymbopogon flexuosus* Steud.	Toxic, insecticidal activity	[80]
Poaceae	*Cymbopogon nardus* L.	Decreased pupa weight, increased larval mortality	[33]
Poaceae	*Zea diploperennis* L.	High larval survival	[81]
Rhamnaceae	*Zizyphus joazeiro* Mart.	Decreased pupa weight, increased larval mortality	[33]
Rubiaceae	*Morinda citrifolia* L.	Decreased pupa weight, increased larval mortality	[33]
Rubiaceae	*Psychotria goyazensis* Mull. Arg.	Reduced hatching rate, Egg mortality	[82]
Rutaceae	*Ruta graveolens* L.	Mortality	[36]
Rutaceae	*Citrus limon* L.	Antifeedant	[83]
Sapindaceae	*Magonia pubescens* A. St.-Hil.	Decreased pupa weight, increased larval mortality	[33]
Sapindaceae	*Talisia esculenta* Rsdlk.	Mortality	[61]
Sapindaceae	*Sapindus saponaria* L.	Mortality	[61]
Solanaceae	*Nicotiana tabacum* L.	High mortality	[23,26]
Verbenaceae	*Lantana camara* L.	High mortality	[26]
Verbenaceae	*Vitex polygama*	High mortality	[84]
Zingiberaceae	*Zingiber officinale* L.	Mortality	[36]

The essential oil of *Cymbopogon flexuosus* was reported to be lethal to FAW (LC_{50} = 1.35 mg Ll^{-1}) when supplemented in to an artificial diet at 2.25, 2.5 and 4 mg Ll^{-1} concentrations and 18.85 h median lethal time (LT_{50}). The insecticidal activity of citral was not significantly different to the essential oil, suggesting that citral, a compound of this essential oil caused insecticidal effects of the *Cymbopogon flexuosus* essential oil to FAW [80].

Moringa oil induced a lower feeding ratio expressed as the ratio of consumed area of treated leaf discs to consumed area of untreated (control) leaf discs and highest total corrected mortality percentage of FAW. This study concluded that at 10% concentration, *Moringa* oils can be used as a botanic insecticide in the management of FAW. Saponifiable components of the *Moringa* oils comprised of oleic acid (74.2%) and palmitic acid (7.16%). However, the LC$_{50}$ of moringa oil, unsaponifiable and saponifiable matters were 1.9%, 3.4% and 7.6% respectively, indicating that saponifiable matter was less effective against FAW larvae [77]. This therefore, means that there is no need for separation and identification of the moringa oil components for application in FAW control. Farmers should be advised to apply whole *Moringa* oil to benefit from the synergistic effects of the components therein.

Linalool showed potential in controlling FAW through non-preference, knockdown and toxicity effects on FAW larvae [56,57]. More than 80% of the essential oil of *Ocimum basilicum* consisted of linalool suggesting that this is the main active component [86]. More recently this species has been evaluated against FAW in Africa as part of a study focused on plants that were either native or widely grown in Malawi [23]. Another species investigated in this study included *Tephrosia vogelii*, a rotenoid producing and widely used species for pest control in Africa but this was not active suggesting a level of tolerance in FAW to the insecticidal rotenoids occurring in this species [23,87]. Another South American plant which grows widely as an invasive weed in Africa where it has been shown to have biological activity against insects [88] and used widely as a pesticide is *Tithonia diversifolia* but again this species was not active [23]. The most promising plant species based on their low mammalian toxicity, abundance and bioactivity against FAW identified through this work were *Lippia javanica*, *Ocimum basilicum* and *Cymbopogon citratus* which showed various activities including anti-feedancy and increased mortality. These three species are consumed as spices and teas so are far safer than synthetics [23]. *C. citratus* has also been shown in studies elsewhere to be effective against FAW. For example, it was been reported that sub-lethal doses of citronella oil altered the biochemical profile of FAW larvae causing damage to their reproductive histophysiology and resulted in diminished reproduction or reproductive failure [42]. The citronella-treated midgut of FAW larvae displayed modifications to the epithelium such as increased periodic acid-Schiff positive granules, columnar cell extrusion, cytoplasmic protrusions and pyknotic nuclei [42]. This study showed further that there was an increase in regenerative cells, which aided successive renewal of the epithelium. Trophocytes which are the main cell type of the fat body, once exposed to citronella, had reduced amounts of proteins, glycogen, and lipids. The fat bodies also showed distended vacuoles and mitotic bodies. This implies that citronella oil acts by causing changes in the morphology of the midgut and reducing stored resources in the fat body, limiting insect reproduction and survival.

FAW larvae feeding activity was reduced when treated with 1% and 10% methanolic extract of *Melia azedarach* seed. Other effects were slowed caterpillar growth due to ingestion of toxic substances present in *M. azedarach* [52,61], extended pupation time, small pupae and deformed moths. While native to Indomalaya and Australasia *Melia azaderach* grows widely in South America and Africa as well, so this study is highly relevant to the African context although there is some concern about the toxicity of the plant [89].

The ethanolic extract of Poinsettia *(Euphorbia pulcherima)* leaves obtained during the vegetative and reproductive phase was evaluated against FAW. The extracts were fed to the FAW larvae after mixing with artificial diet. Administered at 0.5 and 1% concentrations, the extracts increased the larval period, reduced larval and pupae weight as well as egg viability and resulted in greater larval mortality. It was further noted that the extract prepared from leaves that were in the reproductive phase of the plant effectively reduced the FAW population. Cold aqueous extract of *E. pulcherrima* also resulted in 58.5% mortality of FAW [53; 54] affecting *Neonotonia wightii* (perennial soybean).

Trials were done on the bioactivity of aqueous plant extracts of *Calotropis procera*, *Jatropha curcas*, *Cymbopogon nardus* (citronella), *Zyzyphus joazeiro*, 'noni', *Morinda citrofolia*, *Magonia pubescens* and *Annona squamosa* and showed that the consumption of leaves impregnated with different plant extracts increased larval mortality and significantly decreased pupal weight. The *Annona squamosa* treatment had the most effective insecticide activity against FAW. However, no identification of the phytochemicals responsible for these activities was done making the exploitation of these data difficult [33].

Methanolic extracts of leaves, bark and fruit peel of *Copaifera langsdorffii* resulted in low FAW egg viability [79]. Findings of this study showed that the methanolic extracts from leaves and fruit peel added to the artificial diet of 2nd instar FAW had several effects including reduced larval growth, long development duration, lower fertility and fecundity of adults as well as augmented mortality. The aeropylar and micropylar regions of the eggs had abnormalities. The insect feces were high in protein as reflected by repressed trypsin activity in the in vitro test. They suggested that *C. langsdorffii* presented the greatest potential for use as alternative bioinsecticide for control and management of FAW.

The effects of aqueous extracts of *Talisia esculenta* and *Sapindus saponaria* on the FAW at 8 and 14 days of development led to increased larval mortality at 63.15% and 26.71% for *S. saponaria* and *T. esculenta*, respectively [61]. The extract of *T. saponaria* was the most promising for the control of FAW. This might be because their seeds high in fat content, yielding a similarly fatty extract with adjuvant capacity thus facilitating fixing and distribution of the extract on the leaves of maize hence increasing the insecticide action. There remains, however, the need to determine the insecticidal compounds in the plants on whose basis new natural insecticidal products could be produced or improvements to the extraction could be made.

A study under laboratory conditions and conducted on the biological activity of boldus (*Peumus boldus* Molina) water extract against FAW and *Helicoverpa zea* (Boddie) (Lepidoptera: Noctuidae) [76] showed that FAW was the most susceptible with 75% mortality at seven days when exposed to 8% w/w of *P. boldus* extract and had an LC_{50} value of 2.31 mL kg^{-1}. Again, no chemistry was undertaken making the usefulness of this data questionable.

The bioactivity of *Ipomoea murucoides* methanolic extracts and fractions on FAW were evaluated by incorporating the extracts into a meridic diet at concentration of 2 mg mL^{-1}. These were then fed to FAW larvae (1st instars) [48]. After seven days, crude leaf extracts caused up to 46.16% mortality (leaf extract LC_{50} = 2.692 mg mL^{-1}). Other effects were reduced larval weight, increased pupation time and in the time to attain adulthood. No influence was noted for the number of eggs. Despite the fact that the partly purified fraction caused no toxicity to FAW, the greatest effect was on reduced larval weight, augmented pupation time and time to attain adulthood with an influence on number of eggs.

An earlier study on a methanolic extracts of *I. murucoides* calli [49] reported that it induced a higher (95%) neonate larvae mortality than was reported by [48]. This difference was explained by the fact that in [48], the leaf extracts contained a large amount of chlorophyll (that is lacking in calli) which masked the compounds and therefore inhibited their activity.

A study reported in [37] investigated how tocotrienols and hydroquinones from *Roldana barba-johanis* affected the growth of insects. The major compounds obtained from the aerial parts methanol extract were sargachromenol, sargahydroquinoic acid and sargaquinoic acid. These compounds and their associated methylated and acetylated derivatives exhibited insect growth regulatory and insecticidal activities against the FAW. The most biologically active phytochemicals were sargachromenol, sargahydroquinoic acid and sargaquinoic acid in the order of abundance. These compounds and the acetylated form of this mixture resulted in negligible effects. When used at 5.0 and 20.0 ppm in diets, they caused substantial inhibitory effects on FAW larvae with insecticidal activity ranging between 20 and 35 ppm.

Eucalyptus citriodora Hook (Myrtaceae) contained eucalyptin in methanol extract of leaves along with naringenin, chrysin, apigenin, quercetin, and luteolin, oleanolic acid, ursolic acid, betulinic acid and composite mixtures of flavonoids and triterpenes that were not identified [29]. These compounds exhibited insecticidal and insect growth regulatory and antifeedant activities, against FAW and the Yellow Mealworm (*Tenebrio molitor*) (Coleoptera:Tenebrionidae).

The sublethal effects of the essential oils of *Foeniculum vulgare*, *Ocimum gratissimum* and *Eucalyptus staigeriana* on FAW have been reported [34]. The essential oils caused reduced larval and pupal weights, increased larval and pupation periods, reduced oviposition period and adult survival although there were variations in effects. The essential oil of *O. gratissimum* had the greatest effects across the tested doses. These insecticidal effects could have been as a result of essential oil components like limonene, geranial, (E)-anethole, eugenol and α-pinene in the essential oils. This provides an opportunity for researchers to explore other plants with these compounds with the aim of incorporating them in the pool of plants that provide promising outcomes for managing FAW in Africa.

Methanol extracts of *Yucca periculosa* bark yielded 4, 4′-dihydroxystilbene, resveratrol and 3, 3′, 5, 5′-tetrahydroxy-4-methoxystilbene. These compounds showed growth regulatory effects against the FAW. The most active compound was 3, 3′, 5, 5′-tetrahydroxy-4-methoxystilbene which was active at 3 µg g^{-1} in diets [35]. However, the utilization of *Y. periculosa* (Agavaceae) is limited due to its local use as a source of firewood. In addition, the leaves of this plant are used for making handicrafts while the flowers are utilized as food. At 25.0 ppm concentration, the methoxy stilbene and methanolic extract of *Y. periculosa* caused 100% mortality of larvae. Most importantly, the methoxy stilbene and methanolic extract of *Y. periculosa* proved to be more active than gedunin and the methanolic extract of *Cedrela salvadorensis* with LC$_{50}$ values of 5.4 ppm and 7.18 ppm, respectively. They also indicated that there was a decrease in the percentage of larvae attaining pupation across treatments as compared to the control. Survival of the pupae was reduced to 0.05 at 25 and 50 ppm for the methoxy stilbene and methanolic extract, respectively. The percentage of adult emergence showed further impacts at the pupal stage with resveratrol, the methoxy stilbene, methanolic extract of *Y. periculosa*, gedunin and methanolic extract of *C. salvadorensis* with 0.0%, 27.0%, 18.0%, 13.0%, and 8.0% of emergence, respectively at 25, 10, 10, 25 and 25 ppm. The methoxy stilbene and the methanolic extract of *Y. periculosa* with Relative Growth Index (RGI) values of 0.25 and 0.45 at 10 and 15 ppm gave the greatest outcome. The effects of resveratrol, the methoxy stilbene and methanolic extract of *Y. periculosa* did not differ from that of gedunin but had greater potency than the methanolic extract of *C. salvadorensis* [64]. This finding presents these plants as having potential for further development for use against FAW.

The aerial portions of *Gutierrezia microcephala* yielded four oxyflavones, which were tested for activity against neonate larvae of FAW [38]. The flavone, a clerodane, its methyl ester, methanolic and n-hexane extracts caused a major delay in the time taken to attain pupation and adult emergence. Severe toxicity against FAW adults and insect growth inhibition were also reported [38].

Maytenus disticha aerial parts and *Maytenus boaria* seeds were evaluated to determine their effects on the FAW [47]. Several β-dihydroagarofurans were isolated including 9-benzoyloxy-1,2,6,8,15-pentaacetoxy-dihydro-β-agarofuran-(1) and 9-furanoxy-1,6,8-triacetoxy-dihidro- β-agarofuran and their insecticidal activities compared to ethanol extracts from *A. indica* and *M. azedarach* [65]. There was a 58% and 100% growth inhibition at 16 and 80 ppm, respectively. This suggested that agarofurans and MeOH and hexane/EtOAc extracts from *M. disticha* and *M. boaria*, respectively, have potential for use as a biopesticide against FAW.

Extracts of *A. indica* and *M. azedarach* caused significant larval deaths, slowed the growth rate of larva and lengthened pupation time. The influence of 9-furanoxy-1,6,8-triacetoxy-dihidro-β-agarofuran and hexane/EtOAc extract on FAW was comparable to that of limonoids such as gedunin and cedrelone [90]. The action of these compounds was comparable to toosendanin, which is a commercially available biopesticide, suggesting that there is potential for researchers to harness these plants and produce products that can assist in controlling FAW.

Ricinus communis has been identified as a potentially important pesticidal plant owing to its insecticidal properties. Some fatty acids obtained from the aqueous extracts of caster plant have shown insecticidal and insectistatic activity against FAW. For example, linoleic acid, palmitic acid and stearic acid show biological activity against FAW [46] while linolenic acid was reported to have insecticidal and insectistatic activities against FAW [51].

Castor oil and vicinine which can be extracted from seeds or leaves of *R. communis* were active against FAW, however, the seed extract was more potent [50]. The two test substances were associated with the effects observed for FAW. The half maximum larvae viability concentration (LVC$_{50}$) was 0.38 × 103 ppm for the vicinine, 0.75 × 103 ppm for methanol extract of seeds, 1.97 × 103 ppm for ethyl acetate seed extract, 2.69 × 103 ppm for castor oil, 4.83 × 103 ppm for a methanol extract of leaves and 10.01 × 103 ppm for a hexane extract of leaves. Bioactivity in castor plants is particularly relevant to the African context as this plant is cosmopolitan and grows abundantly adjacent to farmland in many parts of Africa.

Trichilia pallida leaf and branch extracts when applied at very low concentrations of ≤ 0.0008% were shown to have no effects on eggs and larvae of FAW [63]. Although less diverse species of the genus *Trichilia* also occur in Africa such as *T. emetica* indicating the potential for using knowledge from American studies to inform the use of African species.

Insecticidal activity of *Salvia* spp. has also been reported on FAW and *Spodoptera littoralis* (Lepidoptera:Nuctuidae) [59]. The extracts from *Salvia keerlii* and *Salvia ballotiflora* were shown to have modest insecticidal action (LV50= 1527 and 1685 µg mL^{-1}, respectively. On the other hand, the extract of *S. ballotiflora* increased the larval and pupal stages by 5.2 and 2.9 days, respectively and caused a decrease in the pupal weight by 13.2%. Furthermore, *Salvia microphylla* showed insecticidal activity against FAW (LC_{50} = 919 ppm) [59]. The bioactivity of the essential oil of *S. ballotiflora* at 1000, 600, 400, 120 and 80 µg mL^{-1} led to reduced viability of larva which was 0%, 5%, 10%, 10%, and 20%, respectively [59]. They also reported extended duration of the larval stage by 30.5, 8.0, 5.5 and 5.5 days at 600, 400, 120 and 80 µg mL^{-1} compared with the control. The pupation period was extended by 1.6 days at 400 µg mL^{-1}. Moreover, the reduction in weight of the pupae decreased by 52%, 39%, 29% and 29% at 600, 400, 120 and 80 µg mL^{-1}, respectively, in relation to the control.

S. microphylla contains palmitic acid, oleic acid and Y-sitosterol which have been associated with its activities against FAW [40]. Furthermore, they pointed out that there was a possibility to use of *Senecio salignus* and *Salvia microphylla* extracts for controlling FAW as they produce bioactive compounds that are antifeedants [91]. Salvia species are abundant in Africa including the South American exotic species *Salvia suaveolens* thus this species may be worthy of investigation to identify similar activities.

FAW eggs died at a rate of 97.7% one day after being exposed to extracts of *Lychnophora ericoides* and *Trichogonia villosa* [39]. Thus only 2.3% of the eggs hatched being a very low percentage to sustain populations that can cause damage.

Citrus-derived limonoids have been implicated in reduced feeding activity in insect pests. They include limonin, nomlin and abacunone and their semisynthetic products. Limonoids from *Citrus limon* have exhibited similar effects on FAW [83]. Citrus crops are also grown widely in Africa so further work on by-products of the peel from the fruit processing sector may provide opportunities for bioactive plant compounds in Africa.

The biological activity of extracts from various plant parts of wild and in-vitro plants of *Piper tuberculatum* on the 3rd instars of FAW in Brazil have been studied [78]. The dichloromethane (DCM): methanol (2:1) and ethanol extracts of leaves and stems and boiling water extracts of leaves, stems and spikes of *P. tuberculatum* showed no effects on FAW 3rd instars across the dosage. However, the DCM: methanol (2:1) and ethanol extracts of mature spikes from wild and DCM: methanol (2:1) extract of in vitro plants were reported to have exhibited potential insecticidal activity on the 3rd instars of FAW. This result suggests that there is a potential for direct use of *P. tuberculation* mature spike of EtOH extracts that would allow farmers to utilize their locally brewed alcoholic drinks as extraction solvents. It would also mean that using in vitro techniques, the respective bioactive compounds can be biologically synthesized in large quantities using in-vitro cell suspension cultures [92]. This may require adequate and well-equipped laboratories most of which are out of reach for the farming support and commercial systems in Africa. *P. tuberculatum* has palmitic and oleic acids which could be responsible for the reduced viability of the larvae at 33.3% and 48.5%, respectively with a concentration of 1600 ppm.

The main components identified in *Carica papaya* seed were oleic acid (45.97%), palmitic acid (24.1%) and stearic acid (8.52%) [44]. When evaluated against FAW the viability of the larvae was reduced to 33.3% for oleic acid, 48.5% for palmitic acid and 62.5% for stearic acid at 1600 ppm. Single fatty acids in *C. papaya* possessed greater potential to kill the insect pest compared to the chloroform extract. Amongst the three, palmitic acid was the most active.

A high mortality of FAW was reported with extracts of Jatropha curcas, Militia ferruginea, Phytolacca dodecandra, Scinus molle, Melia abyssinica, Nicotiana tabacum, Lantana camara, Chenopodium ambroides, Azadirachta indica and Jatropha gossypifolia [26]. This is the first report where these plant species were evaluated against FAW in Africa-Ethiopia. Similar activities were reported for A. indica and N. tabacum against FAW supporting these earlier findings [23].

The neem tree, *A. indica* can control many pest species including FAW [24,70,93,94]. The deleterious properties of neem oils and extracts on pests are associated with the content of limonoids like azadirachtin which is a highly complex and effective molecule [69]. Azadirachtin, is freely decomposable, selective, non-mutagenic causing minimal harm to mammals and the environment and could present an excellent option for controlling FAW [67]. For example, egg laying by female FAW was about 50% lower on the neem treated than on untreated cloth [67]. However, this substance has limitations such as being highly costly, it cannot be synthesized chemically and has to be purified using expensive and sophisticated methods. It can be produced from large quantities of seasonally available seeds [94] so may not be so well suited to small holders in Africa. The main components occur in the seeds and even for a "low-tech" processing method require considerable effort to extract them. One additional problem with the use of neem is that the main active components including the various azadirachtin related structures are highly UV labile so may low residual effects in the field [95]. There is no standardization and control of quality in neem-based preparations manufactured in Brazil an indicator that it may not possible to reproduce the desired effects of the insecticide [96]. To increase effectiveness, controlled-release preparations of insecticides by polymeric encapsulation [97,98] has been done. Encapsulation of neem oil and extracts into films or polymeric walls shelters the active component and permits controlled release stopping the loss of unstable compounds and increasing their stability in the environment [95]. Although again this approach may be beyond the needs and scope of smallholders but illustrates technologies in development to improve persistence in the field for botanicals.

In another study neem seed cake extract was more active (LC_{50} = 0.13%) than leaf extract (LC_{50} = 0.25%) [73]. This was because of the higher amounts of azadirachtin the most effective of the toxic tetranortriterpenoids, because 90% of azadirachtin is more intense in the neem cake after pressing the seeds [74]. Farmers often use Neem leaves when seed is unavailable. However, the concentration of the active constituents is very low in leaves such that it has low and potentially no efficacy so promoting the use of Neem leaves should be discouraged as poor efficacy may negatively influence farmer opinions about the value of plants as alternatives to synthetics. Additionally, the bioassay indicated a static effect on the growth of FAW caterpillars, as most of them exhibited their exuviae in the terminal part of the body, incompletely releasing them as expressed by [69] as it limits the ability of the insects to feed by affecting the physiological functioning of ecdysis and in cellular processes, eventually causing insect death. This process takes some time and that is why comparatively, there is low larval mortality and high pupal mortality [99].

Zea diploperennis was evaluated against FAW and indicated that methanol extract and residual fiber of the plant adversely influenced the size of pupae. The aqueous extract caused 100% of larval cumulative mortality [81].

An extract was obtained from the roots and aerial parts of *M. geometrizans* using methanol as a solvent. Its components were peniocerol, macdougallin and chichipegenin and the mixtures of peniocerol and macdougallin. They all exhibited insect growth regulatory and insect killing activities against FAW [100].

3. Future Prospects

The plant species reviewed above provide an illustration of the extent of work undertaken to identify new pest management options from plants. These plants have been shown to have biological activity against FAW through various modes of action If these initial indications of activities are to be translated to the African context then not only are the bioactivity of extracts in the laboratory required but also the chemistry of these activities needs to be determined and the materials tested in field conditions using tailored approaches to extraction that are appropriate for small scale farmers.

The plants reviewed had a variety of modes of action in controlling FAW including induction of low feeding ratio through the action of oleic acid (74.2%) and palmitic acid (7.16%) [44], repellent effects, severe toxicity, non-preference and knockdown effects by linalool from *Ocimum basilicum*. Citronella oil changes the chemical profile of FAW larvae, affecting reproductive and cell physiological parameters causing reduced reproduction and sometimes reproductive failure. It is also associated with changed epithelium that has cytoplasmic projections, columnar cell extrusion, pyknotic nuclei and increased periodic acid-schiff positive granules. Citronella oil caused morphological changes of the midgut and reduction of stored resources in the fat body, which may adversely affect insect reproduction and survival. It has been further reported that reduced feeding after ingestion of *Melia azedarach* caused starvation. This in addition to ingestion of toxic substances from *M. azedarach* [60]. Leaf extracts at vegetative and reproductive phase of Poinsettia (*Euphorbia pulcherrima*) increased larval period, reduced the weight of larvae and pupae egg viability [54]. The methanolic extracts of *Copaifera langsdorffii* leaves, bark of fruits and fruit peels resulted in low egg viability, reduction in larval growth, prolonged period of development, increased mortality, lowered fertility and fecundity of adults, abnormalities in the aeropylar and micropylar regions, increased excretion of protein in the insect feces and invitation of trypsin activity [79].

There is adequate evidence as indicated by the research findings presented in this review, that there are numerous opportunities for the use of botanical extracts in the management of FAW. However, exploitation of these opportunities is limited because the potential for use may face challenges attributed to the following:

1. Despite there being numerous plant products many are unstable upon application because they are UV labile. This means they may need more frequent application incurring greater costs in time. However, as they are non-persistent, they are potentially less damaging to the environment particularly non-target insects [17,18,99,101].
2. African smallholder farmers are not economically endowed to buy the botanical pesticides as has been the case for other farm inputs [102,103]. This therefore means that farmers will be encouraged to self-harvest these plant materials [104,105] and use them as crudely produced products as reported earlier [21,22].
3. There are different modes of action, which are determined by the stage of growth of both maize and the FAW raising the issue of exposure period, effectiveness, mode of application and method of extraction. However different modes of action could help to reduce the build-up of resistance in the pest where used in combination.
4. The opportunities maybe limited in scope where the products are not standardized for reproducibility and scale-up and this will require uniformity of the chemistry for the plant material and they likely need propagation [87,106]. More so, surprisingly few have been evaluated under field conditions [26]. This is a major oversight of the work as it means there is little evidence that any of the biological activities translate to a real-world setting. Field evaluations provide options to engage with farmers and determine effects on yield and determine non-target effects as undertaken recently in the African context [17,85,101].
5. Some of the plant materials tested may not be available for use by the farmers, for example the use of citrus seeds at farm level may not be attainable because it may not be feasible in terms of availability of the seeds. However, there may be good opportunities for propagation, and this would likely overcome some of the challenges of chemical variation across plant populations and provide consistency, which may otherwise be lacking when plants are harvested from the wild [106].
6. All the studies conducted failed to include economic viability for the tested plant extracts. This is closely linked to sustainable availability of the plant materials which is key driver for farmer

adoption. Most of the plant materials used in these studies were wild harvested and this may not be self-sustaining unless efforts are made to commercialize the promising plant products or at least determine the economic viability of their use compared to alternatives including the use of synthetics [18,85].

4. Recommended Research Areas for Further Studies

1. Evaluate different extraction methods with the aim of documenting the most appropriate for adoption by small-scale maize farmers.
2. Investigate modes of action of different products based on part used, pure compounds and mixtures across the different growth stages of maize plant and FAW.
3. Conduct field evaluations of these plants and potentially determine any benefits of combining materials that could deliver different mechanisms of activity to address issues of insect resistance.
4. Investigate standardization to increase the scope of reproducibility and adoption especially through propagation.
5. Conduct research on approaches for upscaling, commercialization and sustainability of the botanical extracts and products
6. Testing activity of pure compounds from extracts to determine which components are active allowing the evaluation of variability across materials and improving methods for optimizing extraction and [87,107,108].

Author Contributions: Original draft preparation, N.B.R.; review and editing, P.C.S. All authors have read and agreed to the published version of the manuscript.

Funding: The contribution from PCS was funded by a grant from the McKnight foundation to PCS Grant No: 17-070.

Acknowledgments: Authors acknowledge Peter Opala who assisted in editing the manuscript for his valuable input.

Conflicts of Interest: The authors declare no conflict of interest.

References

1. Pashley, D.P. Current status of fall armyworm host strains. *Fla. Entomol.* **1988**, *71*, 227–234. [CrossRef]
2. Luttrell, R.G.; Mink, J.S. Damage to cotton fruiting structures by the fall armyworm *Spodoptera frugiperda* (Lepidoptera: Noctuidae). *J. Cotton Sci.* **1999**, *3*, 35–44.
3. Georgen, G.; Kumar, P.L.; Sankung, S.B.; Togola, A.; Tamo, M. First report of outbreaks of the fall armyworm (*Spodoptera frugiperda* (J.E. Smith) (Lepidoptera, Nuctuidae), a new alien invasive pest in West and Central Africa. *PLoS ONE* **2016**, *11*, 10.
4. FAO. Fall armyworm continues to spread in Ethiopia's maize fields. *Facilitates National Awareness Training for key Partners and Field Offices*. 2017. Available online: http://www.fao.org/ethiopia/news/detail-events/en/c/1028088/ (accessed on 5 February 2020).
5. Chapman, J.W.; Williams, T.; Martinez, A.M.; Cisneros, J.; Caballero, P.; Cave, R.D.; Goulson, D. Does cannibalism in *Spodoptera frugiperda* (Lepidoptera: Noctuidae) reduce the risk of predation? *Behav. Ecol. Sociobiol.* **2000**, *48*, 321–327. [CrossRef]
6. De Almeida Sarmento, R.; de Souza Aguiar, R.W.; de Almeida Sarmento de Souza Aguiar, R.; Vieira, S.M.J.; de Oliveira, H.G.; Holtz, A.M. Biology review, occurrence and control of *Spodoptera frugiperda* (Lepidoptera, Noctuidae) in corn in Brazil. *Biosci. J.* **2002**, *18*, 41–48.
7. Cook, D.R.; Leonard, B.R.; Gore, J. Field and Laboratory performance of novel insecticides against armyworms (Lepidoptera: Noctuidae). *Flav. Entomol.* **2004**, *87*, 433–439. [CrossRef]
8. Williams, W.P.; Sagers, J.B.; Hanten, J.A.; Davis, F.M.; Buckley, P.M. Transgenic corn evaluated for resistance to fall armyworm and southwestern corn borer. *Crop Sci.* **1997**, *37*, 957–962. [CrossRef]

9. Williams, W.P.; Davis, F.M.; Buckley, P.M.; Hedin, P.A.; Baker, G.T.; Luther, D.S. Factors associated with resistance to fall armyworm (Lepidoptera: Noctuidae) and Southwestern corn borer (Lepidoptera: Crambidae) in corn at different vegetative stages. *J. Econ. Entomol.* **1998**, *91*, 1472–1480. [CrossRef]
10. Wiseman, B.R.; Widstrom, N.W. Resistance of corn populations to larvae of the corn earworm (Lepidoptera: Noctuidae). *J. Econ. Entomol.* **1992**, *85*, 601–605. [CrossRef]
11. Ashley, T.R. Classification and distribution of fall armyworm parasites. *Fla. Entomol.* **1979**, *62*, 114–122. [CrossRef]
12. Sparks, A.N. Fall armyworm (Lepidoptera: Noctuidae): Potential for area-wide management. *Fla. Entomol.* **1986**, *69*, 603–614. [CrossRef]
13. Midega, C.A.O.; Pittchar, J.O.; Pickett, J.A.; Hailu, G.W.; Khan, Z.R. A climate-adapted push-pull system effectively controls Fall Armyworm, *Spodoptera frugiperda* (J E Smith), in maize in East Africa. *Crop Protect.* **2018**, *105*, 10–15. [CrossRef]
14. Baudron, F.; Zaman-Allah, M.A.; Chaipa, I.; Chari, N.; Chinwada, P. Understanding the factors conditioning fall armyworm (*Spodoptera frugiperda* J.E. Smith) infestation in African smallholder maize fields and quantifying its impact on yield: A case study in Eastern Zimbabwe. *Crop Prot.* **2019**, *120*, 141–150. [CrossRef]
15. Abate, T.; van Huis, A.; Ampofo, J.K.O. Pest management strategies in traditional agriculture: An African perspective. *Annu. Rev. Entomol.* **2000**, *45*, 631–659. [CrossRef]
16. Abudulai, M.; Shepard, B.M.; Mitchell, P.L. Parasitism and predation on eggs of *Leptoglossus phyllopus* (Hemiptera: Coreidae) in cowpea: Impact of endosulfan sprays. *J. Agric. Urban Entomol.* **2001**, *18*, 105–115.
17. Tembo, Y.; Mkindi, A.G.; Mkenda, P.A.; Mpumi, N.; Mwanauta, R.; Stevenson, P.C.; Ndakidemi, P.A.; Belmain, S.R. Pesticidal plant extracts improve yield and reduced insect pests on legume crop without harming beneficial arthropods. *Front Plant Sci.* **2018**, *9*, 1425. [CrossRef]
18. Mkenda, P.; Mwanauta, R.; Stevenson, P.C.; Ndakidemi, P.; Mtei, K.; Belmain, S.R. Extracts from field margin weeds provide economically viable and environmental benign pest control compared to synthetic pesticides. *PLoS ONE* **2015**, *10*, e0143530. [CrossRef]
19. Mkindi, A.G.; Tembo, Y.L.B.; Mbega, E.R.; Kendal-Smith, A.; Farrell, I.W.; Ndakidemi, P.A.; Stevenson, P.C.; Belmain, S.R. Extracts of Common Pesticidal Plants Increase Plant Growth and Yield in Common Bean Plants. *Plants* **2020**, *9*, 149. [CrossRef]
20. Stevenson, P.C.; Isman, M.B.; Belmain, S.R. Pesticidal plants in Africa: A global vision of new biological control products from local uses. *Ind. Crop. Prod.* **2017**, *110*, 2–9. [CrossRef]
21. Kamanula, J.F.; Sileshi, G.; Belmain, S.R.; Sola, P.; Mvumi, B.; Nyirenda, G.K.C.; Nyirenda, S.P.N.; Stevenson, P.C. Farmers' Pest management practices and pesticidal plant use for protection of stored maize and beans in Southern Africa. *Int. J. Pest Manag.* **2011**, *57*, 41–49. [CrossRef]
22. Nyirenda, S.P.N.; Sileshi, G.; Belmain, S.R.; Kamanula, J.F.; Mvumi, B.; Sola, P.; Nyirenda, G.K.C.; Stevenson, P.C. Farmers' Ethno-Ecological Knowledge of Vegetable Pests and their Management Using Pesticidal Plants in Northern Malawi and Eastern Zambia. *Afr. J. Agric. Res.* **2011**, *6*, 1525–1537.
23. Phambala, K.; Tembo, Y.; Kasambala, T.; Kabambe, V.H.; Stevenson, P.C.; Belmain, S.R. 2020 Bioactivity of common pesticidal plants on fall armyworm larvae (*Spodoptera frugiperda*). *Plants* **2020**, *9*, 112. [CrossRef]
24. Lima, R.K.; Cardoso, M.D.; Moraes, J.C.; Andrade, M.A.; Melo, B.A.; Rodrigues, V.G. Chemical characterization and insecticidal activity of the essential oil leaves of *Ageratum conyzoides* L. on fall armyworm (*Spodoptera frugiperda* (Smith, 1797) (Lepidoptera: Noctuidae). *Biosci. J.* **2010**, *26*, 1–5.
25. Rioba, N.B.; Stevenson, P.C. *Ageratum conyzoides* L. for the management of pests and diseases by small holder farmers. *Ind. Crop. Prod.* **2017**, *110*, 22–29.
26. Sisay, B.; Tefera, T.; Wakgari, M.; Ayalew, G.; Mendesil, E. The efficacy of selected synthetic insecticides and botanicals against Fall armyworm, *Spodoptera frugiperda* in maize. *Insects* **2019**, *10*, 45. [CrossRef] [PubMed]
27. Trindade, R.C.P.; Ferreira, E.S.; Gomes, I.B.; Silva, L.; Santana, A.E.G.; Broglio, S.M.F.; Silva, M.S. Aqueous extracts of yam (*Dioscorea rotundata* Poirr.) and Chenopodium (*Chenopodium ambrosioides* L.) in *Spodoptera frugiperda* (J.E. Smith, 1797). *Rev. Bras. Plantas Med.* **2015**, *17*, 291–296. [CrossRef]
28. Skenjana, N.L.; Poswal, M.A.T. The use of *Chenopodium ambrosioides* (Chenopodiceae) in insect pest control in the Eastern Cape Province. *South Afr. J. Bot.* **2017**, *109*, 370. [CrossRef]
29. Salazar, J.R.; Torres, P.; Serrato, B.; Dominguez, M.; Alarcon, J.; Cespedes, C.L. Insect Growth Regulator (IGR) effects of *Eucalyptus citriodora* Hook (Myrtaceae). *Boletín Latinoam. y del Caribe de Plantas Med. y Aromáticas* **2015**, *14*, 403–422.

30. Singh, M.; Khokhar, S.; Malik, S.; Singh, R. Evaluation of Neem (*Azadirachta indica* A. Juss) extracts against American Bollworm, *Helicoverpa armigera* (Hubner). *J. Agric. Food Chem.* **1997**, *45*, 3262–3268.
31. Isman, M.B. Botanical Insecticides in the Twenty-First Century—Fulfilling their Promise? *Ann. Rev. Entomol.* **2020**, *65*, 233–249. [CrossRef]
32. Isman, M.B.; Grieneisen, M.L. Botanical insecticide research: Many publications, limited useful data. *Trends Plant Sci.* **2014**, *19*, 140–145. [CrossRef] [PubMed]
33. Santos, B.A. Bioactivity of Plant Extracts on Spodopterea frugiperda (J.E. Smith) (Lepidoptera: Noctuidae). Master' Dissertation, State University of Montes Claros, Januaba, Brazil, 2012.
34. Cruz, G.S.; Wanderley-Teixeira, V.; Oliveira, J.V.; Lopez, F.S.C.; Barbosa, D.R.S.; Breda, M.O.; Dutra, K.A.; Guedes, C.A.; Navarro, D.M.A.F.; Teixeira, A.A.C. Sublethal effects of *Eucalyptus staigeriana* (Myrtales: Myrtaceae), *Ocimum gratissimum* (Lamiales: Laminaceae) and *Foeniculum vulgare*) (Apiales: Apiaceae) on the biology of *Spodoptera frugiperda* (Lepidoptera, Noctuidae). *J. Econ. Entomol.* **2016**, *109*, 660–666. [CrossRef] [PubMed]
35. Torres, P.; AÄvila, J.G.; Romo de Vivar, A.; Garcıa, A.M.; Marin, J.C.; Aranda, E.; Cespedes, C.L. Antioxidant and insect growth regulatory activities of stilbenes and extracts from *Yucca periculosa*. *Phytochemistry* **2003**, *64*, 463–473. [CrossRef]
36. Tagliari, M.S.; Knaak, N.; Fiuza, L.M. Efeito de extratos de plantas na mortali-dade de lagartas de *Spodoptera frugierda* (J. E. Smith) (Lepidoptera: Noctuidae). *Arq. do Inst. Biológico São Paulo.* **2010**, *77*, 259–264.
37. Cespedes, C.L.; Torres, P.; Marin, J.C.; Arciniegas, A.; Perez-Castorena, A.L.; Romo de Vivar, A.; Aranda, E. Insect growth inhibition by tocotrienols and hydroquinones from *Roldanabarba-johannis* (Asteraceae). *Photochemistry* **2004**, *65*, 1963–1975. [CrossRef]
38. Calderon, J.S.; Cespedes, C.L.; Rosaura, R.; Gomez-Garibay, F.; Salaza, J.R.; Lina, L.; Aranda Eduardo Kubo, I. Acetylcholinesterase and insect growth inhibitory activities of *Gutierrezia microcephala* on Fall Armyworm (*Spodoptera frugiperda* (J.E.Smith). *Z Naturforschung.* **2001**, *56c*, 382–394. [CrossRef]
39. Tavarez, W.S.; Cruz, I.; Petacci, F.; Assis Jonior, S.L.; Freitas, S.S.; Zanuncio, J.C.; Serrao, J.E. Potential uses of Asteraceae extracts to control *Spodoptera frugiperda* (Lepidoptera:Noctuidae) and selectivity to their parasitoids *Trichogramma pretiosumus* (Hymenoptera:Trichogrammatidae) and *Telenomus remus* (Hymenoptera:Scelionidae). *Ind Crop Prod.* **2009**, *30*, 384–388. [CrossRef]
40. Romo-Asncion, D.; Avila-Calderon, M.A.; Ramos-Lopez, M.A.; Barranco-Florido, J.E.; Rodriquez-Navarro, S.; Romero-Gomez, S. Juvenomimetic and insecticidal activities of *Senecio saliginus* (Asteraceae) and *Salvia mycrophylla* (Lamiaceae) on *Spodoptera frugiperda* (Lepidoptera:Noctuidae). *Fla. Entomol.* **2016**, *99*, 345. [CrossRef]
41. Salinas-Sánchez, D.O.; Aldana-Llanos, L.; Valdés-Estrada, M.E.; Gutiérrez-Ochoa, M.; Valladares-Cisneros, G.; Rodríguez-Flores, E. Insecticidal activity of *Tagetes erecta* extracts on *Spodoptera frugiperda* (Lepidoptera: Noctuidae). *Fla. Entomol.* **2012**, *95*, 428–432. [CrossRef]
42. Silva, C.T.S.; Wanderley-Teixeira, V.; Cunha, F.M.; Oliveira, J.V.; Dutra, K.A.; Navarro, D.M.A.F.; Teixeira, A.A.C. Biochemical parameters of *Spodoptera frugiperda* (J. E. Smith, 1979) treated with citronella oil (*Cymbopogon winterianus* Jowitt ex Bor) and its influence on reproduction. *Acta Histochem.* **2016**, *118*, 347–352. [CrossRef]
43. Silva, S.M.; Rodrigues da Cunha, J.P.A.; Malfitano de Carvalho, S.; Souza Zandonadi, C.H.; Martins, R.C.; Chang, R. *Ocimum basilicum* essential oil combined with deltamethrin to improve the management of *Spodoptera frugiperda*. *Ciênc Agrotec.* **2017**, *41*, 665–675. [CrossRef]
44. Perez-Gutierrez, S.; Zavala-Sanchez, M.A.; Gonzalez-Chavez, M.M.; Cardenas-Ortega, N.C.; Ramos-Lopez, M.A. Bioactivity of *Carica papaya* (Caricaceae) against *Spodoptera frugiperda* (Lepidoptera: Noctuidae). *Molecules* **2011**, *16*, 7505–7509. [CrossRef] [PubMed]
45. Franco, A.S.L.; Jiménez, P.A.; Luna, L.C.; Figueroa-Brito, R. Efecto tóxico de semillas de cuatro variedades de *Carica papaya* (Caricaceae) en *Spodoptera frugiperda* (Lepidoptera: Noctuidae). *Folia Entomol. Mex.* **2006**, *45*, 171–177.
46. Figueroa-Brito, R.; Camino, M.; Pérez-Amador, M.C.; Muñoz, V.; Bratoeff, E.; Labastida, C. Fatty acid composition and toxic activity of the acetonic extract of *Carica papaya* L. (Caricaceae) seeds. *Phyton. Inter. J. Exp. Bot.* **2002**, *69*, 97–99.
47. Cespedes, C.L.; Alarcon, J.; Aranda, E.; Becerra, J.; Silva, M. Insect growth regulatory and insecticidal activity of β-di-hydroagarofurans from *Maytenus* spp. (Celastraceae). *Z. Naturforsch.* **2001**, *56c*, 603–613. [CrossRef]
48. Curzio, L.G.V.; Velazquez, V.M.H.; Rivera, I.L.; Fefer, P.G.; Escobar, E.A. Biological activity of methanolic extracts of *Ipomoea murucoides* (Roem & Schult) on *Spodoptera frugiperda* (J. E. Smith). *J. Entomol.* **2009**, *6*, 109–116.

49. Vera-Curzio, L.G.; Aranda, E.E.; Castillo, E.P. *A Study of Cytotoxicity and Morphogenetic Responses of Calli of Ipomoea Murucoides Roem & Schults (Convolvulaceae) and Its Potential in Insecticidal Activity*; Memorias del X Congreso Nacional de Biotecnologia y Bioingenieria: Pueto Vallarta, Mexico, 2003.
50. Ramos-Lopez, M.A.; Perez, S.; Rodriguez-Hernandez, G.C.; Guevara_Fefer, P.; Zavala-Sanchez, M.A. Activity of *Ricinus communis* (Euphorbiaceae) against *Spodoptera frugiperda* (Lepidoptera: Noctuidae). *Afr. J. Biotechnol.* **2010**, *9*, 1359–1365.
51. Ramos-Lopez, M.A.; Gonzalez-Chavez, M.M.; Cardenas-Ortega, N.C.; Zavala-Sanchez, M.A.; Perez, G.S. Activity of the main fatty acid components of the hexane leaf extract of *Ricinus communis* against *Spodoptera frugiperda*. *Afri. J. Biotechnol.* **2012**, *11*, 4274–4278.
52. Bullangpoti, V.; Wajnberg, E.; Audant, P.; Feyereisen, R. Antifeedant activity of *Jatropha gossypifolia* and *Melia azedarach* senescent leaf extracts on *Spodoptera frugiperda* (Lepidoptera: Noctuidae) and their potential use as synergists. *Pest Manag. Sci.* **2012**, *68*, 1255–1264. [CrossRef]
53. D'incao, M.P.; Quadros, B.; Fiuza, L. Efeito agudo e crônico de três diferentes extratos de *Euphorbia pulcherrima* sobre *Spodoptera frugiperda* (J.E.SMITH, 1797) (Lepidoptera, Noctuidae). In I Simpósio de Integração das Pós-Graduações do CCB/UFSC. *Florianópolis* **2012**, 28.
54. Almeida, V.T.; Ramos, V.M.; Saqueti, M.B.; Gorni, P.H.; Pacheco, A.C.; Marcos de Leão, R. Bioactivity of ethanolic extracts of *Euphorbia pulcherrima* on *Spodoptera frugiperda* (J.E. Smith) (Lepidoptera: Noctuidae). *Afr. J. Biotechnol.* **2017**, *16*, 615–622.
55. Soares, C.S.A.; Silva, M.; Costa, M.B.; Bezerra, C.E.S. Ação inseticida de óleos essenciais sobre a lagarta desfolhadora *Thyrinteina arnobia* (Stoll) (Lepidoptera: Geometridae). *Rev. Verde* **2011**, *6*, 154–157.
56. Praveena, A.; Sanjayan, K.P. Inhibition of Acetylcholinesterase in Three Insects of Economic Importance by Linalool, a Monoterpene Phytochemical. In *Insect Pest Management, A Current Scenario*; Ambrose, D.P., Ed.; Entomology Research Unit, St. Xavier's College: Palayamkottai, India, 2011; pp. 340–345.
57. Labinas, A.M.; Crocomo, W.B. Effect of Java grass (*Cymbopogon winterianus* Jowitt) essential oil on fall armyworm *Spodoptera frugiperda* (J. E. Smith, 1797) (Lepidoptera, Noctuidae). *Acta Sci.* **2002**, *4*, 1401–1405.
58. Zavala-Sánchez, M.A.; Pérez-Gutierrez, S.; Romo-Asunción, D.; Cárdenas-Ortega, N.C.; Ramos-López, M.A. Activity of four *Salvia* species against *Spodoptera frugiperda* (J. E. Smith) (Lepidoptera: Noctuidae). *Southwest. Entomol.* **2013**, *38*, 67–73. [CrossRef]
59. Cardenas-Ortega, N.C.; Gonzalez-Chavez, M.M.; Figueroa-Brito, R.; Flores-Macias, A.; Romo-Asuncion, D.; Martinez-Gonzalez, D.E.; Perez-Moreno, V.; Ramos-Lopez, M.A. Composition of the essential oil of *S. ballotiflora* (Lamiaceae) and its insecticidal activity. *Molecules* **2015**, *20*, 8048–8059. [CrossRef] [PubMed]
60. Breuer, M.; Schmidt, G.H. Influence of a short period treatment with *Melia azedarach* extract on food intake and growth of the larvae of *Spodoptera frugiperda* (J. E. Smith) (Lepidoptera: Noctuidae). *J. Plant Dis. Prot.* **1995**, *102*, 633–654.
61. Dos Santos, W.; Freire, M.; Bogorni, P.C.; Vendramim, J.D.; Macedo, M.L. Effect of the aqueous extracts of the seeds of *Talisia esculenta* and *Sapindus saponaria* on fall armyworm. *Braz. Arch. Biol. Technol.* **2008**, *51*, 373–383. [CrossRef]
62. Bogorni, P.C.; Vendramim, J.D. Sublethal effect of aqueous extracts of *Trichilia* spp. on *Spodoptera frugiperda* (J.E. Smith) (Lepidoptera:Noctuidae) development on maize. *Neotrop. Entomol.* **2005**, *34*, 311–317.
63. Roel, A.R.; Vendramim, J.D.; Frighetto, R.T.S.; Frighetto, N. Efeito do extrato acetato de etila de *Trichilia pallida* (Swartz) (Meliaceae) no desenvolvimento e sobrevivência da lagarta-do-cartucho. *Bragantia* **2000**, *59*, 53–58. [CrossRef]
64. Cespedes, C.L.; Calderón, J.S.; Lina, L.; Aranda, E. Growth inhibitory effects on fall armyworm *Spodoptera frugiperda* of some limonoids isolated from *Cedrela* spp. (Meliaceae). *J. Agric. Food Chem.* **2000**, *48*, 1903–1908. [CrossRef]
65. Mikolajczak, K.L.; Zilkowski, B.W.; Bartelt, R.J. Effect of meliaceous seed extracts on growth and survival of *Spodoptera frugiperda* (J.E. Smith). *J. Chem. Ecol.* **1989**, *15*, 121–128. [CrossRef] [PubMed]
66. Hellpap, C.; Mercado, J.C. Effect of neem on the oviposition behaviour of the fall armyworm *Spodoptera frugiperda* Smith. *J. Appl. Entomol.* **1986**, *105*, 463–467. [CrossRef]
67. Campos, A.P.; Boica Junior, A.L.; Lagartas, D.C. Spodoptera frugiperda (J. E. Smith) (Lepidoptera: Noctuidae) submetidas a diferentes concentrações de óleo de nim. *Rev. Bras. Milho Sorgo* **2012**, *11*, 137–144.
68. Mordue, A.J.; Nisbet, A.J. Azadirachtin from the Neem tree (*Azadirachta indica*): Its actions against insects. *Anais da Soc. Entomol. Bras.* **2000**, *29*, 615–632.

69. Viana, P.A.; Prates, H.T.; Ribeiro, P.E.A. Efeito de extratos de nim e de metodos de aplicaçao sobre o dano foliar e o desenvolvimento da lagarta-do-cartucho, *Spodoptera frugiperda*, em milho. *Rev. Bras. Milho Sorgo* **2007**, *6*, 17–25. [CrossRef]
70. Viana, P.A.; Prates, H.T. Desenvolvimento e mortalidade larval de *Spodoptera frugiperda* em folhas de milho tratadas com extrato aquoso de folhas de Azadirachta indica. *Bragantia* **2003**, *62*, 69–74. [CrossRef]
71. Viana, P.A.; Prates, H.T. Mortalidade de lagarta de *Spodoptera frugiperda* alimentadas com folhas de milho tratadas com extrato aquoso de folhas de nim (*Azadirachta indica*). *Rev. Bras. Milho Sorgo Sete Lagoas* **2005**, *4*, 316–322. [CrossRef]
72. Mordue, A.J.; Blackwell, A. Azadirachtin: An update. *J. Insect. Physiol.* **1993**, *39*, 903–924. [CrossRef]
73. Silva, S.M.; Broglio, S.M.F.; Trindade, R.C.P.; Ferreira, E.S.; Gomes, I.B.; Micheletti, L.B. Toxicity and application of neem in fall armyworm. *Comun. Sci.* **2015**, *6*, 359–364. [CrossRef]
74. Gutierrez-Garcia, S.D.; Sanchez-Escudero, J.; Perez-Dominguez, J.F.; Carballo-Carballo, A.; Bergvinson, D.; Aguilera-Pena, M.M. Effect of neem on damage caused by fall armyworm *Spodoptera frugiperda* (Smith) (Lepidoptera:Noctuidae) and three agricultural variables on resistant and susceptible maize. *Acta Zool. Mex.* **2010**, *26*, 1–6.
75. Brechelt, A.; Fernandez, C.L. *El nim. Un arbol para la Agricultura y el Medio Ambiente*; Experienses en La Republica Dominicana. San Cristobal, Rep. Dom.; Fundacion Agricola Y Meio Ambiente: San Domingo, Dominican Republic, 1995; p. 133.
76. Silva, G.; Rodriguez, J.C.; Blanco, C.A.; Lagunes, A. Bioactivity of a water extract of boldus (*Peumus boldus* Molina) against *Spodoptera frugiperda* (J.E. Smith) and *Helicoverpa zea* Boddie (Lepidoptra: Noctuidae). *Chil. J. Agr. Res.* **2013**, *73*, 135–141. [CrossRef]
77. Kamel, A.M. Can we use the moringa oil as botanical insecticide against *Spodoptera frugiperda*? *Acad. J. Entomol.* **2010**, *3*, 59–64.
78. Soberon-Risco, G.V.; Rojas, C.; Kato, M.J.; Diaz, J.S. Larvicidal activity of *Piper tuberculatum* on *Spodoptera frugiperda* (Lepidoptera: Noctuidae) under laboratory conditions. *Rev. Colomb. Entomol.* **2012**, *38*, 35–40.
79. Alves, D.S.; Carvalho, G.A.; Oliveira, D.F.; Samia, R.R.; Villas-Boas, M.A.; Carvalho, G.A.; Correa, A.D. Toxicity of copaiba extracts to armyworm (*Spodoptera frugiperda*). *Afr. J. Biotechnol.* **2012**, *11*, 6578–6591.
80. Oliveira, E.R.; Alves, D.S.; Aazza, S.; Bertolucci, S.K.V. Toxicity of *Cymbopogon flexuosus* essential oil and Citral for *Spodoptera frugiperda*. *Ciencia e Agrotech.* **2008**, *42*, 408–419. [CrossRef]
81. Farias-Rivera, L.A.; Hernandez-Mendoza, J.L.; Molina-Ochoa, J.; Pescador-Rubio, A. Effect of leaf extracts of teosinte, *Zea diploperrennis* L. and a Mexican maize variety, Criollo "uruapeno" on the growth and survival of the Fall armyworm (Lepidoptera:Noctuidae). *Fla. Entomol.* **2002**, *86*, 239–343. [CrossRef]
82. Tavares, W.S.; Grazziotti, G.H.; De Souza Juniour, A.A.; Freitas, S.S.; Consolaro, H.N.; Ribeiro, P.E.A.; Zanuncio, J.C. Screening of extracts of leaves and stems of *Psychotria* spp. (Rubiaceae) against *Sitophillus zeamais* (Coleoptera: Curculionidae) and *Spodoptera frugiperda* (Lepidoptera:Noctuidae) for maize protection. *J. Food Prot.* **2013**, *76*, 1892–1901. [CrossRef]
83. Ruberto, G.; Renda, A.; Tringali, C.; Napoli, E.M.; Simmonds, M.S. Citrus limonoids and their semisynthetic derivatives as antifeedant agents against *Spodoptera frugiperda* larvae. A structure-activity relationship study. *J. Agric. Food Chem.* **2002**, *50*, 6766–6774. [CrossRef]
84. Gallo, M.B.C.; da Rocha, W.C.; Cunha, U.S.; Diogo, F.A.; da Silva, F.C.; Vieira, P.C.; Vendramim, J.D.; Fernandes, J.B.; da Silva, M.F.; Batista-Pereira, L.G. Bioactivity of extracts and isolated compounds from *Vitex polygama* (Verbenaceae) and *Siphoneugena densiflora* (Myrtaceae) against *Spodoptera frugiperda* (Lepidoptera:Noctuidae). *Pest Manag. Sci.* **2006**, *62*, 1072–1081. [CrossRef]
85. Amoabeng, B.W.; Gurr, G.M.; Gitau, C.W.; Nicol, H.I.; Munyakazi, L.; Stevenson, P.C. Tri-Trophic Insecticidal Effects of African Plants against Cabbage Pests. *PLoS ONE* **2013**, *8*, 10. [CrossRef]
86. Blank, A.F.; Souza, E.M.O.; Arrigoni-Blank, M.D.F.; Paula, J.W.; Alves, P.B. Maria Bonita: Cultivar de manjericão tipo linalol. *Pesqui Agropecu Bras.* **2007**, *42*, 1811–1813. [CrossRef]
87. Stevenson, P.C.; Kite, G.C.; Lewis, G.P.; Forest, F.; Nyirenda, S.P.; Belmain, S.R.; Sileshi, G.W.; Veitch, N.C. Distinct chemotypes of *Tephrosia vogelii*: Implications for insect pest control and soil enrichment. *Phytochemistry* **2012**, *78*, 135–146. [CrossRef] [PubMed]
88. Green, P.W.C.; Belmain, S.R.; Ndakidemi Patrick, A.; Farrell, I.W.; Stevenson, P.C. Insecticidal activity in Tithonia diversifolia and Vernonia amygdalina. *Ind. Crop Prod.* **2017**, *110*, 15–21. [CrossRef]

89. Phua, D.H.; Tsai, W.; Ger, J.; Deng, J.; Yang, C. Human Melia azedarach poisoning. *J. Clin. Toxicol.* **2009**, *46*, 1067–1107.
90. Govindachari, T.R.; Narasimhan, N.S.; Suresh, G.; Partho, P.D.; Gopalakrishnan, G.; Krishna-Kumari, G.N. Structure-related insect antifeedant and growth regulating activities of some limonoids. *J. Chem. Ecol.* **1995**, *21*, 1585–1601. [CrossRef]
91. Tomas-Barberan, F.A.; Wollenweber, E. Flavonoid aglycones from the leaf surfaces of some Labiatae species. *Plant Syst. Evol.* **1990**, *173*, 109–118. [CrossRef]
92. Danelutte, A.P.; Costantin, M.B.; Delgado, G.E.; Braz-Filho, R.; Kato, M.J. Divergence of secondary metabolism in cell suspension cultures and differentiated plants of *Piper cernuum* and *P. crassinervium*. *J. Braz. Chem. Soc.* **2005**, *16*, 1425–1430. [CrossRef]
93. Gupta, P.K. Pesticide exposure- Indian scene. *Toxicology* **2004**, *198*, 83–90. [CrossRef]
94. Allan, E.J.; Stuchbury, T.; Mordue, A.J. Azadirachta indica A. Juss. (Neem tree): In vitro culture, micropropagation and the production of Azadirachtin and other secondary metabolites. In *Medical Aromatic Plants*; Biotechnology in Agriculture and forestry science series; Bajaj, Y.P.S., Ed.; Springer: Berlin/Heidelberg, Germany, 1999; Volume 43, pp. 11–41.
95. Riyajan, S.; Sakdapipanich, J.T. *Development of a Controlled Release Neem Capsule with a Sodium Alginate Matrix, Cross Linked by Glutaraldehyde and Coated with Natural Rubber*; Polymer Bulletin: Berlin, Germany, 2009.
96. Forim, R.M.; Matos, A.P.; Silva, M.F.G.F.; Cass, Q.B.; Vieira, P.C.; Fernandes, J.B. Uso de CLAE no controle de qualidade em produtos comerciais de nim: Reproductividade da acao inseticida. *Quim. Nova.* **2010**, *33*, 1082–1087. [CrossRef]
97. Perlatti, B.; Fernandes, J.B.; Silva, M.F.; Forim, M.R.; de Souza Bergo, P.L. Polymeric nanoparticle-based insecticides: A controlled release purpose for agrochemicals. In *Insecticides-Development of Safer and More Effective Technologies*; Trdan, S., Ed.; INTECH Open Access Publisher: Rijeka, Croatia, 2013; pp. 523–550.
98. Das, R.K.; Sarma, S.; Brar, S.K.; Verma, M. Nanoformulation of insecticides: Novel products. *Biofertil Biopestic.* **2014**, *5*, 2.
99. Martinez, S.M.; Emdem, H.F. Growth disruption, abnormalities and mortality of *Spodoptera littoralis* (Baisduval) (Lepidoptera: Noctuidae) caused by azadirachtin. *Neotrop. Entomol.* **2001**, *30*, 113–125. [CrossRef]
100. Cespedes, C.L.; Salazar, J.R.; Martinez, M.; Aranda, E. Insect growth regulatory effects of some extracts and sterols from *Myrtillocactus geometrizans* (Cactaceae) against *Spodoptera frugiperda* and *Tenebrio molitor*. *Phytochemistry* **2005**, *66*, 2481–2493. [CrossRef] [PubMed]
101. Mkindi, A.; Mpumi, N.; Tembo, Y.; Stevenson, P.C.; Ndakidemi, P.; Mtei, K.; Machunda, R.L.; Belmain, S.R. Invasive weeds with pesticidal properties as potential new crops. *Ind. Crops Prod.* **2017**, *110*, 113–122. [CrossRef]
102. Kassie, M.; Jalta MShiferaw, B.; Mmbando, F. *Plot and Household Level Determinants of Sustainable Agricultural Practices in Rural Tanzania*; Environ Dev.; Resources for the Future: Washington, DC, USA, 2012.
103. Erenstein, O.; Samaddar, A.; Teufel, N.; Blümmel, M. The paradox of limited maize stover use in India's smallholder crop-livestock systems. *Exp. Agric.* **2011**, *47*, 677–704. [CrossRef]
104. Grzywacz, D.; Stevenson, P.C.; Belmain, S.R.; Wilson, K. The Use of Indigenous Ecological Resources for Pest Control in Africa. *Food Secur.* **2014**, *6*, 71–86. [CrossRef]
105. Belmain, S.R.; Stevenson, P.C. Ethnobotanicals in Ghana: Reviving and modernising an age-old practise. *Pestic. Outlook* **2001**, *6*, 233–238.
106. Sarasan, V.; Kite, G.C.; Sileshi, G.W.; Stevenson, P.C. The application of phytochemistry and invitro tools to the sustainable utilization of medicinal and pesticidal plants for income generation and poverty alleviation. *Plant Cell Rep.* **2011**, *30*, 1163–1172. [CrossRef]
107. Belmain, S.R.; Amoah, B.A.; Nyirenda, S.P.; Kamanula, J.F.; Stevenson, P.C. Highly variable insect control efficacy of *Tephrosia vogelii* chemotypes. *J. Agric. Food Chem.* **2012**, *60*, 1055–1066.
108. Mkindi, A.G.; Tembo, Y.; Mbega, E.R.; Medvecky, B.; Kendal-Smith, A.; Farrell, I.W.; Ndakidemi, P.A.; Belmain, S.R.; Stevenson, P.C. Phytochemical analysis of *Tephrosia vogelii* across East Africa reveals three chemotypes that influence its use as a pesticidal plant. *Plants* **2019**, *8*, 597. [CrossRef]

© 2020 by the authors. Licensee MDPI, Basel, Switzerland. This article is an open access article distributed under the terms and conditions of the Creative Commons Attribution (CC BY) license (http://creativecommons.org/licenses/by/4.0/).

Review

The Phytochemical Composition of *Melia volkensii* and Its Potential for Insect Pest Management

Victor Jaoko [1,2,3,*], Clauvis Nji Tizi Taning [1], Simon Backx [2], Jackson Mulatya [3], Jan Van den Abeele [4], Titus Magomere [5], Florence Olubayo [5], Sven Mangelinckx [2,*], Stefaan P.O. Werbrouck [1] and Guy Smagghe [1]

[1] Department of Plants and Crops, Ghent University, Coupure Links 653, B-9000 Ghent, Belgium; tiziclauvis.taningnji@ugent.be (C.N.T.T.); stefaan.werbrouck@ugent.be (S.P.O.W.); guy.smagghe@ugent.be (G.S.)
[2] SynBioC, Department of Green Chemistry and Technology, Ghent University, Coupure Links 653, B-9000 Ghent, Belgium; simon.backx@ugent.be
[3] Kenya Forestry Research Institute, P.O. Box 20412-00200 Nairobi, Kenya; jmulatya@kefri.org
[4] Better Globe Forestry, P.O Box 823-00606 Nairobi, Kenya; jan@betterglobeforestry.com
[5] Department of Plant Science and Crop Protection, University of Nairobi, P.O. Box 30197-0010 Nairobi, Kenya; magomere.titus@ku.ac.ke (T.M.); olubayo@uonbi.ac.ke (F.O.)
* Correspondence: victor.jaoko@ugent.be (V.J.); sven.mangelinckx@ugent.be (S.M.); Tel.: +254-722157414 (V.J.); +32-9-264-59-51 (S.M.)

Received: 31 December 2019; Accepted: 20 January 2020; Published: 22 January 2020

Abstract: Due to potential health and environmental risks of synthetic pesticides, coupled with their non-selectivity and pest resistance, there has been increasing demand for safer and biodegradable alternatives for insect pest management. Botanical pesticides have emerged as a promising alternative due to their non-persistence, high selectivity, and low mammalian toxicity. Six Meliaceae plant species, *Azadirachta indica*, *Azadirachta excelsa*, *Azadirachta siamens*, *Melia azedarach*, *Melia toosendan*, and *Melia volkensii*, have been subject to botanical pesticide evaluation. This review focuses on *Melia volkensii*, which has not been intensively studied. *M. volkensii*, a dryland tree species native to East Africa, has shown activity towards a broad range of insect orders, including dipterans, lepidopterans and coleopterans. Its extracts have been reported to have growth inhibiting and antifeedant properties against *Schistocerca gregaria*, *Trichoplusia ni*, *Pseudaletia unipuncta*, *Epilachna varivestis*, *Nezara viridula*, several *Spodoptera* species and other insect pests. Mortality in mosquitoes has also been reported. Several limonoids with a wide range of biological activities have been isolated from the plant, including volkensin, salannin, toosendanin, trichilin-class limonoids, volkendousin, kulactone among others. This paper presents a concise review of published information on the phytochemical composition and potential of *M. volkensii* for application in insect pest management.

Keywords: Meliaceae; *Melia volkensii*; botanical pesticide; limonoid; insect pest; antifeedant; growth inhibitor

---

1. Introduction

The continuous and indiscriminate use of synthetic pesticides in crop protection has led to an increase in pest resistance, health and environmental concerns [1]. This has led to a renewed interest in natural products as alternative sources for insect pest control [1]. One of the most promising options is the use of secondary metabolites produced by plants, many of which are toxic to a wide spectrum of insect pests [2]. Plant extracts can offer a solution to insect pest control because they are environmentally friendly, easily biodegradable, and are target-specific [3].

The *Meliaceae* plant family has been reported to produce a wide range of compounds, including flavonoids, chromones, coumarins, benzofurans, mono-, sesqui-, di-, and triterpenoids, but tetranortriterpenoids with a β-substituted furanyl ring at C17α are the best known for the production of limonoids [4]. Limonoids are known for a range of biological activities, including insect antifeedant and growth-regulating properties and antibacterial properties [4]. Alkaloids are rarely isolated from Meliaceae [4]. Reviews on the Meliaceae plant family have been reported in the literature. The use of Meliaceae plant extracts as potential mosquitocides have been reviewed, and *Azadirachta indica* A. Juss (Indian neem tree) is reported as a potential plant for the control of vector mosquitoes [5]. Reviews on the chemical constituents of the genus Melia reported the isolation of terpenoids, steroids, alkaloids, flavonoids, anthraquinones with a wide range of biological activities including antiviral, pesticidal, inhibition of iNOS, antitumor, antibacterial and antifungal activities [6,7]. A phytopharmacological review of *Melia azedarach* Linnaeus (chinaberry) has been reported outlining its use in folk medicine having antifertility, antiviral, cytotoxic, antibacterial, immunomodulatory, repellent, antifeedant, antilithic and anthelmintic activity from various parts of the plant [8,9]. A review on *A. indica* has reported its use in agriculture for application as manure, fertilizer, soil conditioner, fumigant, and as botanical pesticide [10]. *Melia volkensii* (Gurke) has also been identified as one of the pesticidal plants in Africa [11]. Another review has explored the phytochemical and antimicrobial activities of the Meliaceae family [12]. Detailed information on commercially available neem products developed for agricultural pest control has also been reviewed [13].

Several plant species of the Meliaceae have shown promising bioactivity against a variety of insects [3]. Their insect growth regulatory and antifeedant properties against many insect pests have made them emerge as a potent source of insect control products [14]. Six species have been subjected to botanical pesticide evaluation; these include *A. indica* (Indian neem tree), *Azadirachta excelsa* Jack (Philippine neem tree), *Azadirachta siamens* Valeton (Siamese neem tree), *M. azedarach* (chinaberry), *Melia toosendan* Siebold and Zucc., and *M. volkensii* [13]. However, research has concentrated mostly on *A. indica* (neem tree) and *M. azedarach* (chinaberry) [15]. Azadirachtin, a commercial biopesticide, and other limonoids isolated from *A. indica*, have been effective growth regulators and feeding deterrents for a wide range of insect species [16]. Azadirachtin targets the corpus cardiacum in insects, which in turn affects neuroendocrine activity and turnover of neurosecretion [17]. Extracts from *M. azedarach* have also shown antifeedant activity against the juvenile and adult *Xanthogaleruca luteola* Muller (elm leaf beetles) and mortality against its larvae [16]. Fruit extracts from *M. azedarach* are also effective against *Napomyza lateralis* Fallen (agromyzid leafminers) and *Trialeurodes vaporariorum* Westwood (whiteflies) [16]. Toosendanin, a limonoid constituent of *M. azedarach* which has been commercialized in China, is an effective growth inhibitor against *Ostrinia nubilalis* Hübner (European corn borer), effective repellent against *Pieris brassicae* Linnaeus (cabbage moth) and an oviposition deterrent against *Trichoplusia ni* Hübner (cabbage looper) [16]. Toosendanin is reported to be mainly active against lepidopteran pests and is less active than azadirachtin [18].

M. volkensii, a dryland tree species native to East Africa has, however, not been intensively studied [16]. It is a tall tree (15–25 m), shown in Figure 1, which grows in semi-arid areas of Kenya, Tanzania, Ethiopia, and Somalia at altitudes of between 350 to 1700 m above sea level [19]. The tree, like other meliaceous plants, is fast growing and produces fruits after 4–5 years [19]. It remains green for most of the year and is prized by farmers for its termite-resistant timber. It is intercropped with food crops, used for shade, firewood, and livestock fodder [19]. Several chemical compounds occur only in *M. volkensii*. These include: 1-O-cinnamoyltrichilin, meliavolkinin, 1,3-diacetylvilasinin, meliavolkin, volkensin, volkensinin, 12β- and 6β-hydroxykulactone, meliavolkenin, meliavolin, meliavolen, meliananinone, meliavolkensin A and B, melianin C, (E)- and (Z)-volkendousin, meliavosin, 2-9-epoxymeliavosin [6]. *M. volkensii* seed kernel extracts have more insect growth inhibitory and acute lethal toxicity than azadirachtin-containing fractions from neem seed kernel extracts [20]. It has been reported that when *M. volkensii* dried fruit powder and residual fruit cake obtained after extraction with ethanol are used as goat feed, their growth and performance are not negatively affected, indicating that both fruit powder and its cake could be used as safe ruminant feed supplement [21]. Its use as a fodder crop underscores its safety in mammals [20], and traditionally, it is used for the treatment of diarrhea, pain, skin rashes, and eczema [22]. Aqueous extracts of *M. volkensii* have also traditionally been used to control ticks and fleas in goats [19]. *M. volkensii* offers a key indigenous tree species that can be used to mitigate against desertification in arid and semi-arid lands [23], while also offering a high economic potential for the rural community in these regions [24]. This paper presents a concise review of published information on the phytochemical composition and potential application of *M. volkensii* in insect pest management.

Figure 1. *Melia volkensii* and its various parts: (**a**) 10-year old *M. volkensii* plantation, (**b**) leaves, (**c**) seeds, (**d**) fruits and (**e**) nuts [23].

2. Biological Activity of *Melia volkensii* Extracts Against Insects

Crude fruit extracts from *M. volkensii* have been reported to pose activity towards a broad range of insect orders including Diptera, Lepidoptera, Coleoptera among others [19] as shown in Table A1 (Appendix A). The methanolic fruit extracts were first reported to have antifeedant effects against *Schistocerca gregaria* Forsk. (desert locusts) [25]. Repellency effect, decreased mobility, retarded development and reduced fecundity were observed against *S. gregaria* when seed extract was applied to their preferred host plants mainly *Schouwia thebaica* Webb, *Fagonia olivieri* DC (fagonbush plant) and *Hyoscyamus muticus* Linnaeus (Egyptian henbane) in a field trial experiment [26]. Although the mode of action of the extracts is still unknown, it is postulated that the active compounds in *M. volkensii* extracts could affect hormone levels in *S. gregaria* larvae [27]. In fifth-instar nymphs of *S. gregaria*, 80% mortality was recorded 48 hours after injection with crude ethanolic and methanolic extracts at a concentration of 30 µg/g of the insect [19]. When sprayed on third- to fifth-instar *S. gregaria*, *M. volkensii* and neem oil have been reported to cause mortality of up to 91% and 92%, respectively, after 14 days in a comparative study [26]. In contrast to synthetic pesticides, these botanicals do not have a knock-down effect, but their slow response is similar to inhibitors of chitin synthesis [26].

Antifeedant and larval growth inhibitory effects of fruit extracts have been observed in *Trichoplusia ni* Hübner (cabbage looper) and *Pseudaletia unipuncta* Haworth (true armyworm) [25,28]. Crude seed extracts are also an effective growth inhibitor against *T. ni* (dietary EC_{50} = 7.6 ppm) and feeding deterrent (DC_{50} = 0.9 µg/cm^2) [29]. Prolonged exposure to *M. volkensii* extracts has been observed to lead to a decrease in antifeedant response when tested against *T. ni* implying that the insect could develop tolerance to the extracts [30]. However, when tested against *Plutella xylostella* Linnaeus (diamondback moth) and *P. unipuncta*, there was no significant decrease in feeding deterrent response to the extracts following continuous exposure [31]. It has been postulated that triterpenoids from seed kernels of *M. volkensii* are responsible for the insecticidal activity in *T. ni* [11]. Comparative efficacy has been observed with *M. volkensii* extracts, other Meliaceae plant extracts (*A. indica, A. excelsa, M. azedarach,* and *Trichilia americana* Sessé & Mocino) and commercial botanical insecticides (ryania, pyrethrum, rotenone and essential oils of rosemary and clove leaf) when tested against *T. ni* and *P. unipuncta* [32].

M. volkensii fruit extracts when tested at concentrations ranging from 1 to 50 µg/µL showed feeding deterrence, growth disruption and mortality against *Nezara viridula* Linnaeus (stink bug), a polyphagous pest which attacks a variety of crops, including nuts, corn, cotton, grains and tomatoes [16]. The disruption of the molting process led to eventual mortality in *N. viridula* [16]. Furthermore, deformities and malfunctions like shortened or missing antennae, legs failing to detach from the exuvium, absent or shortened hemelytra, notching, and lack of symmetry have been observed in *N. viridula* when exposed to fruit extracts, with 10 µg/µL causing malformation in up to 85.70% of surviving adults [16]. A delay of the imaginal molt was observed in immature *Coranus arenaceus* Walker even though there were no deformities in resultant adults after topical application of the *M. volkensii* extracts at 1, 5, and 10 µg/µL [16].

When applied to cabbage leaf disks in a choice bioassay, *M. volkensii* fruit extract showed potent antifeedant properties against *Epilachna varivestis* Mulsant (Mexican bean beetle) [16]. Growth inhibition has also been observed in *P. unipuncta* (dietary EC_{50} = 12.5 ppm) with refined seed extracts to the leaf discs in a choice bioassay [29]. The seed extracts also showed feeding deterrent effects on third-instar larvae of *P. unipuncta* and *P. xylostella*, and adults of *E. varivestis* (DC_{50} = 10.5, 20.7 and 2.3 µg/cm^2, respectively) [29]. In fact, *M. volkensii* seed extracts have been recorded to have stronger antifeedant activity compared to pure allelochemicals: digitoxin, cymarin, xanthotoxin, toosendanin, thymol and *trans*-anethole against *P. unipuncta*, *P. xylostella* and *E. varivestis* [29]. When applied to *Spodoptera litura* Fabricius, neem, rotenone, *M. volkensii* extract, toosendanin, *Annona squamosal* L. extract and pyrethrum at 1% concentration recorded larval growth (% relative to control) of 4.1, 97.5, 26.2, 48.3, 61.4, and 56.6%, respectively after 96 h in a comparative study [1].

Dried *M. volkensii* fruit extracts have shown growth-inhibiting activity against *Aedes aegypti* Linnaeus (yellow fever mosquito) larvae at 2 µg/mL in water, whilst recording high mortality during the molting and melanization process with LC_{50} of 50 µg/mL in 48 h [13]. At a high dose (100 µg/mL), the extracts caused acute toxicity, while at a low dose, the lethal effect took a long time, indicating the presence of compounds with an acute toxic effect at a high concentration and a growth-inhibiting effect at a low concentration [20]. Growth inhibiting and disrupting effects in *A. aegypti* could be a result of synergistic effects of a plethora of limonoid compounds or a single active compound exerting these effects [20].

A column chromatography-purified fraction of *M. volkensii* fruit kernel extract showed growth-inhibiting activity against *Anopheles arabiensis* Giles with an LC_{50} of 5.4 µg/mL in 48 h [13]. Mortality (LC_{50} of 34.72 µg/mL in 48 h) and oviposition deterrence was observed in second-instar larvae of *Culex quinquefasciatus* Say (Southern house mosquito) when treated with refined methanolic fruit extracts [33]. The granular formulation of *M. volkensii* fruit acetone extract showed S- and U-shaped postures and frequent stretching in *C. quinquefasciatus*; such postures and stretching are a characteristic of mosquito larvae reared in *M. volkensii* fruit extract [34]. The test granules also caused 86% mortality in third- and fourth-instar larvae of *C. quinquefasciatus* within 36 h [34]. Acetone extracts from *M. volkensii* seeds have recorded growth inhibitory effects and equal toxicity (LD_{50} of 30 µg/mL) for larvae and pupae of *C. pipiens* f. molestus Forskål (London underground mosquito) [17]. *M. azedarach* seed extracts recorded lower toxicity (LD_{50} of 40 µg/mL) while pure azadirachtin A recorded higher toxicity (LD_{50} of 1–5 µg/mL) against *C. pipiens* when compared with *M. volkensii* extracts [17]. The water solubility of the acetone seed extract from *M. volkensii* may indicate the presence of saponins as toxic principles thus making it an interesting candidate for application against aquatic insects such as mosquitoes and other vectors of diseases [17].

3. Phytochemistry and Insect Bioactivity of *Melia volkensii*

Insect antifeedants have been found in major classes of secondary metabolites—alkaloids, phenolics, and terpenoids [35]. However, it is in the terpenoids that the greatest number and diversity of antifeedants, and the most potent, have been found. Most well-documented antifeedants are triterpenoids [35]. Effective insect antifeedants have been isolated from various parts of *M. volkensii*, as shown in Figure 2 and Table A2 (Appendix B), although azadirachtin, the major ingredient in neem seeds, does not occur in *M. volkensii*. This indicates that insect control bioactivity is, therefore, based on other compounds than azadirachtin [25]. It is postulated that the major active compound in *M. volkensii* fruit is more lipophilic than azadirachtin [20]. Botanical antifeedants are easily degraded after application thereby causing little environmental impact [36].

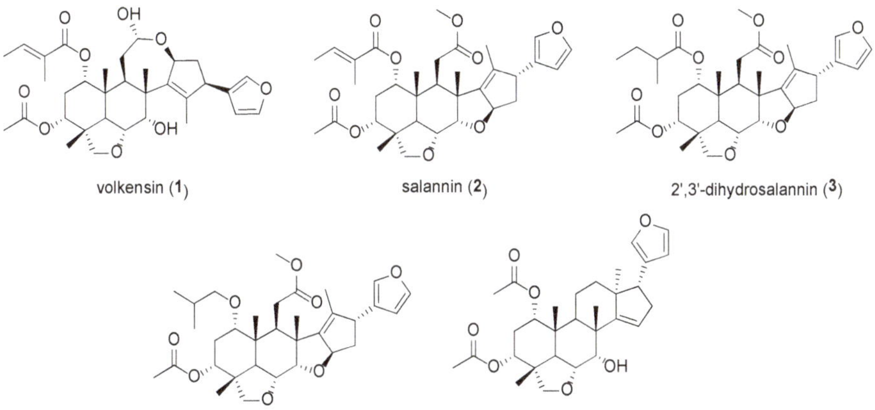

Figure 2. Chemical structures of compounds isolated from *Melia volkensii* with antifeedant and growth-inhibition activity against insects.

The insect antifeedants volkensin (**1**) and salannin (**2**) have been isolated from seed extracts of *M. volkensii* [37]. Additionally, volkensin (**1**) and salannin (**2**) were isolated from the whole fruits of *M. volkensii* [37]. Volkensin (**1**) has shown antifeedant activity against *Spodoptera frugiperda* Smith (fall armyworms) larvae with an ED_{50} of 3.5 µg/cm² [19]. Salannin (**2**) has also shown antifeedant activity against insect pests such as *Acalymma vittata* Fabricius (striped cucumber beetle), *Musca domestica* Linnaeus (housefly), *Epilachna varivestis* Mulsant (Mexican bean beetle), *Heliothis virescens* Fabricius (tobacco budworm), *S. frugiperda* and *Spodoptera littoralis* Boisduval (cotton leafworm) [38]. Salannin (**2**) has also been reported to cause feeding suppression against larvae of *Earias insulana* Boisduval (Egyptian stemborer), weight reduction (59%–89%) in *Cnaphalocrocis medinalis* Guenee (rice leafroller) and reduction in activities of acid phosphatases (ACP), alkaline phosphatases (ALP) and adenosine triphosphatases (ATPase), implying that gut enzyme activities were affected. 2',3'-Dihydrosalannin (**3**), 1-detigloyl-1-isobutylsalannin (**4**) and 1α,3α-diacetylvilasinin (**5**) have also been isolated from the plant [7].

M. volkensii seed extracts, extracted in cold water, have been reported to contain unsaturated fatty acids (oleic acid (**6**), linoleic acid (**7**) and gadoleic acid (**8**)) and saturated fatty acids (palmitic acid (**9**), stearic acid (**10**) and arachidic acid (**11**)) as shown in Figure 3 [39]. Fatty acids with at least 18 carbon atoms have been found to synergistically enhance insecticidal activity of insecticides [40]. Oleic acid (**6**), linoleic acid (**7**), linolenic acid, and ricinoleic acid have enhanced insecticidal activity of organophosphates and carbamates when applied against sucking insects and defoliating insects [40].

Other chemical compounds that have been isolated from various parts of *M. volkensii* are shown in Figure 4. Toosendanin (**12**), which has been isolated from the root bark of *M. volkensii* [22], has been reported to be an effective growth inhibitor against *O. nubilalis*, an effective repellent against *P. brassicae* and an oviposition deterrent against *T. ni* [16]. 1-Cinnamoyltrichilinin (**13**) has shown antifeedant activity towards *S. littoralis* having minimum antifeedant concentration (MAC) value of 1000 mg/L and a significant antibacterial activity against *Porphyromonas gingivalis* ATCC 33277 with minimum inhibitory concentration (MIC) value of 15.6 µg/mL [7]. Nimbolin B (**14**) has been reported to have antifeedant activity against several *Spodoptera* species (*S. exigua*, *S. eridania* and *S. littoralis*) [7]. There was a clear-cut structure-activity relationship when trichilin-class limonoids (1-cinnamoyltrichilinin **13**, 1-acetyltrichilinin **15**, 1-tigloyltrichilinin **16**) were tested against *Spodoptera eridania* Stoll (Southern armyworm) where the 12α-OH function was the most potent, followed by 12β-OH, 12-desoxy, and 12α-acetoxy groups in order of decreasing potency [7]. The 12-OH functionality could be necessary for maximum bioactivity in trichilin-class limonoids (**13**, **15**, **16**) [7]. 2,19-oxymeliavosin **17**, which has weak activity with marginally significant selectivity for breast cancer cell line (MCF-7), has also been isolated from the root bark of *M. volkensii* [41]. Ohchinin-3-acetate (**18**), isolated from methanolic extract of *M. volkensii* fruits [42], and meliantriol (**19**), both insect antifeedants have also been reported [15]. Meliantriol has exhibited moderate cytotoxicity against human epidermoid carcinoma of the nasopharynx (KB), multidrug-resistant (KB-C2), and breast cancer cell line (MCF-7) [43].

Figure 3. Chemical structures of saturated and unsaturated fatty acids isolated from *Melia volkensii*.

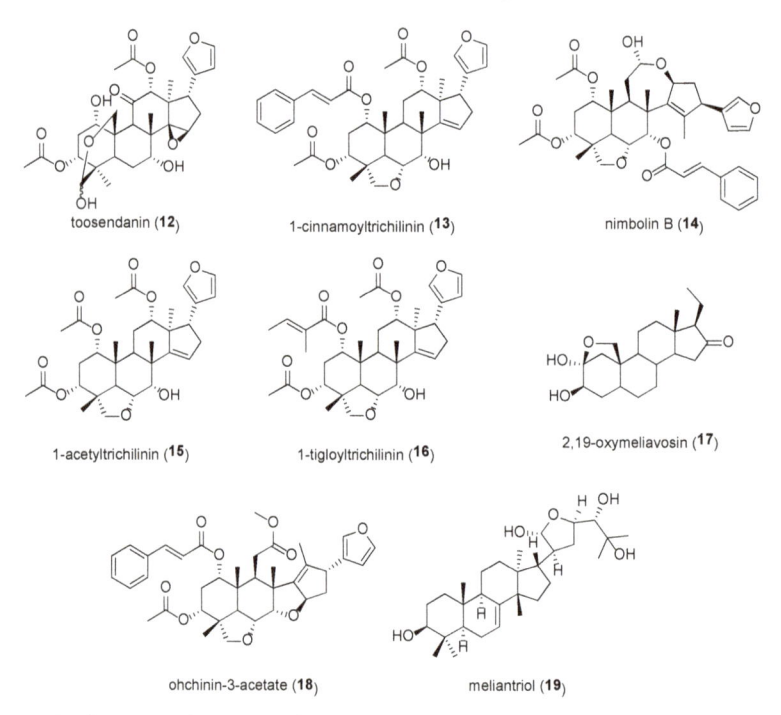

Figure 4. Further chemical structures of compounds isolated from *Melia volkensii* with antifeedant and growth-inhibition activity against insects.

4. Further Phytochemical Composition and Biological Activity of *Melia volkensii*

Other compounds have also been isolated from *M. volkensii* with different biological activities. These include volkensinin, as isolated from ethanolic extracts of *M. volkensii* root bark [44], which showed weak bioactivity in the brine shrimp lethality test BST (LC_{50} = 57 µg/mL) and weak cytotoxicity against six human tumor cell lines with ED_{50} values of 27.90, 28.35, 33.56, 29.55, 8.43, and 28.51 µg/mL in A-498 (human kidney carcinoma), PC-3 (prostate adenocarcinoma), PACA-2 (pancreatic carcinoma), A-549 (human lung carcinoma), MCF-7 (human breast carcinoma), and HT-29 (human colon adenocarcinoma), respectively [44]. Toosendanin has activity against *Escherichia coli* Migula and *Aspergillus niger* Tiegh. with respective minimum inhibitory concentration (MIC) values of 12.5 and 6.25 µg/mL [22]. Melianin B, isolated from the root bark of *M. volkensii*, showed cytotoxicity against six human solid tumor cell lines: A-549, MCF-7, HT-29, A-498, PACA-2, and PC-3 [45]. Bioactivity-guided fractionation of *M. volkensii* root bark led to the isolation of meliavolkenin which showed moderate cytotoxicity against three human tumor cell lines with a respective ED_{50} value of 10.33 µg/mL, 4.30 µg/mL, and 0.67 µg/mL in A-549, MCF-7, and HT-29 cells [46]. The bioactive apotirucallane triterpenes meliavolkensin A and meliavolkensin B, both isolated from the root bark of *M. volkensii* [47], have shown cytotoxicity against human colon tumor cell lines H-29 (human colon adeno-carcinoma) with ED_{50} values of 0.49 µg/mL and 0.25 µg/mL, respectively [47]. (*E*)-volkendousin, isolated from *M. volkensii* root bark, also showed activity against six human tumor cell lines (A-549, MCF-7, HT-29, A-498, PACA-2 and PC-3) [48]. Meliavolin, marginally cytotoxic against human tumor cell lines with an ED_{50} of 11.25 µg/mL, 0.57 µg/mL and 6.65 µg/mL in A-549, MCF-7 and HT-29 cells, respectively [49], has been isolated from *M. volkensii* root bark following activity-directed fractionation with brine shrimp test [49]. Kulactone was isolated from root bark of *M. volkensii* and exhibited significant activity against *E. coli* and *A. niger* with a respective minimum inhibitory concentration (MIC) value of 12.5 and 6.25 µg/mL [22]. Bioactivity-guided antimycobacterial investigations against *Mycobacterium tuberculosis* Zopf resulted in the isolation of 12β-hydroxykulactone, 6β-hydroxykulactone and kulonate from *M. volkensii* seeds with MIC values of 16 µg/mL, 4 µg/mL, and 16 µg/mL, respectively [50]. Meliavolkin has shown anticancer activity against three human tumor cell lines: A-549 (ED_{50} = 0.57 µg/mL), MCF-7 (ED_{50} = 0.26 µg/mL), and HT-29 (ED_{50} = 0.12 µg/mL) [7]. Other limonoids isolated from *M. volkensii* include 3-episapelin, meliavolen, melianinone [4], and nimbolin B [51] and all have shown selectivity for the colon cell line HT-29 [51]. Other compounds, which have been isolated from *M. volkensii* include scopoletin [22], melianin C and meliavolkinin [7], methyl kulonate and 2,19-epoxymeliavosin [6], nimbolidins C-E [12]. However, their activity against insects has not been reported in literature.

5. Conclusions

Extracts and pure compounds isolated from *M. volkensii* have proved to be effective insect antifeedants and growth inhibitors. Extensive research has been done on mosquito control using *M. volkensii*; however, more research needs to be done on insect pests of agricultural importance. *M. volkensii* has no reported adverse effect on the environment or mammals, making it a potential botanical pesticide for the biosafe application in integrated pest management. The availability of renewable resources from the tree, such as fruits, stem bark, and leaves makes this plant a potential candidate for insect control with minimal interference on the plant. In this regard, *M. volkensii* could be further exploited as a source of natural insecticide.

Author Contributions: Conceptualization—G.S., S.P.O.W., J.M., T.M., F.O. and J.V.d.A.; investigation—V.J., S.B., C.N.T.T., G.S., S.M., S.P.O.W., F.O.; resources—S.P.O.W. and G.S.; writing—original draft preparation—V.J.; writing—review and editing—G.S., S.P.O.W., S.M., C.N.T.T., S.B., J.M., T.M., F.O. and J.V.d.A.; supervision—G.S., S.M., S.P.O.W., F.O., C.N.T.T.; project administration—S.P.O.W., F.O., T.M.; funding acquisition—G.S., S.P.O.W. All authors have read and agreed to the published version of the manuscript.

Funding: This research was funded by VLIR-UOS. Grant number KE2018TEA465A103.

Acknowledgments: The authors thank VLIR-UOS for the financial support.

Conflicts of Interest: The authors declare no conflict of interest. The funders had no role in the design of the study; in the collection, analyses, or interpretation of data; in the writing of the manuscript, or in the decision to publish the results.

Appendix A

Table A1. *Melia volkensii* as a botanical pesticide for insect pest control.

Target Insect *	Order	Biological Activity	Plant Part Used	Reference
Desert locust, *Schistocerca gregaria*	Orthoptera	Antifeedant, repellency, growth inhibition, mortality	Fruit	[19,25,26]
Cabbage looper, *Trichoplusia ni*	Lepidoptera	Antifeedant, growth inhibition, mortality	Fruit, seed	[25,28–30]
True armyworm, *Pseudaletia unipuncta*	Lepidoptera	Antifeedant, growth inhibition	Fruit, seed	[11,25,28,29,31]
Diamondback moth, *Plutella xylostella*	Lepidoptera	Antifeedant	Fruits	[29,31]
Stink bug, *Nezara viridula*	Hemiptera	Antifeedant, growth disruption, mortality	Fruit	[16]
Coranus arenaceus	Hemiptera	Growth inhibition	Fruit	[16]
Mexican bean beetle, *Epilachna varivestis*	Coleoptera	Antifeedant, growth inhibition	Seed	[16,29]
Yellow fever mosquito, *Aedes aegypti*	Diptera	Growth inhibition, mortality	Fruit	[13,20]
Anopheles arabiensis	Diptera	Growth inhibition	Fruit kernel	[13]
Southern house mosquito, *Culex quinquefasciatus*	Diptera	Oviposition deterrence, mortality	Fruit	[13,33,34]
London underground mosquito, *Culex pipiens molestus*	Diptera	Growth inhibition, mortality	Seed	[17]

* Non exhaustive list of potential target insect pests.

Appendix B

Table A2. Phytochemical investigation of *Melia volkensii*.

Compound *	Plant Part Isolated From	Biological Activity	Reference
Volkensin	Seed, fruit	Antifeedant against fall armyworms, *Spodoptera frugiperda*	[19,37]
Salannin	Seed, fruit	Antifeedant and weight reduction against *Acalymma vittata*, *Musca domestica*, *Epilachna varivestis*, *Heliothis virescens*, *Spodoptera frugiperda*, *Earias insulana*, *Cnaphalocrocis medinalis* and *Spodoptera littoralis*	[7,37,38]
Toosendanin	Root bark	Growth inhibitor and oviposition deterrent against *Ostrinia nubilalis*, *Pieris brassicae*, *Trichoplusia ni*	[16,22]
Meliantriol	Not reported	Antifeedant	[15]
Unsaturated fatty acids (oleic acid, linoleic acid and gadoleic acid); saturated fatty acids (palmitic acid, stearic acid and arachidic acid)	Seed	Synergistic enhancement of insecticidal activity	[39,40]
1-cinnamoyltrichilinin	Not reported	Antifeedant against *Spodoptera littoralis*	[7]
1-tigloyltrichilinin	Not reported	Antifeedant against *Spodoptera eridania*	[7]
1-acetyltrichilinin	Not reported	Antifeedant against *Spodoptera eridania*	[7]
Nimbolin B	Not reported	Antifeedant against *Spodoptera* species. (*exigua*, *eridania* and *littoralis*)	[7,51]
Ohchinin-3-acetate	Fruit	Antifeedant	[42]

* Non exhaustive list of compounds present in *M. volkensii*.

References

1. Bhuiyan, K.R.; Hassan, E.; Isman, M.B. Growth inhibitory and lethal effects of some botanical insecticides and potential synergy by dillapiol in *Spodoptera litura* (Fab.) (*Lepidoptera*: *Noctuidae*). *J. Plant Dis. Prot.* **2001**, *108*, 82–88.
2. Rai, M.; Carpinella, M.C. *Naturally Occurring Bioactive Compounds*, 3rd ed.; Elsevier: Amsterdam, The Netherlands, 2006.
3. Agbo, B.E.; Nta, A.I.; Ajaba, M.O. Bio-pesticidal properties of Neem (*Azadirachta indica*). *Adv. Trends Agric. Sci.* **2019**, *1*, 17–26.
4. Mulholland, D.A.; Parel, B.; Coombes, P.H. The chemistry of the Meliaceae and Ptaeroxylaceae of Southern and Eastern Africa and Madagascar. *Curr. Org. Chem.* **2000**, *4*, 1011–1054. [CrossRef]
5. Baskar, K.; Mohankumar, S.; Sudha, V.; Mahetswaran, R.; Vijayal.akshmi, S.; Jayakumar, M. Meliaceae plant extracts as potential mosquitocides—A review. *Entomol. Ornithol. Herpetol.* **2016**, *5*, 1–4. [CrossRef]
6. Zhao, L.; Huo, C.-H.; Shen, L.-R.; Yang, Y.; Zhang, Q.; Shi, Q.-W. Chemical constituents of plants from the Genus Melia. *Chem. Biodivers.* **2010**, *7*, 839–859. [CrossRef] [PubMed]
7. Tan, Q.-G.; Luo, X.-D. Meliaceous limonoids: Chemistry and biological activities. *Chem. Rev.* **2011**, *111*, 7437–7522. [CrossRef]
8. Rana, A. Melia azedarach: A phytopharmacological review. *Pharmacogn. Rev.* **2008**, *2*, 173–179.
9. Al-Rubae, A.Y. The potential uses of *Melia azedarach* L. as pesticidal and medicinal plant, review. *Am. Eurasian J. Sustain. Agric.* **2009**, *3*, 185–194.
10. Jagannathan, R.; Ramesh, R.V.; Kalyanakumar, V. A review of neem derivatives and their agricultural applications. *Int. J. Pharm. Technol.* **2015**, *6*, 3010–3016.
11. Stevenson, P.C.; Isman, M.B.; Belmain, S.R. Pesticidal plants in Africa: A global vision of new biological control products from local uses. *Ind. Crops Prod.* **2017**, *110*, 2–9. [CrossRef]
12. Paritala, V.; Chiruvella, K.K.; Thammineni, C.; Ghanta, R.G.; Mohammed, A. Phytochemicals and antimicrobial potentials of mahogany family. *Rev. Bras. Farmacogn.* **2015**, *25*, 61–83. [CrossRef]
13. Su, M.; Mulla, M.S. Activity and biological effects of neem products against arthropods of medical and veterinary importance. *J. Am. Mosq. Control Assoc.* **1999**, *15*, 133–152.
14. Nathan, S.S.; Savitha, G.; George, D.K.; Narmadha, A.; Suganya, L.; Chung, P.G. Efficacy of *Melia azedarach* L. extract on the malarial vector *Anopheles stephensi* Liston (*Diptera*: *Culicidae*). *Bioresour. Technol.* **2006**, *97*, 1316–1323. [CrossRef] [PubMed]
15. Mwangi, R.W. Locust antifeedant activity in fruits of *Melia volkensii*. *Entomol. Exp. Appl.* **1982**, *32*, 277–280. [CrossRef]
16. Mitchell, P.L.; Thielen, J.B.; Stell, F.M.; Fescemyer, H.W. Activity of *Melia volkensii* (*Meliaceae*) extract against Southern green stink bug (*Hemiptera*: *Heteroptera*: *Pentatomidae*). *J. Agric. Urban Entomol.* **2004**, *21*, 131–141.
17. Al-Sharook, Z.; Balan, K.; Jiang, Y.; Rembold, H. Insect growth inhibitors from two tropical *Meliaceae*: Effect of crude seed extracts on mosquito larvae. *J. Appl. Entomol.* **1991**, *111*, 425–430. [CrossRef]
18. Koul, O.; Jain, M.P.; Sharma, V.K. Growth inhibitory and antifeedant activity of extracts from *Melia dubia* to *Spodoptera litura* and *Helicoverpa armigera* larvae. *Indian J. Exp. Biol.* **2000**, *38*, 63–68.
19. Sombatsiri, K.; Ermel, K.; Schmutterer, H. Other Meliaceous Plants Containing Ingredients for Integrated Pest Management and Further Purposes. In *The Neem Tree*; VCH: Weinheim, Germany, 1995; pp. 642–666.
20. Mwangi, R.W.; Mukiama, T.K. Studies of insecticidal and growth-regulating activity in extracts of *Melia volkensii* (*Gurke*), an indigenous tree in Kenya. *East Afr. Agric. For. J.* **1989**, *54*, 165–171. [CrossRef]
21. Wycliffe, W. Toxicological studies of fruit powder and extracted cake of *Melia volkensii* Guerke (Family: *Meliaceae*) on Maasai goats in Kenya. *Int. J. Pharm. Chem.* **2017**, *3*, 82–85. [CrossRef]
22. Kamau, R.W.; Juma, B.F.; Baraza, L.D. Antimicrobial compounds from root, stem bark and seeds of *Melia volkensii*. *Nat. Prod. Res.* **2016**, *30*, 1984–1987. [CrossRef]
23. Kenya Forest Service. *Guidelines to On-Farm Melia volkensii Growing in the Dryland Areas of Kenya*; CADEP-SFM: Nairobi, Kenya, 2018.
24. Wekesa, L.; Muturi, G.; Mulatya, J.; Esilaba, A.; Keya, G.; Ihure, S. Economic viability of *Melia volkensii* (Gurkii) production on smallholdings in drylands of Kenya. *Int. Res. J. Agric. Sci. Soil Sci.* **2012**, *2*, 364–369.
25. Isman, M.B. Tropical forests as sources of natural insecticides. In *Chemical Ecology and Phytochemistry of Forest Ecosystems*; University of British Columbia: Vancouver, BC, Canada, 2009; pp. 145–160.

26. Diop, B.; Wilps, H. Field trials with neem oil and *Melia volkensii* extracts on *Schistocerca gregaria*. In *New Strategies in Locust Control*; Birkhäuser: Basel, Switzerland, 1997; pp. 201–207.
27. Nasseh, O.; Wilps, H.; Rembold, H.; Krall, S. Biologically active compounds in *Melia volkensii*: Larval growth and phase modulator against the desert locust *Schistocerca gregaria* (Forskal) Orth., Cyrtacanthacrinae. *J. Appl. Entomol.* **1993**, *116*, 1–11. [CrossRef]
28. Gokce, A.; Stelinski, L.L.; Whalon, M.E.; Gut, L.J. Toxicity and antifeedant activity of selected plant extracts against larval obliquebanded leafroller, *Christoneura rosaceana* (Harris). *Open Entomol. J.* **2010**, *4*, 18–24. [CrossRef]
29. Akhtar, Y.; Isman, M.B. Comparative growth inhibitory and antifeedant effects of plant extracts and pure allelochemicals on four phytophagous insect species. *J. Appl. Entomol.* **2004**, *128*, 32–38. [CrossRef]
30. Akhtar, Y.; Rankin, C.H.; Isman, M.B. Decreased response to feeding deterrents following prolonged exposure in the larvae of a generalist herbivore, *Trichoplusia ni* (*Lepidoptera*: Noctidae). *J. Insect Behav.* **2003**, *16*, 811–831. [CrossRef]
31. Akhtar, Y.; Isman, M.B. Feeding responses of specialist herbivores to plant extracts and pure allellochemicals: Effect of prolonged exposure. *Entomol. Exp. Appl.* **2004**, *111*, 201–208. [CrossRef]
32. Akhtar, Y.; Yeoung, Y.R.; Isman, M.B. Comparative bioactivity of selected extracts from *Meliaceae* and some commercial botanical insecticides against two noctuid caterpillars, *Trichoplusia ni* and *Pseudaletia unipuncta*. *Phytochem. Rev.* **2008**, *7*, 77–88. [CrossRef]
33. Irungu, L.W.; Mwangi, R.W. Effects of a biologically active fraction from *Melia volkensii* on *Culex quinquefasciatus*. *Int. J. Trop. Insect Sci.* **1995**, *16*, 159–162. [CrossRef]
34. Awala, P.; Mwangi, R.W.; Irungu, L.W. Larvicidal activity of a granular formulation of a *Melia volkensii* (Gurke) acetone extract against *Aedes aegypti* L. *Int. J. Trop. Insect Sci.* **1998**, *18*, 225–228. [CrossRef]
35. Isman, M. Insect antifeedants. *Pestic. Outlook* **2002**, *13*, 152–157. [CrossRef]
36. Pan, L.; Ren, L.; Chen, F.; Feng, Y.; Luo, Y. Antifeedant activity of *Ginkgo biloba* secondary metabolites against *Hyphantria cunea* larvae: Mechanisms and application. *PLoS ONE* **2016**, *11*, e0155682. [CrossRef] [PubMed]
37. Rajab, M.S.; Bentley, M.D.; Alford, A.R.; Mendel, M.J. A new limonoid insect antifeedant from the fruits of *Melia volkensii*. *J. Nat. Prod.* **1988**, *511*, 168–171. [CrossRef]
38. Champagne, D.E.; Koul, O.; Isman, M.B.; Scudder, G.G.; Towers, G.N. Biological activity of limonoids from the rutales. *Phytochemistry* **1992**, *31*, 377–394. [CrossRef]
39. Milimo, P.B. Chemical composition of *Melia volkensii* Gurke: An unrealised browse potential of semi-arid agroforestry systems. *J. Sci. Food Agric.* **1994**, *64*, 365–370. [CrossRef]
40. Puritch, G.S.; Condrashoff, S.F. Insecticide Mixtures Containing Fatty Acids. U.S. Patent 4,861,762, 29 August 1989.
41. Rogers, L.L.; Zeng, L.; McLaughlin, J.L. New bioactive steroids from *Melia volkensii*. *J. Org. Chem.* **1998**, *63*, 3781–3785. [CrossRef]
42. Rajab, M.S.; Bentley, M.D. Tetranotriterpenes from *Melia volkensii*. *J. Nat. Prod.* **1988**, *51*, 840–844. [CrossRef]
43. Kurimoto, S.I.; Takaishi, Y.; Ahmed, F.A.; Kashiwada, Y. Triterpenoids from fruits of *Azadirachta indica* (Meliaceae). *Fitoterapia* **2014**, *92*, 200–205. [CrossRef]
44. Rogers, L.L.; Zeng, L.; McLaughlin, J.L. Volkensinin: A new limonoid from *Melia volkensii*. *Tetrahedron Lett.* **1998**, *39*, 4623–4626. [CrossRef]
45. Rogers, L.L.; Zeng, L.; Kozlowski, J.F.; Shimada, H.; Alali, F.Q.; Johnson, H.A.; McLaughlin, J.L. New bioactive triterpenoids from *Melia volkensii*. *J. Nat. Prod.* **1998**, *61*, 64–70. [CrossRef]
46. Zeng, L.; Gu, Z.M.; Chang, C.J.; Wood, K.V.; McLaughlin, J.L. Meliavolkenin, a new bioactive triterpenoid from *Melia volkensii* (Meliacae). *Bioorg. Med. Chem.* **1995**, *3*, 383–390. [CrossRef]
47. Zeng, L.; Gu, Z.M.; Chang, C.J.; Smith, D.L.; McLaughlin, J.L. A pair of new apotirucallane trirerpenes, meliavolkensins A and B, from *Melia volkensii* (Meliaceae). *Bioorganic Med. Chem. Lett.* **1995**, *5*, 181–184. [CrossRef]
48. Di Filippo, M.; Fezza, F.; Izzo, I.; De Riccardis, F.; Sodano, G. Novel syntheses of (E)- and (Z)-Volkendousin, cytotoxic steroid from the plant *Melia volkensii*. *Eur. J. Org. Chem.* **2000**, *18*, 3247–3252. [CrossRef]
49. Zeng, L.; Gu, Z.M.; Fang, X.P.; Fanwick, P.E.; Chang, C.J.; Smith, D.L.; McLaughlin, J.L. Two new bioactive triterpenoids from *Melia volkensii* (Meliaceae). *Tetrahedron* **1995**, *51*, 2477–2488. [CrossRef]

50. Cantrell, C.L.; Franzblau, S.G.; Fischer, N.H. Antimycobacterial plant terpenoids. *Plant Med.* **2001**, *67*, 685–694. [CrossRef]
51. Romeo, J.T. *Phytochemicals in Human Health Protection, Nutrition, and Plant Defense*; Springer: Boston, MA, USA, 1999.

© 2020 by the authors. Licensee MDPI, Basel, Switzerland. This article is an open access article distributed under the terms and conditions of the Creative Commons Attribution (CC BY) license (http://creativecommons.org/licenses/by/4.0/).

plants

Review

Botanicals Against *Tetranychus urticae* Koch Under Laboratory Conditions: A Survey of Alternatives for Controlling Pest Mites

Ricardo A. Rincón [1,2], Daniel Rodríguez [1,*] and Ericsson Coy-Barrera [2,*]

1. Biological Control Laboratory, Universidad Militar Nueva Granada, Cajicá 250247, Colombia
2. Bioorganic Chemistry Laboratory, Universidad Militar Nueva Granada, Cajicá 250247, Colombia
* Correspondence: daniel.rodriguez@unimilitar.edu.co (D.R.); ericsson.coy@unimilitar.edu.co (E.C.-B.); Tel.: +57-6500000 (ext. 3269) (D.R.); +57-6500000 (ext. 3270) (E.C.-B.)

Received: 1 July 2019; Accepted: 3 August 2019; Published: 7 August 2019

Abstract: *Tetranychus urticae* Koch is a phytophagous mite capable of altering the physiological processes of plants, causing damages estimated at USD$ 4500 per hectare, corresponding to approximately 30% of the total cost of pesticides used in some important crops. Several tools are used in the management of this pest, with chemical control being the most frequently exploited. Nevertheless, the use of chemically synthesized acaricides brings a number of disadvantages, such as the development of resistance by the pest, hormolygosis, incompatibility with natural predators, phytotoxicity, environmental pollution, and risks to human health. In that sense, the continuous search for botanical pesticides arises as a complementary alternative in the control of *T. urticae* Koch. Although a lot of information is unknown about its mechanisms of action and composition, there are multiple experiments in lab conditions that have been performed to determine the toxic effects of botanicals on this mite. Among the most studied botanical families for this purpose are plants from the Lamiaceae, the Asteraceae, the Myrtaceae, and the Apiaceae taxons. These are particularly abundant and exhibit several results at different levels; therefore, many of them can be considered as promising elements to be included into integrated pest management for controlling *T. urticae*.

Keywords: *Tetranychus urticae*; resistance; botanical pesticides; acaricide; integrated pest management

1. Introduction

One of the most important pests in commercial crops worldwide is the polyphagous, two-spotted spider mite, *Tetranychus urticae* Koch. This mite is able to alter the physiological processes of plants, reducing the area of photosynthetic activity and causing the abscission of leaves in severe infestations [1]. The cost of damages caused by this pest in crops such as beans, citrus, cotton, avocado, apples, pears, plums, and many other horticultural and ornamental crops are estimated at over USD$ 4500 per hectare. Such costs correspond to 30% of the total cost of pesticides in crops of ornamental flowers. This constitutes a spending of almost 62% of the global market value on *T. urticae* Koch control based on data of 2008 [2]. The main tools used to control this pest are chemically synthesized acaricides. However, this mite is known to generate a resistance to these chemicals in a short period of time [3]. In addition, when the *T. urticae* Koch is exposed to sublethal pesticide levels, this mite has the ability to increase its reproduction rate, thus its populations increase in a shorter time [4]. Furthermore, many of the active ingredients in pesticide formulations are incompatible with the *T. urticae* Koch's natural predators; consequently, when they are applied to crops, they suppress populations of predators that can contribute to the decrease of phytophagous mites [5].

Courtesy of the above-mentioned issues—together with problems related to environmental contamination, the risk for human and animal health, and phytotoxicity—it is necessary to complement the control of *T. urticae* Koch with tools other than chemically-synthesized acaricides, such as biological control and the use of botanical pesticides (plant extracts), a growing alternative for the control of this pest. From the perspective of locating new options for the control of two-spotted spider mites, the use of botanical pesticides represents a useful tool with minimal detrimental effects on the environment, a low residuality, a slight induction of resistance due to its complex matrix, and with fewer harmful effects on human health when compared to those of the chemically-synthesized acaricides. Therefore, in the present review, a survey is presented based on some characteristics of *T. urticae* Koch behavior in the presence of toxic substances. In addition, this review builds upon other studies in order to determine the biological activity of some botanical pesticides on the phytophagous mite *T. urticae* Koch under laboratory conditions.

2. Characteristics of *T. urticae*

The *T. urticae* Koch is the most abundant and the most widely distributed species of the genus *Tetranychus*. This genus presents a confusing taxonomy due to partial reproductive incompatibilities that have been found in some populations. It is known that, in certain cases, these incompatibilities are caused by species of bacteria from the genus *Wolbachia* [6].

The individuals of the *T. urticae* Koch are characterized by having two spots on their back (dorsal idiosome), green or brown coloration, and white or yellow colored legs [4]. They present sexual dimorphism, as males are smaller than females [4]. An important feature of this species is that it is is able to form a web on the plants in which it grows [4]. These mites feed initially on the leaves of the lower part of the plants, but they can later colonize the rest of them as the population grows. The damage they cause is observed in the form of chlorotic spots and, in some cases, the tanning of leaves and defoliation [4].

2.1. The Biology of T. urticae Koch

The life cycle of the family Tetranychidae includes the stages of egg, larva, protonymph, deutonymph, and adult [7], between each of which a quiescent state usually occurs. Their eggs are round, white, or translucent, and the duration of their cycle depends on the temperature, the relative humidity, and the host plant in which they develop. Under temperature conditions between 25 and 30 °C, the *T. urticae* Koch can complete its cycle between three and five days [8,9]. The eggs are approximately 0.13 mm in diameter. The larvae are spherical or oval in shape, generally greenish yellow with three pairs of legs, and their size is approximately 0.16 mm in length. The protonymphs have an oval shape and a pale green color. They are distinguished from larvae by having four pairs of legs, and their length is approximately 0.2 mm. In the case of deutonymphs, they reach a length of approximately 0.3 mm and have a yellow or light brown color. At this stage, two dark brown spots usually appear on the dorsal level. On the other hand, adults have a globular or oval shape and range from pale green to reddish yellow in color; adults present two red or dark brown spots on the idiosome. Males are smaller than females, with lengths of 0.4 and 0.5 mm, respectively [7,10]. This species is arrhenotokous [8], which increases the probability that a female will mate with her offspring. According to some authors, its high genetic variability allows it to adapt quickly and decreases its probability of expressing deleterious mutations [9].

2.2. Characteristics of Resistance of T. urticae Koch to Acaricides

The *T. urticae* Koch is a widespread polyphagous pest that attacks more than 1100 different plant species [9,11], making it one of the main phytosanitary problems for many crops. This trait is owing (among other reasons) to its capacity for quickly generating resistance to synthetic acaricidal products [12]—from two to four years of new active ingredients [9]—even after a few applications of the active ingredient [11].

This resistance capacity to pesticides of the *T. urticae* Koch has encouraged some researchers to carry out several studies regarding their genetic characteristics in response to the pressure generated by the use of acaricides. Such is the case of Grbić et al. (2011) [9], who carried out a deep analysis of the *T. urticae* Koch genome. They found that more than 10% of their genome comprises transposable elements (9.09 Mb). In the same study, they also observed the presence of several families of genes involved in digestion, detoxification, and transport of xenobiotic compounds with a unique composition. Eighty-six genes encode for cytochromes P450, a group of 32 genes encode for glutathione S-transferases (GST) (12 of these are believed to be unique to vertebrates), and 39 genes encode for drug-resistant proteins of the ABC transporters type (ATP-binding cassette). This repertoire of transporter proteins greatly exceeds the number presented by crustaceans, insects, vertebrates, and nematodes.

All these detoxifying enzymes are closely related to the resistance of *T. urticae* Koch, but this is not the only mechanism used by these mites to counteract the effect of xenobiotics. A set of mutations in the action points of pesticides is another way they are able to mitigate the effect of these compounds. Demaeght et al. (2014) [13] reported a resistance case for this species when there was a mutation in quitin synthase 1, which is the target enzyme of etoxazole. Additionally, because of its similarity to the mechanism of action of hexythiazox and clofentezine, this mutation can cause a cross-resistance to these products. Table 1 shows an example of the effects of 10 different acaricides on four different populations of *T. urticae* Koch in the state of Pernambuco (Brazil) [14]. This information demonstrates the ability of this pest to counteract the effects of different active ingredients, showing variable responses to the same compounds in different regions.

Table 1. The resistance of different populations of *T. urticae* Koch—from the state of Pernambuco (Brazil)—to 10 different acaricides. Adapted from Ferreira et al. [14].

Acaricide	Population	N°	LC_{50} (mg/L) *	LC_{95} (mg/L) *	RR_{50}
Diafenthiuron	Petrolina II	426	6.6	70	1
	Piracicaba	484	10.7	105	1.6
	Brejão	340	4053	93,708	619
	Bonito	401	7732	133,440	1180
Milbemectin	Piracicaba	484	0.6	8.3	1
	Petrolina II	455	5.4	101	9.9
	Bonito	373	357	3726	650
	Brejão	380	384	2386	700
Fenpyroximate	Piracicaba	315	22	341	1
	Petrolina II	469	87	1929	4
	Bonito	378	3246	10,014	200
	Brejão	387	4343	16.234	150
Clorfenapyr	Piracicaba	424	1.3	9.3	1
	Petrolina II	481	2.8	20.1	2.2
	Brejão	477	735	4157	570
	Bonito	524	4652	94.598	3600
Spirodiclofen	Piracicaba	401	16.4	1590	1
	Petrolina II	547	37.5	370.870	2.3
	Bonito	538	6401	127.750	390
	Brejão	414	6586	56.390	400
Fenbutatin oxide	Piracicaba	465	0.83	436	1
	Petrolina II	538	1.72	1093	2.1
	Bonito	477	293	52.892	350
	Brejão	459	1705	197.990	2048
Propargite	Piracicaba	472	6.5	28	1
	Petrolina II	397	15	66	2.3
	Bonito	395	291	990	45
	Brejão	391	622	4410	96

Table 1. Cont.

Acaricide	Population	N°	LC$_{50}$ (mg/L) *	LC$_{95}$ (mg/L) *	RR$_{50}$
Hexythiazox	Piracicaba	416	2938	64.871	1
	Petrolina II	404	4370	100.510	1.5
	Brejão	440	12,700	605.400	4.3
	Bonito	415	1384	381.630	4.7
Spiromesifen	Piracicaba	418	373	18.404	1
	Petrolina II	487	487	17.752	1
	Brejão	467	1388	42.781	3.7
	Bonito	470	3201	90.424	8.6
Abamectin	Petrolina II	613	0.0011	0.033	1
	Piracicaba	714	0.0084	0.066	8
	Petrolina I	676	0.036	0.205	34.4
	Gravatá	787	0.041	1.66	39.3
	Goiânia	584	1.79	27.9	1716
	Brejão	610	118	3000	113.532
	Bonito	693	326	3397	295.270

* LC$_{50}$: the mortality-causing concentration of 50% of the test population. LC$_{95}$: the mortality-causing concentration of 95% of the test population. N°: the number of mites used in the trial. RR$_{50}$: the resistance proportion between the resistant population and the susceptible one at LC$_{50}$.

Function of Detoxifying Enzymes

Cytochrome P450 has been extensively investigated, as it is the most important group of detoxifying proteins in arthropods [15]. This enzyme group has been linked to cases of resistance in the common fly, *Musca domestica* L., with resistance to those furanocoumarins produced by a host plant in *Papilio polyxenes* Fabricius [15], and in cases of resistance to abamectins in *T. urticae* Koch [16]. One of the characteristics of this group of proteins in arthropods is their inducibility over time, which is proportional to the consumption of certain toxic compounds from the plants that serve them as food. Such is the case of the *Spodoptera frugiperda* Smith. In this species, it was demonstrated that, when consuming a diet containing indole-3-carbinol, in time and with the increase in the concentration of this compound, there was an increase in the production of P450 enzymes [15].

Another group of proteins that is important in the response to xenobiotics is the Glutathione-S-Transferases (GST) family. Among these proteins, two enzymes belonging to the delta class—tuGSTd10 and tuGSTd14—and one of the mu class—tuGSTm09—are present in *T. urticae* Koch. They are strongly associated with mite resistance to the active ingredient abamectin [17]. Similarly, Pavlidi et al. (2017) [18], through molecular docking analysis and implementation of HPLC-MS, deduced that the active ingredient cyflumetofen and its de-esterified metabolite could be transformed by the enzyme TuGSTd05 in the same mite species.

On the other hand, Merzendorfer (2014) [19] and Dermauw and Van Leeuwen (2014) [20] mentioned the presence of 104 genes belonging to subfamilies of ABC genes in *T. urticae* Koch. This number is higher than that of other different species such as *Homo sapiens* L., *Apis mellifera* L., *Drosophila melanogaster* Meigen, *Anopheles gambiae* Giles, *Bombyx mori* L., *Tribolium castaneum* Herbst, *Pediculus humanus* L., *Daphnia pulex* Leydig, *Caenorhabditis elegans* Maupas, and *Saccharomyces cerevisiae* Meyen ex EC Hansen, demonstrating its importance within this species. This group of genes has also been related to the development of elytra and wings in some insects and to the transport of certain drugs of hydrophobic origin. The type of transport of compounds of these proteins has been elucidated through models constructed by crystallography, for which it is known that they act as proteins of import, export, or as flipases [19].

2.3. Relationship Between Resistance and the Host Plant

The different mechanisms of resistance presented by the *T. urticae* Koch suggest that these adaptations may not be due exclusively to the pressure generated from the use of pesticides.

This question was asked by Dermauw et al. (2012) [21], who made an interesting finding when studying the transcriptome of resistant and susceptible strains of the *T. urticae* Koch in the presence of different host plants.

In that study, they demonstrated that a susceptible strain of this phytophagous mite was capable of expressing diverse deactivated genes when it was relocated from a bean to a tomato as its host plant [21]. In addition, the number of expressed genes that are related to the generation of resistance increased considerably, going from 13 genes—expressed after two hours from host plant change—to 1206 genes after five generations. On the other hand, they compared the transcriptome of two resistant strains and that of the susceptible strain developed in the tomato. They also found that both mite strains shared the expression of a significant number of genes related to resistance (Figure 1). This seems to indicate that there is a strong relationship between the resistance mechanisms developed by the *T. urticae* Koch and its host plants. These mechanisms may be similar to those developed by this species to face exposure to different pesticides.

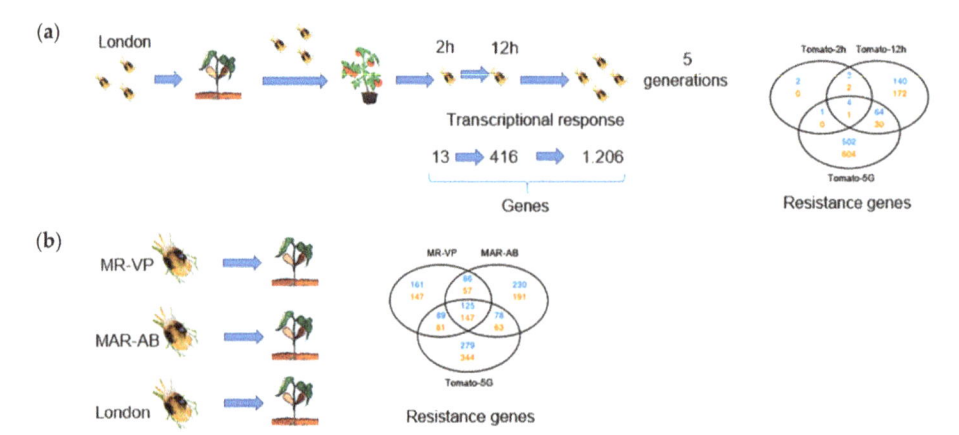

Figure 1. A graphical representation of the study performed by Dermauw et al. (2012) [21]. (**a**) represents the transcriptional changes in the susceptible London strain of the *T. urticae* Koch when changing host plant. (**b**) represents the number of genes expressed in two resistant strains and the susceptible London strain of *T. urticae* Koch after 5 generations from the relocation to another host plant. The scheme was constructed by R.A. Rincón for this review from the data published by Dermauw et al. (2012) [21].

Evidence of the resistance capacity of this phytophagous mite is shown in Table 2. A list of important pest arthropod species is shown, reporting the number of active ingredients to which they developed resistance until the year 2012 [22,23]. The list is led by the *T. urticae* Koch, a species that showed a reported resistance to 93 active ingredients until that moment.

Owing to the large number of reports of resistance existing for the *T. urticae* Koch, some studies have provided important information and promising aspects in terms of understanding the resistance with promising results. Such is the case of the research conducted by Demaeght et al. (2013) [24] concerning cross-resistance. They studied two *T. urticae* Koch strains that were resistant to Spirodiclofen—an active ingredient belonging to group 23 of the IRAC (i.e., inhibitors of acetyl CoA carboxylase). Although strains appeared to be strongly resistant to this ingredient, they had a very low cross-resistance to spirotetramat and spirodiclofenenol. This information could serve as a base for the understanding of some routes of resistance-generation in this phytophagous mite, because they demonstrated that the spirodiclofen detoxification route affects—at least partially—all of the tetranic and the tetronic acid derivatives in the *T. urticae* Koch.

In the same study, Demaeght et al. (2013) discarded resistance to spirodiclofen by active site mutations after aligning the sequences of active sites from target proteins with BlastP [24]. However, when microarrays were made to express the genome of the studied strains and subsequently compared, they found similarities in several genes expressed among the spirodiclofen resistant strains, which were identified as P450 family proteins, carboxylesterases, glutathione S-transferases, transport proteins, lipocalins, and several proteins without homology in the available databases. This fact demonstrated that this detoxifying route is strongly related to the response of the *T. urticae* Koch to this ingredient.

On the other hand, Kwon et al. (2012) [25] detected a fitness decrease of *T. urticae* Koch strains that demonstrated Monocrotophos resistance. Although the presence of more than one mutation increased the resistance up to 1165-fold, these modifications in genes significantly decreased the catalytic capacity of acetyl cholinesterase, thus gene overexpression seems to be necessary in order to compensate for deficiency acquired by resistance-conferring mutations to the acaricide.

Table 2. A list of pest arthropods based on the reported number of active ingredients resistance and the number of reported cases per species—adapted from Van Leeuwen et al. (2010, 2012) [22,23]. The information for the species *Plutella xylostella* L., *Myzus persicae* Sulzer, *Leptinotarsa decemlineata* Say, *Blatella germanica* L., and *Panonychus ulmi* Koch correspond to the cases reported up to 2010.

Species	Taxonomy	Kind of Pest	Number of Active Ingredients	Cases of Resistance
Tetranychus urticae Koch	Acari: Tetranychidae	Crop	93	389
Plutella xylostella L.	Lepidoptera: Plutellidae	Crop	81	437
Myzus persicae Sulzer	Hemiptera: Aphididae	Crop	73	320
Leptinotarsa decemlineata Say	Coleoptera: Chrysomelidae	Crop	51	188
Musca domestica L.	Diptera: Muscidae	Urban	53	266
Blatella germanica L.	Blattodea: Blatellidae	Urban	43	213
Rhipicephalus microplus Canestrini	Acari: Ixodidae	Cattle	43	158
Helicoverpa armigera Hubner	Lepidoptera: Noctuidae	Crop	43	639
Bemisia tabaci Gennadius	Hemiptera: Aleyrodidae	Crop	45	428
Panonychus ulmi Koch	Acari: Tetranychidae	Crop	42	181
Varroa destructor Anderson y trueman	Acari: Varroidae	Bees parasite	2	10
Ixodes scapularis Say	Acari: Ixodidae	Cattle	0	0
Culex pipiens L.	Diptera: Culicidae	Disease vector	36	161
Culex quinquefasciatus Say	Diptera: Culicidae	Disease vector	32	256
Tribolium castaneum Herbst	Coleoptera: Tenebrionidae	Stored-grain pest	32	113
Aedes egypti egypti L.	Diptera: Culicidae	Disease vector	24	267
Spodoptera frugiperda Smith	Lepidoptera: Noctuidae	Crop	16	25
Pediculus humanus L.	Phthiraptera: Pediculidae	Disease vector	9	59
Anopheles gambiae Giles	Diptera: Culicidae	Disease vector	3	39
Manduca sexta L.	Lepidoptera: Sphingidae	Crop	3	4
Rhodnius prolixus Stal	Hemiptera: Reduviidae	Disease vector	3	3
Anopheles darlingi Root	Diptera: Culicidae	Disease vector	1	2
Linepithema humile Mayr	Hymenoptera: Formicidae	Urban	2	2

3. Control Strategies for *T. urticae* Koch

In agricultural crops, the main pest control method used is the spraying of solutions based on chemically synthetic products such as insecticides and acaricides [26]. Although this method has been effective in some cases for *T. urticae* Koch control, it has also demonstrated serious limitations and disadvantages, especially due to *T. urticae* Koch's high reproductive potential. This peculiarity encourages farmers to use acaricides in larger volumes and doses, causing high levels of toxic waste in fruits, the development of resistant populations, the intoxication of mammals, and the destruction of beneficial organisms [23,27,28].

Another strategy used for *T. urticae* Koch management is biological control. Among the predators of this pest are some mites of the family Phytoseiidae. Within this family, two predators stand out—the *Neoseiulus californicus* McGregor and the *Phytoseiulus persimilis* Athias-Henriot. These mites are characterized by consuming a large number of prey at adequate conditions and having high reproductive rates and a capacity for rapid development [4]. Other natural predators that are less commonly used for the control of this mite are the beetle *Stethorus punctillum* Weise (Coccinellidae) and the *Conwentzia psociformis* Curtis (Neuroptera: Coniopterygidae)—which are found naturally in Spain [29]—purely to mention some of the predators of this phytophagous species. Additionally, the fungus *Neozygites floridana* Weiser and Muma has also exhibited significant control over the *T. urticae* Koch, but some difficulties in cultivation have hampered its use [30]. However, other fungi such as the *Lecanicillium lecanii* Zimmermann and the *Beauveria bassiana* Bals.-Criv., as well as the bacterium *Bacillus thuringiensis* Berliner, have been commercially used for the management of the two-spotted spider mite with positive effects.

3.1. Other Methods for T. urticae Koch Control

As complementary strategies, these mites are controlled in some crops through the application of water washings and the manual massaging of the affected leaves using water and soap in order to remove the mites from the plant, kill them mechanically, and break their webs. Within the strategies used for controlling the *T. urticae* Koch, biopesticides based on plant extracts or phytochemicals are considered to be another alternative to chemically-synthesized acaricides [31,32], which have also emerged as a complement to traditional management. This has allowed the development of commercial products with formulations based on substances of natural origin, such as CinnAcar®, Biodie® and PHC Neem®, which are produced from compounds and mixtures isolated from plant extracts. As an example for this case, they have demonstrated compatibility with the natural predator *Tamarixia radiata* Waterston (Hymenoptera: Eulophidae)—parasitoid of the *Diaphorina citri* Kuwayama (Hemiptera: Psyllidae)—thus these formulations may constitute excellent alternatives to be included into integrated management programs (IPM) of the so-called "Asian citrus psyllid" [33]. Therefore, the fact that these botanical pesticides are compatible with natural predators becomes an advantage in the control of pests and constitutes an additional tool that can be used in integrated pest management strategies.

An essential prerequisite for success when using extracts as a control strategy for pests is their compatibility with other management strategies. Within the context of the IPM, a relevant issue is the evaluation of how this type of product can affect biological control agents. In the particular case of the *T. urticae* Koch, a question arises about how phytoseiid mites that have been successfully used as a control strategy could be affected—a topic that has been explored by different researchers. Among the botanical pesticides, probably the most used are the neem derivatives, a trend that is also present in the case of the *T. urticae* Koch. A moderate reduction in female survival and fecundity in response to Azadirachtin use on *P. persimils* Athias-Henriot was reported by Duso et al. (2008) [34], although a positive shift in favor of the predator in terms of the predator–prey interaction can be inferred, since azadirachtin was more toxic to the *T. urticae* Koch. A moderate effect was also reported by Spollen and Isman (1996) [35], who found a maximum mortality of 14% in *P. persimils* Athias-Henriot adults sprayed with neem extract. Although variables such as egg eclosion, the mean number of eggs laid per female, and differences in preference between treated or untreated leaves were not found, the authors concluded that neem-derived insecticides could be effective and safe. Neem pesticides have exhibited few negative impacts on the phitoseids *N. californicus* McGregor [36], *Euseius alatus* De Leon [37], and *Phytoseiulus macropilis* Banks [36,37]. On the other hand, the negative effects of NeemAzal-T/S in terms of its potential impact on populations of the predator *Metaseiulus occidentalis* Nesbitt were reported by Yanar (2019) [38], who recommended the use of low concentrations of this product in cases where the *M. occidentalis* is a relevant component of IPM. Regarding another plant species used to obtain botanicals, crude extracts of *Artemisia judaica* L. exhibited acaricidal bioactivity against *T. urticae* Koch in terms of LC_{50}, while its negative impacts upon *P. persimils* Athias-Henriot were clearly lower, suggesting that such extracts are compatible with the predaceous mite [39]. Similarly, a promising toxic effect of the *Melissa officinalis* L. on *T. urticae* Koch has been reported, along with an LC_{50} for the *N. californicus* McGregor, which is comparatively extremely lower. Undertaking compatibility evaluations between extracts and natural predators is essential, since there is no reason to generalize slight or innocuous effects of these products on said beneficial organisms. Commercial formulations and application rates similar to those used by farmers are needed in order to obtain results with more predictive value in respect of those expected in the field. Sublethal effects will also be a subject of relevant research in the future, because, although many of the evaluations that demonstrate little or no effect on the natural predators have been carried out in adults, the sublethal effects could raise compatibility issues that are not evident when restricting evaluations to adult individuals [40].

3.2. The Use of Plant Extracts for the Control of T. urticae Koch in The Field

There are many studies regarding the use of plant extracts for the control of the *T. urticae* Koch. Although many of these trials have delivered successful results, others have not demonstrated the level of expected control over this mite species. For that reason, a greater understanding of the mechanisms of action presented by molecules that demonstrate biological activity on these mites and the way in which these molecules interact is highly required. In addition, the toxic effects of such molecules are generated in many cases by the presence of several toxic compounds contained in the same extract, which act in a synergistic manner. An understanding of these factors will help to foster a broader understanding of the use of this tool in the control of the two-spotted spider mite. Further studies must take into account the results of the studies developed thus far, which have delivered promising results, not just in terms of the toxic effects demonstrated on these mites, but also in terms of sublethal effects such as low fecundity and repellency.

3.2.1. Methods for the Evaluation of Extracts under Laboratory Conditions

The methods of testing the effects of extracts are very much the same as those used thus far for evaluations of chemical compounds. An important prerequisite for making appropriate evaluations is having a susceptible population of individuals. Generally, this population can be obtained by rearing individuals that have not been exposed to any type of chemical substance with a possible acaricidal effect for a considerable number of generations [5,41–43]. Additionally, the origin and the type of the selected plant material must be clearly defined in order to ensure the repeatability of results. Hence, correct taxonomic classification, location, season of the year, time of day, phenological stage, and organs to be collected and processed in order to obtain the extracts affect the particular composition of the tested botanical and influence the acaricidal activity [5,44–46]. Finally, the type of preparation and the extracting protocol are also crucial steps for obtaining a standardized mixture of plant-based compounds, which would be the source of effective botanical-based acaricidal or repellent agents. Thus, solid–liquid (S–L) extractions—i.e., the selected plant material directly enters into contact with the extracting solvent during a defined period through a continuous (maceration) or discontinuous (percolation or Soxhlet) procedure—are the most commonly used method for obtaining different types of extracts, depending on the polarity of the extracting solvent. In order of polarity, water, water/ethanol mixture (hydroalcoholic), ethanol/methanol, chloroform, ethyl acetate, and hexane are the most commonly used solvents for extractions. Other types of preparations are essential oil (usually obtained by steam distillation) or low-polar/volatile extracts (afforded by hydrodistillation, supercritical fluid extraction, microwave or ultrasound-assisted hydrodistillation, among others) [47]. The physicochemical nature of these naturally occurring compounds, which are present in the preparation (extract or essential oil), is the critical prerequisite information required to identify the extracting procedure. The purification or the isolation of the active principles requires several steps, usually using preparative techniques such as column chromatography under a bioguided fractionation strategy—although the isolated compounds might be separately assessed after a conventional purification protocol. In any case, these efforts could affect upon acaricidal rather than repellent activities to facilitate mite control, but this choice depends of the aims of use. Essential oils often exhibit repellent activity in comparison to extracts, owing to their volatile nature.

3.2.2. Bioassays

The purpose of bioassays is to determine the effect of a given agent on the physiology of an organism, which, in the context of acari research, is generally associated with determining the toxicity of a chemical compound—or resistance to it—either in the field or in laboratory conditions [48]. Repeatability of results, practical facilities, and conditions as similar as possible to those under which the acaricide will be used are desirable [49]. In the case of mites, a small size and fast movement are aspects conditioning the bioassay design. The main aspects of some common bioassays used for the evaluation of botanicals on *T. urticae* Koch adults (generally females) and their advantages and disadvantages are described below.

Slide Dip Methods

An initial method was described by Voss (1961) [50] as part of an acaricide screening procedure. Double-sided Scotch® tape is adhered on one of its sides to a microscope slide. It is important to avoid bubbles or empty spaces between the tape and the glass, because deposits of the substance under evaluation could be formed, which could affect the test results [51]. After this, the mites must be affixed to the other side of the tape by the dorsal part of the hysterosoma. A fine brush is usually used to transfer individuals to the tape. The slides are then dipped into the solution containing the toxicant for 5 s [48,51] and, after this, are placed on a paper towel. It is important to remove any excess of liquid with filter paper. After this, the slides are placed on trays covered with slightly moistened disposable towels, which must then be taken into controlled conditions. The mortality criterion in the different methods is usually an absence of movement when the individual is gently prodded with a fine brush. High control mortalities due to desiccation and an absence of food are common in this method for times of evaluation greater than 24 h, limiting the accuracy of the response parameters. Furthermore, individuals are exposed to toxicants in an artificial substrate, and in some cases, problems distinguishing alive and dead mites arise [48]. Despite such problems, this method has been used repeatedly to determine the effect of botanicals on adults of the *T. urticae* Koch, in some cases considering evaluation times of 24 h [52], but in other cases employing higher evaluation times [41,52–57]. An advantage of this method is that the results obtained are highly reproducible. It can also be modified by employing a spray tower to supply the toxicant, which allows for a better coverage [58].

Petri Dish Methods

The main variant is the Petri Dish Residue-Potter Tower Method (PDR-PT), in which the bottom and the top inner surfaces of a Petri dish are sprayed with the toxicant using a Potter Tower and allowed to dry for around 30 min at room temperature. After this, the individuals are transferred to the dishes using a fine brush. For this method, high mortalities after 48 h have been reported, thus it is advisable to restrict the evaluation time to shorter periods, such as 24 h [59]. The petri dish methods have been used in some cases for the evaluation of botanicals against *T. urticae* Koch [3,60].

Leaf Disc Methods

For this kind of bioassay, leaf discs of variable diameter (approximately 20 mm) are cut from leaves of several plant species, such as beans [1,43,46], peaches [48], or roses [3], and placed upside down in a Petri dish containing moistened cotton wool when bean or rose leaf discs are used or a semi-solid agar pad in the case of peach leaf discs. A variable number of adults (between five and 20) must be transferred to each leaf disc using a fine brush. The experimental assembly is maintained without disturbance for at least one hour before the spraying of the toxicant is performed. This spraying can be performed using an airbrush, provided the required distance—as well as the number of drops/cm^2 and the pressure—may be adequately standardized [61], which corresponds to the basic Leaf Disc Direct Method (LDD). This method can be improved employing the Potter spray tower, derivating in the so-called "Leaf Disc Direct-Potter Tower Method" (LDD-PT) [48]. This device was developed by C.

Potter from the Rothamsted Experimental Station [62], and it is recognized as a reference standard for making chemical sprays under laboratory conditions, since it enables the achievement of an even deposition of spray in the target area. The LDD and the LDD-PT methods can also be used to evaluate the effect of a residual film of the toxicant on adults placed on a sprayed surface (such as a leaf disc in this instance). In this case, the procedures are named "Leaf Disc Residue Method" (LDR) and "Leaf Disc Residue-Potter Tower Method" (DR-PTM) [48]. Both a direct spray and the residual film are intended to evaluate the toxic effects generated by contact between the individual mites and the test substance. After the spray, Petri dishes are kept uncovered for around 30 min, which allows for the drying of the leaf disc surface. They are then covered and placed under controlled conditions. Generally, mites that cannot walk a distance equivalent to their body length are considered dead. Since the leaf disc method implies the presence of the natural substrate of spider mites, it can be considered to have a greater similarity with field conditions than the slide dip or the petri dish methods. However, one drawback is the escape of individuals. This problem intensifies when the toxicant requires a prolonged time to act or when it should be ingested in the feeding process. The fate of individuals that escape is uncertain, thus the most advisable procedure is to discard them in the analysis; to consider them as part of the mortality rate would not be justifiable [48]. An alternative approach is the development of methods that do not allow for the escape of individuals, as proposed by Bostanian et al. [63]. In their setup, a large leaf disc (50 mm in diameter) is placed upside down and tightly fitted to the bottom of a plastic Petri dish of the same diameter, thus it occupies the whole dish. The base of each petri dish contains thinly moistened cotton wool (1.5 mm in thickness) to prevent desiccation. A circular window of 28 mm is cut in the top of the Petri dish to facilitate air circulation and avoid condensation. For bioassays involving tetranychyds, they recommend covering the window with a 40 μm polyester mesh screen to avoid run-off. The edges of the Petri dish bottoms are wrapped with masking tape to ensure a strong grip on the top, preventing the escape of individuals. A small hole in the lower half of the Petri dish allows the petiole to protrude outwards, where it must be covered with a wet cloth. This method enables observations for a period as long as nine days, which makes it suitable for slow- and fast-acting reduced-risk toxicants. A different variant of the leaf disc method is the Leaf Disc-Residue Dipping Method (LDR-D), in which the leaf discs are dipped into the solution containing the toxicant [64]. Although estimations of lethal concentrations obtained by this method are less precise when compared to the LDD-PT, this fact could be explained by an uneven distribution of residues on the leaf surface. The leaf disc methods have been widely used in several trials of botanicals against *T. urticae* Koch [5,40,54–62].

Leaf Absorption Method

In this method, the leaf is placed in some kind of recipient containing the toxicant solution in order to allow the absorption by the leaf for an adequate period (usually around 72 h). The leaf is then located in a Petri dish containing agaropectin to prevent desiccation, and the mites are transferred onto the leaf, where they are allowed to feed by 24 h, and mortality is then evaluated. The design of the experimental unit should consider alternatives to prevent the escape of mites, as discussed in the leaf disc method [42,63].

Whole Plant Direct Method

The purpose of this kind of bioassay is to evaluate the direct effect of toxicants under conditions as similar as possible to those of the field. Young bean plants with 2–3 leaves can be used, and the apical part of the plant must be removed to prevent the appearance of new leaves, which have not received the treatment. Adult females are then placed in the plants long enough before performing the application in order to allow oviposition. Alternatively, a specific number of immature stages and adults can be placed in the plants. The spray of the toxicant is made using an atomizer, considering an application volume similar to the required under crop conditions. The number of eggs, nymphs, and adults is recorded at predefined evaluation times, usually between 5 and 10 days [65,66]. This method has been also used to evaluate repellency [67].

Filter Paper Difussion Methods (Fumigant Bioassays)

This kind of assay is designed to evaluate the fumigant action of toxictants, thus it is essential to avoid any direct contact between individuals and the toxicant. The setup is similar to that employed in the leaf disc method, but the bottoms of the Petri dishes are covered using a tight-fitting lid with a fine wire sieve. The toxicant is applied to filter papers, which should be allowed to dry before being placed over the wire sieve [68]. In some cases, the paper is attached to the downside of the lid with a small quantity of solid glue that should not affect individuals [42].

3.3. Studies Carried Out for the Control of T. urticae Koch from Plant Extracts Grouped by Plant Families

The investigations carried out, which focused on the effects of biopesticides on *T. urticae* Koch, have led to the identification of a large number of plant extracts with acaricidal, repellent, and deterrent properties. Below are descriptions of some species—grouped by plant families—whose plant extracts have been used in laboratory studies that have exhibited their biological activity on the two-spotted spider mite (the information is summarized and complemented in Table A1 in the Appendix A).

3.3.1. Family Amaranthaceae

This family has aroused interest in different areas such as traditional and alternative medicine, given the properties that have been identified in some of the species that comprise it. Such is the case of *Achyranthes aspera* L., whose secondary metabolites have antinociceptic activity [69], or *Chenopodium ambrosioides* Mosyakin et Clemants, which has toxic effects that have been studied in some human parasites [70].

Due to these toxic effects, Hiremath et al. (1995) [71] evaluated the acaricidal effect of the extract of this plant. They compared the activity of the methanolic extracts obtained from 21 different species of African plants against *T. urticae* Koch adults using the leaf immersion method. Among the most active extracts, the whole plant of *Celosia trigyna* Linn. exhibited the highest biological activity, causing mortality rates between 40% and 60% of evaluated mites.

Chiasson et al. (2004) [41] also evaluated the acaricidal effects of a species of this family. They studied the effect of an emulsifiable concentrate—obtained from *Chenopodium ambrosioides* Mosyakin et Clemants essential oil—on adults and eggs of the *T. urticae* Koch and the *Panonychus ulmi* Koch and compared it with the effect obtained from the use of commercially available products. The products were applied with an airbrush on females that were placed on microscope slides with glue. In the case of eggs, the application was made on the eggs previously laid by females on leaf discs located within Petri dishes. Thus, a dose of 0.5% produced a mortality of 94.7% in females, which was higher than that obtained from the Neem extract (22.1%). Otherwise, hatching was diminished on days five and nine after application. This hatching effect was lower in treatments with Neem, Abamectin, and insecticide soap. A lower effect was observed for an ethanolic extract from seeds of *Chenopodium quinoa* Willd. on adult females and nymphs of this mite, exhibiting an LD_{50} of 1.24% w/v [72].

Two years later, Shi et al. (2006) [52] evaluated the effect of *Kochia scoparia* (L.) Schrad extract on *T. urticae* Koch, *T. cinnabarinus* Boisdu-Val, and *T. viennensis* Zacher using three different solvents for extracting the compounds contained in the plant material: methanol, chloroform, and petroleum ether. The mortality trials were carried out using three different methodologies: (1) the slide dip method measuring mortality after 24 h of immersion, (2) the LDD-PT, and (3) the leaf absorption method. Using these methodologies, the highest mortality of the *T. urticae* Koch was obtained with the chloroform-soluble extract, which exhibited a 78.86% average mortality and an LC_{50} of 0.88 using the dipping method, in which mites were glued to an adhesive tape.

3.3.2. Family Amaryllidaceae

This family is studied widely due to its potential uses in the control of human diseases [73] as well as its antitumor [74] and insecticides properties [75]. Abbassy et al. (1998) [76] determined the LC_{50} of the alkaloidal extract, the ethanolic extract, and the essential oil of the bulb of the ornamental plant *Pancratium maritimum* L. (Amaryllidaceae) on the *T. urticae* Koch, whose values were 0.2%, 0.36%, and 1.5%, respectively.

The insecticidal properties demonstrated by various studies led Attia et al. (2011) [77] to expose adult *T. urticae* Koch females to different concentrations of garlic extract (*Allium sativum* L.). These concentrations ranged between 0.46 and 14.4 mg/L using the Potter Tower application. After the bioassays, they determined the LD_{50} and the LD_{90}, whose values were 7.49 and 13.5 mg/L, respectively. On the other hand, they concluded that fecundity was reduced by using the concentrations of 0.36 and 0.74 mg/L. Geng et al. (2014) [78] measured the toxicity by the contact and the repellency of the garlic extract at 20, 10, 5, 2.5, and 1.25 g/L. From these tests, they found that treatment with 20 g/L caused a 76.5% mortality rate on mites at 48 h after its application. Additionally, with the obtained data, they calculated the regression equation of toxicity as $Y = 1.3 x + 3.9$. They were also able to determine the LD_{50} value, which corresponded to 7.2 g/L. Furthermore, the repellencies were found to be 95.6% and 65.2% at extract concentrations of 10 g/L and 20 g/L, respectively.

3.3.3. Family Annonaceae

Within this group of plants, the presence of several important secondary metabolites involved in the communication of arthropods and plants' defenses against the attack of pests has been identified [79]. However, Ohsawa et al. (1991) [80] obtained negative results when using *Annona glabra* L. seed extract on *T. urticae* Koch eggs. During their experiment, they dissolved 10 mg of the extract in acetone (1 mL) and applied 0.5 mL of the solution to a bean leaf where the eggs were laid. After this, they noticed that the extract demonstrated no impact on mortality rates, deterrence in feeding, or mite growth.

Pontes et al. (2007) [44] also demonstrated the acaricidal activity of the essential oils of this family of plants, but in this case, they used the species *Xilopia serícea* A.St.-Hil., which was evaluated on *T. urticae* Koch. Using gas chromatography–mass spectrometry (GC–MS), they identified the compounds present in both leaves and fruits, finding mostly monoterpenes and sesquiterpenes. When comparing their acaricidal activity, they concluded that the essential oils of the leaves exhibited a greater toxicity than those obtained from the fruits.

3.3.4. Family Apiaceae

Plants of this family are widely used within the diet of different human communities [81], although their nature is so varied that many species have been used as pesticides and repellents [82]. For example, Choi et al. (2004) [42] tested the essential oils of 53 plants to determine their acaricidal potential on *T. urticae* Koch eggs and adults. Among these oils, the highest toxicity was exhibited by species of the family Apiaceae—i.e., *Carum carvi* L.—since a 100% mortality rate of adult mites was obtained. To carry out this study, the researchers conducted bioassays by diffusion on filter paper, avoiding any direct contact between the oil and the mites. The tests were developed in a plastic container (4.5 × 9.5 cm) at a concentration of 14×10^{-3} µL/mL air.

Tsolakis and Ragusa (2007) [83] studied the effect of a mixture of essential oils from the *C. carvi* L. with potassium salts of fatty acids on the *T. urticae* Koch and one of its predators, *Phytoseiulus persimilis* Athias-Henriot. This combination proved to be very selective, since it generated a mortality rate of 83.4% in *T. urticae* Koch females compared to a 24% mortality rate in *P. persimilis* Athias-Henriot females. Besides, the product also caused a decrease in the intrinsic growth rate of the phytophagous mite while having no effect on the growth rate of the predator. Approximately four years later, this same essential oil was tested by Han et al. (2010) [68] on the same species of mite. In this case, by using mortality bioassays by vapor phase to evaluate fumigant effect (see section of Myrtaceae Family), they established an LD_{50} of 22.4 µg/cm^3 air.

Among works carried out with plants of this family, Attia et al. (2011) [43] showed that the *Deverra scoparia* Coss. & Durieu essential oil has an acaricidal effect and decreases the fecundity of the *T. urticae* Koch. In the same study, they isolated the components of the oil and tested them individually on the pest, obtaining the highest toxicities with the compounds α-pinene, Δ^3-carene, and terpinen-4-ol. Amizadeh et al. (2013) [84] also decided to evaluate the effect on two species of this family of the inhalation of essential oils. For this purpose, they carried out tests to determine the fumigant activity of *Heracleum persicum* Desf. Ex. Fisch. essential oils and *Foeniculum vulgare* Mill. seeds on adult females and eggs of the *T. urticae* Koch. The LD_{50}s were 3.15 µL/L and 1.53 µL/L for females and eggs treated with *Heracleum persicum* Desf. Ex. Fisch. essential oil, respectively, and 5.75 µL/L and 1.17 µL/L for females and eggs treated with *Foeniculum vulgare* Mill. essential oil, respectively. Other essential oils obtained from Apiaceae plants having acaricidal activity on *T. urticae* Koch were *Cuminum cyminum* L. (seeds) and *Ferula gumosa* Boiss (leaves), showing LD_{50} values of 3.74 and 6.52 µL/L air, respectively [85,86].

On the other hand, Pavela (2015) [65] tested acaricidal and ovicidal effects of the methanolic extract of *Ammi visnaga* (L.) Lamarck seeds on *T. urticae* Koch. The efficacy in terms of adult mortality rates increased over time, with LD_{50}s (after 72 h from the time of application) estimated at 17, 10, and 98 µg/cm^2 for the extract and its two major compounds, khellin and visnagin (furanochromenes), respectively. Moreover, the extract and the two isolated furanochromenes inhibited the development of eggs and caused their mortality, with LD_{50}s of 13.3, 0.5, and 1.8 µg/cm^2 for the extract, the visnagin, and the khellin, respectively. The application of the extract to leaves infested with *T. urticae* Koch achieved a reduction of the number of individuals in all stages of development. The concentration of 10 mg/mL showed the highest efficacy, which was 98.5% on the tenth day since the application. The terpenes isofuranodiene and germacrone, isolated from *Smyrnium olusatrum* L. inflorescences, also exhibited toxicity on this mite (LD_{50}s = 1.9 and 42.7 µg/mL, respectively) [87].

3.3.5. Family Asteraceae

There have been numerous studies carried out with species from this group to evaluate their acaricidal activity. First, Chiasson et al. (2001) [45] evaluated the essential oils of two plant species known as potential pesticides—*Artemisia absinthium* L. and *Tanacetum vulgare* L.—to determine their acaricidal activity against the *T. urticae* Koch. The oils were obtained via a microwave-assisted process (MAP), distillation in water (DW), and by direct distillation with steam (DDS), and their relative toxicities were tested by direct contact. All oils were tested at 1%, 2%, 4%, and 8% as emulsions prepared in water with 9% denatured ethanol and 0.32% Alkamul EL-620 as emulsifier, and mite mortality was evaluated after 48 h.

The three oils of *A. absinthium* L. were toxic to the *T. urticae* Koch; however, there were differences in their levels of toxicity. For example, the oil extracted by MAP and DW methods caused 52.7% and 51.1% mortality in the mites, respectively, while the oil obtained by DDS produced a mortality rate of 83.2%. Consequently, the LC_{50} of the oil extracted by DDS was lower (0.043 mg/cm^2) than those obtained by MAP (0.134 mg/cm^2) and by DW (0.130 mg/cm^2). The extracts of *T. vulgare* L. obtained by DW and DDS exhibited greater acaricidal activity than the extract prepared by the MAP method. At a concentration of 4%, oils delivered mortality rates of 60.4%, 75.6%, and 16.7%, respectively. The chemical analysis of the extracts of *T. vulgare* L. indicated that the compound *p*-thujone is the major compound in the oil (>87.6%) and probably contributes significantly to its acaricidal activity. Additionally, the acetone extract from leaves of *Artemisia judaica* L. exhibited an LD_{50} of 0.56 µg/mL against adult females [39].

Trials that have shown acaricidal activity within this family have also identified important compounds in essential oils that may play a role in the toxic activity against the *T. urticae* Koch. One of these cases was developed by Attia et al. (2012) [46], who identified the terpinen-4-ol compound in the *Santolina africana* Jord. & Fourr. essential oil. This compound was the most abundant component

(54.96%) within the study. They evaluated the acaricidal activity of the *S. africana* Jord. & Fourr. and the *Hertia cheirifolia* (L.) Kuntze essential oils, with positive impacts upon the mortality rates of the *T. urticae* Koch and important effects in the reduction of oviposited eggs.

In another study, this same group of researchers tested the effect of the *Chrysanthemum coronarium* L. essential oil on the *T. urticae* Koch and produced mortality rates of 88% and 93% on larvae and adult females, respectively [88]. In the same year, another paper was published by Afify et al. (2012) [89], who tested the acaricidal activity of *Chamomilla recutita* L. extract on the *T. urticae* Koch. The LD_{50} values obtained for adults and eggs in this study were 0.65% and 1.17%, respectively. In this study, the authors identified the main compounds of *C. recutita* L. by means of gas chromatography–mass spectrometry. The most predominant compounds were α-bisabolol oxide (35.25%) and trans-β-farnesene (7.75%). The essential oil from the aerial part of *Achillea mellifolium* L. showed LD_{50} values of 1.208% v/v and 1.801 µL/L air when evaluated through leaf dipping and fumigation, respectively. The GC–MS chemical profile of this oil was mainly composed of piperitone (12.8%) and *p*-cymene (10.6%) [64].

However, not all studies using species from this plant family obtained satisfactory results in terms of the *T. urticae* Koch. For example, extracts obtained from *Artemisia absinthium* L.—known insecticides and acaricides used throughout the world to control aphids—demonstrated weak activity upon the *T. urticae* Koch, as reported by Aslan et al. (2005) [90]. Similarly, Derbalah et al. (2013) [91] found that the extract of castor leaves (*Artemisia cinae* O. Berg & C.F. Schmidt ex.Plajakov) exhibited low toxicity against the *T. urticae* Koch, with an LD_{50} of 1326.53 ppm. Similarly, Pavela et al. (2016) [92] studied the effect of the methanolic extract taken from leaves of the *Tithonia diversifolia* Hemsl. on *T. urticae* Koch and its ethyl acetate fraction in order to measure acute and chronic toxicity as well as its inhibitory effects on oviposition. In acute toxicity trials, mortality did not exceed 50%, even for the highest dose evaluated (150 µg/cm^3). On the other hand, in the chronic toxicity tests on the fifth day after application, the LD_{50} of the methanolic extract was 41.3 µg/cm^3, and the LD_{90} was 98.7 µg/cm^3. However, the two extracts caused inhibition in the oviposition of mites.

3.3.6. Family Boraginaceae

A low polar extract from roots of *Onosma visianii* Clem. demonstrated significant chronic toxicity and oviposition inhibition on *T. urticae* Koch adult females (LD_{50} = 2.6 µg/mL). Eleven naphthoquinone-type related compounds were isolated and structurally elucidated [93]. Although all isolated derivatives exhibited effects against this mite, isobutylshikonin and isovalerylshikonin were found to be the most active isolated compounds (LD_{50}s = 2.69 and 1.06 µg/mL, respectively).

3.3.7. Family Burseraceae

Several species belonging to this family exhibit anti-inflammatory properties [94]. They are considered to be anticarcinogenic agents with antimalarial, antidiarrheal, and antifever properties and uses as insecticides [95], antimicrobials, and antioxidants [96] (among others) for disease treatment [95]. However, some studies have pursued applications in agriculture, specifically for the management of important pests. In that respect, Pontes et al. (2007) [97] studied the acaricidal and the repellent effects of the *Protium bahianum* Daly plant resin oil on the *T. urticae* Koch by fumigant tests. For this, they kept mites in leaf discs of *Canavalia ensiformis* (L.) DC. inside 9 cm Petri dishes as test chambers. Each chamber had a strip stuck on the inner side that was saturated with different amounts and concentrations of the oil (5, 10, 15, 20, and 25 µL, corresponding to 2, 4, 6, 8, and 10 µL/L of air, respectively). They evaluated the fresh resin oil and the old resin oil separately. Results showed that the fumigant effect of the oil in both cases increased with concentration and exposure times and had mortality rates of 79.6% and 59.0% after 72 h for the old and the new resin oils, respectively. Regarding the deterrent effect of oviposition, the fresh resin oil presented an increased activity, with only 14 eggs oviposited at 72 h at a concentration of 10 µL/L of air. In repellency tests, only fresh resin oil showed positive effect against mites.

3.3.8. Family Cannabaceae

Although this family of plants is recognized for its various pharmaceutical uses, little has been studied about its effects as an insecticidal and an acaricidal agent. Among the studies that have been performed, Yanar et al. (2011) [60] used the extract obtained from the flower buds of *Humulus lupulus* L. on *T. urticae* Koch adults at 5% (adhesive tape method) and at 50% (residual film method). Using the adhesive tape methodology (in which 1 mL of solution was applied to the tape left for 4 to 5 h to dry, and 20 adult females were then placed on it), the mortality rate after 24 h was 67.84% ± 2.52%. On the other hand, with the residual film methodology (in which the extract was applied to a 90 mm Petri dish, distributed homogeneously, and left for 2 to 4 h to dry before the addition of 20 *T. urticae* Koch adult females), the mortality observed after 24 h was 56.37% ± 0.99%. The acaricidal effect against this mite of an essential oil from panicles of hemp (*Cannabis sativa* L.) was also evaluated, exhibiting 83.28% of mortality on adult females at 0.10% [98].

3.3.9. Family Caryophyllaceae

The acaricidal effect of an aqueous extract from roots of *Saponaria officinalis* was evaluated against all developmental stages of *T. urticae* Koch [66]. The lowest sensitivity was found for adults (LD_{50} = 0.31% w/v), while eggs revealed the highest sensitivity (LC_{50} = 1.18% w/v). Oviposition was also inhibited by this extract (LC_{50} = 0.91% w/v).

3.3.10. Family Combretaceae

There are several plant species of this group on which acaricidal activity studies of the *T. urticae* Koch have been carried out—the majority of them successfully. An example of this is the study performed by Hiremath et al. (1995) [71], who compared the activity of the methanolic extracts obtained from 21 different species of African plants against adults of the *T. urticae* Koch using the leaf immersion method. Among the results found, the *Combretum micronthum* G. Don. and the *Piloitigma vetilicolin* whole plant extracts demonstrated effects on the rates of *T. urticae* Koch mortality of between 40% and 60%.

3.3.11. Family Convolvulaceae

There are few studies on the *T. urticae* Koch that involve this plant family, with plants of the genera *Convolvulus* and *Ipomaea* being the most used. Chermenskaya et al. (2010) [99] studied the effect of the species *Convolvulus krauseanus* Regel. and Schmalh. on three species of pest arthropods, among which was the *T. urticae* Koch. From this study, which gathered the effect of extracts from 123 plant species, they concluded that the *C. Kraseanus* Regel. and Schmalh. roots extract was one of the two that showed the highest miticidal effect [together with the *Ailanthus altissima* (Mill.) Swingle leaf extract, Simarubaceae], causing a mortality rate of 95.6% after seven days from the application (using the immersion method).

3.3.12. Family Cupressaceae

Essential oils from two plants of this family were evaluated against adult females of *T. urticae* Koch in the same study [55]. Oil from leaves of *Cupressus macrocarpa* Hartw. ex Gordon had an LD_{50} of 5.69 µL/L air, whereas *Thuja orientalis* L. leaves resulted in an LD_{50} of 7.51 µL/L air. The main compounds in these essential oils were β-citronellol (35.92%) and α-pinene (35.49%), respectively.

3.3.13. Family Euphorbiaceae

The species of this family have not been well studied in terms of their pesticide properties. One of the works carried out in this area was that of Dang et al. (2010) [100], who investigated the effect of the dried root extract of *Euphorbia kansui* S.L. Liou ex S.B. Ho on the *T. urticae* Koch, as well as that of two of its compounds separately: 3-O-(2,3-dimethylbutanoyl)-13-ododecanoilingenol (compound

1) and 3-O-(2'E, 4'Z-decadienoyl)-ingenol (compound 2). Concerning the extract, they found that it generated mortality rates of 27% and 55% at concentrations of 3 and 5 g/L, respectively. When testing the two compounds obtained by fractionation and evaluating them on mites, they determined that compound 1 caused mortality rates of 45% and 59% when applied at 500 and 1000 mg/L, respectively. In contrast, compound 2 showed no acaricidal activity during the study.

On the other hand, in 2015, Numa et al. (2015) [61] published a study in which they tested the susceptibility of *T. urticae* Koch females to the *Cnidoscolus aconitifolius* (Mill) I.M. Johnst. leaf extract using the leaf immersion methodology merged with direct application using an airbrush. In this study, they determined that a dose of 2000 µg/mL was the only one that did not show differences in the positive control (based on chlorfenapyr as the active ingredient). This dose could be the most appropriate for an extract formulation based of this plant during its potential use in the control of pests in agricultural crops, taking into account the fact that it caused a 92% rate of mortality of mite females in the trials.

3.3.14. Family Fabaceae

This family is well known as an aspect of human diets throughout the world. Several studies have been carried out to evaluate the effects of their plant extracts on arthropods with very varied results. These include the study performed by Hiremath et al. (1995) [71], who compared the activity of methanolic extracts obtained from 21 different species of African plants against adults of the *T. urticae* Koch using the leaf immersion method. The most active extracts were those obtained from the leaves, the fruits, and the whole plant of *Prosopis chinensis* (Molina) Stuntz, which caused mortality rates between 61% and 80% for the leaf extract and higher than 80% in the case of the extracts obtained from the fruits and the whole plant. The plant oil of *Millettia pinnata* L. showed an LD_{50} of 0.004% on adult females after four days of testing [101].

3.3.15. Family Gramineae (Poaceae)

Although this family is made up of nearly 10,000 plant species, studies involving the effect of its plant extracts on the *T. urticae* Koch have focused on only some of the 55 species that make up the *Cymbopogon* genus [102]. In one of these cases, Choi et al. (2004) [42] included the oil from *Cymbopogon nardus* (L) Rendle within the 53 essential oils that they evaluated on the *T. urticae* Koch. This oil showed a positive result, causing a mortality rate greater than 90% on adults of this phytophagous mite. In a study of another species of genus *Cymbopogon*, Han et al. (2010) [68] examined the effect of Citronella Java oil on the *T. urticae* Koch, evaluating its fumigant effect. To do this, they took disc-shaped bean leaves and placed them on moistened cotton contained in Petri dishes together with *T. urticae* Koch adult mites. On each Petri dish, a mesh cover was placed and placed over this was filter paper moistened with the essential oil. Under these conditions, the LD_{50} found was 22.5 µg/cm^3.

3.3.16. Family Lamiaceae

The effects of plant extracts and essential oils from the species that make up this family have been the most studied on the phytophagous mite *T. urticae* Koch. Among the studies reported in the literature are, for instance, those from the species *Rosmarinus officinalis* L. and *Salvia officinalis* L. The essential oils of these plants demonstrated effective control over populations of the *T. urticae* Koch and a decrease in the number of oviposited eggs when concentrations increased [53]. In a similar way, Choi et al. (2004) [42] performed trials using the *S. officinalis* L. essential oil on the same species of mite, obtaining an adult mortality rate of 82%. In the same study, they included another species from the family Lamiaceae—*Mentha spicata* L.—from which they obtained the essential oil that was evaluated on the *T. urticae* Koch. As a result, the mortality rate of these arthropods in the adult stage was 81%.

On the other hand, Rasikari et al. (2005) [103] carried out a screening of the leaf extracts of 67 species of plants belonging to the Lamiaceae family. They were evaluated on the *T. urticae* Koch, which were applied by direct contact with the Potter Tower to bean leaves kept in Petri dishes with cotton. From the extracts tested, 14 had a moderate to acute toxic effect on mites. From these, extracts obtained from the plants *Clerodendrum traceyi* F. Muell., *Premna serratifolia* L., *Ceratanthus longicornis* (F.Muell.) G. Taylor, *Plectranthus habrophyllus* P.I. Forst, and *Plectranthus* sp. Hann caused a 100% mortality rate, whereas the extracts of *Gmelina leichardtii* F.Muell. & Benth, *Premna acuminata* R. Br., *Viticipremna queenslandica* Munir, *Plectranthus diversus* S.T. Blake, *Plectranthus glabriflorus* P.I. Forst, and *Plectranthus suaveolens* S.T. Blake caused mortality rates that were between 90% and 99%.

In 2006, a study performed by Miresmailli et al. (2006) [104] was published. In that investigation, they tested the effect of the *R. officinalis* L. essential oil on the *T. urticae* Koch. For that, they took two different populations of mites, one from bean plants and another from tomato plants. For the tests, they used five different concentrations (2.5, 5, 10, 20, 40, and 80 mL/L) of the essential oil diluted in methanol and water (70:30 *v/v*). In order to evaluate the mortality rates of mites, they took 3 mm disc leaves within Petri dishes, to which they applied 20 µL of the treatment solution. Once dried at room temperature, they placed five adult females on the leaves and kept them at a temperature of 26 ± 2 °C, a relative humidity (RH) between 55% and 60%, and a photoperiod of 16:8 (light:dark). From these assays, they determined that the LC_{50} for the females maintained on bean plants was 10 mL/L, while for the females kept on tomato plants, it was 13 mL/L. Moreover, with a concentration of 20 mL/L, a mortality of 100% of females produced in bean plants was obtained, whereas a 40 mL/L concentration was necessary before females on the tomato plants reached total mortality (100%).

Additionally, Miresmailli et al. [104] identified the components of *R. officinalis* L. essential oil using GC–MS by column chromatography and tested them individually on the *T. urticae* Koch. In the case of mites reared on bean plants, two compounds revealed a significant toxicity—1,8-cineol and α-pinene (with 88% ± 4.8% and 32% ± 4.8% mortality, respectively)—whereas for mites raised on tomato plants, the same two compounds were those that revealed a significant toxicity. The resulting values were 80% ± 6.2% and 72% ± 4.8% for 1,8-cineol and α-pinene, respectively.

In a similar study, Çalmaşur et al. (2006) [105] tested the effect of the vapors of three essential oils from *Micromeria fruticosa* L., *Nepeta racemosa* L., and *Origanum vulgare* L. on nymphs and adults of the *T. urticae* Koch and adults of the *Bemisia tabaci* Gennadius, finding the highest mortality rates (96.7%, 95%, and 95%, respectively, for *T. urticae* Koch, and 100% for *B. tabaci* Gennadius) when using doses of 2 µL/L of air at 12 h of exposure. Han et al. (2010) [68] also studied several essential oils obtained from species of this family. To do this, they evaluated its fumigant effects on the *T. urticae* Koch and, as a result, obtained LD_{50}s of 22.7, 22.8, 23.7, 38.8, 39.5, and 63.7 µg/cm^3 for *Thymus vulgaris* L., *Mentha* L. *piperita*, *Mentha pulegium* L., *Mentha spicata* L., *Ocimum basilicum* L., and *Salvia officinalis* L., respectively.

In 2012, Afify et al. (2012) [89] tested the acaricidal activity of *Majorana hortensis* Moench extract on the *T. urticae* Koch. The LD_{50} values obtained for adults and eggs in the trial were 1.84% and 6.26%, respectively. In the study, they identified the main compounds of *M. hortensis* Moench by means of gas chromatography–mass spectrometry as terpinen-4-ol (23.86%), *p*-cymene (23.40%), and sabinene (10.90%)—the main compounds for this species. In the same year, Attia et al. (2012) [88] tested the effect of the essential oil of *Mentha pulegium* L. on the *T. urticae* Koch, obtaining a mortality rate of 91% in larvae and adult females. The same essential oil was evaluated by Choi et al. (2004) [42] on the same mite species, in which a mortality rate higher than 90% was obtained. Within the same experiment, they analyzed the effect of the essential oil of the *Mentha piperita* L., in which the mortality rate also exceeded 90%. On the other hand, Amizadeh et al. (2013) [84] studied the fumigant effect of the essential oil obtained from leaves of the *Satureja sahendica* Bornm. on eggs and adult females. The LD_{50} obtained for females was 0.98 µL/L, while it was of 0.54 µL/L for eggs.

3.3.17. Family Meliaceae

The insecticidal properties of plants belonging to the family Meliaceae have been studied extensively [106]. For this reason, Ismail (1997) [107] evaluated the relative toxicity of the extracts of *Melia azedarach* L. and some synthetic acaricides against recently hatched larvae of the *T. urticae* Koch and third-instar larvae of the predatory beetle, *Stethorus gilvifrons* Mulsant. The methanolic extract of the plant was the most effective among the tested products, followed by the extracts of acetone and petroleum ether. The toxicity of the plant material obtained was less active against the predator compared to the effect it had on the two-spotted spider mite, in which a decrease in fecundity was also observed. The study of the joint action of the products also revealed a strong synergy in the bromopropylate mixture with the methanolic extract of the *M. azedarach* L. Interestingly, this mixture demonstrated no effect on the predator.

In a similar way, Brito et al. (2006) [37] tested the toxicity of different commercial products based on one of the plants with the highest pesticide potential, the Neem (*Azadirachta indica* A. Juss.). It was tested not only on the *T. urticae* Koch but also on its predators, *Euseius alatus* DeLeon and *Phytoseiulus macropilis* Banks. In this study, they found that the formulation of the product Neemseto (1%) was the one that obtained the best result on the *T. urticae* Koch by topical contact. In the same way, they tested the product at different concentrations (0.25%, 0.5%, and 1.0%) and found that the product had a repellent effect on *T. urticae* Koch and *E. alatus* DeLeon; however, it did not affect the *P. macropilis* Banks. Additionally, the Neemseto exhibited an important reduction in *T. urticae* Koch fecundity, but on the predatory mites, a significant decrease was only observed when mites were exposed to the highest concentrations. This shows that this product can be a promising option for the management of the two-spotted spider mites within integrated pest management schemes given its relative compatibility with predatory mites.

3.3.18. Family Myrtaceae

T. urticae Koch toxicity studies involving these plants have had varying results. First, Choi et al. (2004) [42] determined that the *Eucalyptus citriodora* Hook's essential oil is capable of causing a mortality rate of more than 90% on *T. urticae* Koch adults. This essential oil was also tested by Han et al. (2010) [68] on the same mite species using the vapor-phase mortality bioassay; they found similar fumigant activity results to those obtained previously [42]. The test performed consisted of placing 3 cm diameter bean leaf discs on wet cotton inside Petri dishes, each with 20 adult mite individuals [68]. On each Petri dish, they installed a mesh cover on which a filter paper moistened with the essential oil at the evaluated concentrations was placed (after drying for two minutes). From these experiments, they estimated an LD_{50} of 19.3 µg/cm^3.

On the other hand, they also wanted to evaluate the fumigant effect of *Syzygium aromaticum* (L.) Merr. & L.M. Perry essential oil. Within the study, they found an LD_{50} value of 23.6 µg/cm^3 on *T. urticae* Koch adults. In 2011, Afify et al. [108] tested the activity of six extracts of *Syzygium cumini* (L.) Skeels at three different concentrations (75, 150, and 300 µg/mL) on the *T. urticae* Koch. The highest mortality rates were obtained with the ethanolic extract (98.5%), followed by the hexane extract (94%) and the ether-ethyl acetate extract (90%). The LD_{50} values obtained were 85, 101, 102, and 98 µg/mL, respectively. The same group of researchers in 2012 conducted a study to measure the acaricidal activity of *Eucalyptus* sp. on the same mite [89]. The LD_{50} values obtained for adults and eggs in the assay were 2.18 and 7.33 µg/mL, respectively.

In 2013, Amizadeh et al. [84] also tested the fumigant effect of some essential oils of this family, including those obtained from leaves and fruits of the *Eucalyptus microtheca* F. Muell. on both eggs and adult females of the *T. urticae* Koch. For the tests, mites were placed on bean leaf discs laid in plastic containers in which an oil-impregnated filter paper was held without coming into direct contact with leaf discs or mites. The LD_{50}s on the adult females were 1.52 µL/L and 5.7 µL/L for the extracts of leaves and fruits, respectively, while for eggs, they were 0.56 µL/L and 2.36 µL/L for leaf and fruit extracts, respectively.

3.3.19. Piperaceae Family

There have been few studies carried out concerning the effects of extracts of species of the Piperaceae family on the *T. urticae* Koch—particularly considering the fact that they have focused on very few species of the genus *Piper*, which has more than 1000 species [109]. One of those studies was developed by Araújo et al. (2012) [110], who reported acaricidal and repellent activity of the essential oils obtained from *Piper aduncum* L. leaves and its components separately on the *T. urticae* Koch. The repellent activity was attributed to the components (*E*)-nerolidol, α-humulene, and β-caryophyllene, while the toxicity was attributed to β-caryophyllene. The extracts and their components exhibited a better performance in fumigation than in contact.

3.3.20. Family Ranunculaceae

In general terms, the toxicity studies of extracts of these plants used on the *T. urticae* Koch have not been very satisfactory. A case demonstrating this is the study conducted by Derbalah et al. (2013) [91], which found that the black cumin seeds (*Nigella sativum* L.) extract showed a low toxic effect on the *T. urticae* Koch, with an LD_{50} of 708.57 ppm. However, some species of this family—such as *Aconitum soongaricum* Stapf and *Clematis orientalis* L.—have shown toxic effects on the *T. urticae* Koch with mortality rates ranging between 50% and 80% of mites [99].

3.3.21. Family Rutaceae

In 2005, Tewary et al. [111] tested two concentrations (5000 and 10,000 ppm) of the *Zanthoxylum armatum* DC. leaf extract on the arthropods *H. armigera* Hübner, *P. xylostella* L., *T. urticae* Koch, and *A. craccivora* Koch, with mortality rates of 46% at 10,000 ppm in the *H. armigera* Hübner, 42% at 10,000 ppm in the *P. xylostella* L., 36% and 39% at 5000 and 10,000 ppm, respectively, in the *T. urticae* Koch, and 30% and 65% at 5000 and 10,000 ppm, respectively, in the *A. craccivora* Koch. On the other hand, Attia et al. (2012) [88] also included plants of the Rutaceae family, since they proved the effect caused by the essential oil of *Haplophyllum tuberculatum* (Forssk.) A. Juss. on the *T. urticae* Koch, obtaining a mortality of 93%.

Da Camara et al. (2015) [67] demonstrated that essential oils obtained from the epicarp of pear orange fruits (*Citrus sinensis* Osbeck var. Pera) and the lime orange (*Citrus aurantium* L.) had repellent effects against the *T. urticae* Koch, with very similar repellency results to those obtained with eugenol. Using mass spectrometry, 27 compounds were idenitified both in *C. sinensis* Osbeck and in *C. aurantium* L., which corresponded to 98.1% and 98.9% of the total constituents of the two extracts, respectively. This demonstrated that the major compound in the two essential oils was d-limonene. Within this study, the authors determined that all the identified compounds were responsible for the repellency.

3.3.22. Family Santalaceae

Within this family, Roh et al. (2011) [112] studied the effect of *Santalum* L. sp. essential oil on the *T. urticae* Koch using the leaf immersion method. Through this methodology, they found that the mortality rate of mites was 87.2% ± 2.9%. Additionally, they noticed an oviposition decrease of 89.3% on leaves treated with oil. Subsequently, they evaluated a mixture of α and β–Sandalool—the two main compounds of *Santalum* L. sp.—on the *T. urticae* Koch and obtained a mortality of 85.5% ± 2.9% and a decrease of 94.7% in fecundity.

3.3.23. Family Scrophulariaceae

The toxic effects of Scrophulariaceae plants on the *T. urticae* Koch have been less studied than plant species of other groups. Within the investigations carried out in this regard, Khambay et al. (1999) [113] studied the effect of two compounds of *Calceolaria andina* Benth extract with recognized insecticidal activity—2-(1,1-dimethylprop-2-enyl)-3-hydroxy-1,4-naphthoquinone (compound **1**) and 2-acetoxy-3-(1,1-dimethylprop-2-enyl)-1,4-naphthoquinone (compound **2**)—on 29 pest species, including the *T. urticae* Koch. The LD_{50}s for this species were 80 ppm and 30 ppm for each compound, respectively. The two cases were evaluated using the micro-immersion method. Additionally, they performed the same test on individuals from a population that showed resistance to chlorpyrifos and bifenthrin using the same compounds of *C. andina* Benth extract, thus obtaining $LD_{50}s$ of 44 ppm and 33 ppm for compounds **1** and **2**, respectively.

3.3.24. Family Simarubaceae

The toxicity of plant extracts from species of this family on the tetraniquid mite *T. urticae* Koch have not been well studied. Among the studies accomplished, Latif et al. (2000) [114] tested the extract from *Quassia* sp. aerial parts on this mite at a concentration of 10,000 ppm, finding acaricidal activity. Subsequently, they identified the quassinoid Chaparinone compound and tested it separately, obtaining an LC_{50} of 47 ppm. Chermenskaya et al. (2010) [99] evaluated extracts from 123 different plant species on the *T. urticae* Koch, *Frankliniella occidentalis* Pergande and *Shizaphis graminum* Rondani, using the leaf immersion method. Within these extracts, one that demonstrated a high acaricidal effect was obtained from the *Ailanthus altissima* (Mill.) Swingle leaves, which caused a mortality rate of 97.4% after 7 days of evaluation.

3.3.25. Family Solanaceae

Although most studies involving plant extracts tested on the *T. urticae* Koch have focused on assessing the effects on mortality and fecundity, those involving the Solanaceae family have been mostly dedicated to determining the repellent effects of certain extracts. Such is the case of the study conducted by Snyder et al. (1993) [115]. They isolated dihydrofarnesoic acid as one of the phytoconstituents in trichomes of *Lycopersicon hirsutum* Dunal, and its repellent effect on the phytophagous mite was then evaluated. For this purpose, 10 µL of the extract was applied to a filter paper separated by 1.5 cm from another similar filter, which was impregnated with 10 µL of hexane. Once the solvent was evaporated, a strip of filter paper was positioned to connect the two filter papers, and a mite was placed in the middle of the paper bridge to evaluate its displacement preference. This process was performed with approximately 40 adult females. According to the obtained results, they concluded that dihydrofarnesoic acid exhibited a repellent activity against the *T. urticae* Koch. Similarly, Antonious et al. (2006) [116] also evaluated toxic and repellent effect of the fruit extracts of *Capsicum chinense* Jacq., *Capsicum frutescens* L., *Capsicum baccatum* L., *Capsicum annuum* L., and *Capsicum pubescens* Ruíz & Pav. In their results, they determined that the highest mortality rate (45%) occurred when using the extract of the *C. annuum* L., while the extracts of the fruits of the *C. baccatum* L. and *C. annuum* L. caused repellence on mites.

Extracts of leaves and seeds of the *Datura stramonium* L. were used by Kumral et al. (2009) [5] to evaluate their acaricidal, repelling, and deterrent effects on oviposition over *T. urticae* Koch adults at 167.25 mg/L and 145.75 mg/L (for leaves and seeds, respectively). For these tests, they used a Potter Tower in order to place the mites on leaf discs contained in Petri dishes. These concentrations caused 98% and 25% of the mortality, respectively, for the two concentrations after 48 h of application. Through a simple logistic regression analysis, they determined that an increase in the leaf extract dose caused a significant increase in mite mortality, while the effect of increasing the dose of the seed extract was not significant. Based on Probit analysis, they estimated that the lethal dose (LD_{50}) with the leaf extract was 70.59 mg/L. According to the Pearson X^2 test, they concluded that mites showed a strong tendency to flee from areas treated with leaf and seed extracts to untreated areas.

3.3.26. Verbenaceae Family

In this family, a highlighted study was conducted by Cavalcanti et al. (2010) [117], in which they carried out a characterization of the essential oils of the *Lippia sidoides* Cham. (Verbenaceae) by GC–MS and tested their acaricidal activity on *T. urticae* Koch females. They concluded that the compounds thymol and carvacrol—as well as the essential oil of *L. sidoides* Cham.—showed a promising miticidal activity against this mite.

3.4. Additional Studies with Isolated Compounds Obtained after Plant Extract Fractionation

As with essential oils and plant extracts, a considerable number of their isolated constituents have also been tested on the *T. urticae* Koch. For example, Lee et al. (1997) [118] studied the insecticidal and the acaricidal effects of several monoterpenes and their possible phytotoxicity in maize plants that served as hosts of the *Diabrotica virgifera virgifera* LeConte, *T. urticae* Koch, and *Musca domestica* L. Twenty-nine compounds belonging to different chemical classes were tested against the *T. urticae* Koch by means of the leaf immersion method.

These tests used: the alcohols carveol, carvomentenol, citronellol, geraniol, 10-hydroxygeranol, isopulegol, linalool, menthol, perilyl alcohol, aterpineol, and verbenol; the phenols carvacrol, eugenol, and thymol; the ketones (−)-carvone, (+)-carvone, (+)-fenchone, menthone, pulegone, tuyone, and verbenone; the aldehydes citral and citronellal; citronelic acid; ether 1,8-cineol; and the hydrocarbons limonene, α-terpinene, and y-terpinene.

All compounds were tested in water with Triton X-100 as a wetting agent at 10,000 and 1000 ppm, and the activity was evaluated 24, 48, and 72 h after the treatment. The toxicity varied depending on the concentrations and the exposure times. All of the monoterpenes tested—except for 1,8-cineole, 10-hydroxygeraniol, aterpineol, verbenol, and verbenone—caused a 100% mortality rate at the highest concentration after 24 h. However, carvacrol was the most effective compound in the lowest concentrations, followed by citronellol.

On the other hand, geraniol produced a 100% rate of mortality, while its 10-hydroxy geraniol analogue exhibited a 0% mortality rate. During the trial, a longer exposure time increased acaricidal effects. Alternately, the most effective monoterpenoids (carvacrol, carvomenthenol, carvone, citronellol, eugenol, geraniol, perilyl alcohol, 4-terpineol, thymol) were evaluated separately in more detail tests. From these compounds, carvomentenol and 4-terpineol demonstrated greater acaricidal activity (LC_{50}s = 59 and 96 ppm, respectively).

In another study, Martínez et al. (2005) [119] examined the effect of azadirachtin at 64 and 128 ppm on different biological parameters of the *T. urticae* Koch, such as longevity, fecundity, fertility, and offspring development. The tests were performed on bean leaf discs in Petri dishes using the Potter Tower. The results found that this compound affected mortality and fecundity but exhibited no effects on fertility and offspring development. In a later analysis of life table, they determined that, with the application of azadirachtin at 80 ppm, the adult survival rate was reduced to 50%. Duso et al. (2008) [34] also tested the toxicity of Azadirachtin on the *T. urticae* Koch. In that case, the micro-immersion bioassay methodology was implemented using a concentration of 4.5 g of active ingredient/L on *T. urticae* Koch females. For those conditions, the mortality rate obtained was 86.49%.

Similarly, Han et al. (2011) [120] tested some constituent compounds of the *Eucalyptus citriodora* Hook extract and other plants on resistant and susceptible acaricidal *T. urticae* Koch females. Among them, those that showed the highest toxicity were menthol (LD_{50} of 12.9 µg/cm^3) and citronellium acetate (LD_{50} of 16.8 µg/cm^3), evaluated on females susceptible to acaricides. Other compounds such as β-citronellol, citral, geranyl acetate, and eugenol also demonstrated a high toxic activity, with LD_{50}s between 21.7 µg/cm^3 and 24.6 µg/cm^3. When comparing the mortality results obtained for both susceptible and acaricide-resistant mites, the researchers estimated that they were very similar to each other and therefore evidenced that the mechanisms of action of the components of the essential oil and of the synthetic acaricides are different and do not present processes that promote cross-resistance.

One year later, Akhtar et al. (2012) [121] studied the effect of eight quinones on the *T. urticae* Koch—*Myzus persicae* Sulzer, *Myzocallis walshii* Monell, and *Illinoia liriodendri* Monell—using the leaf immersion method. The compound plumbagine was the one that exhibited the greatest activity on the mite, with an LC_{50} of 0.001%. Marčić and Međo (2014) [122] also performed experiments with secondary metabolites from plants. In their study, they tested a combination of oximatrin and psoralen (0.2% and 0.4%, respectively) on the *T. urticae* Koch and measured acute toxicity and repellency. The applications were made on bean leaves with a Potter Tower, and the subsequently calculated LD_{50}s were 55.49, 52.68, 6.88, 13.03, and 8.8 µL/L for eggs, females that had not oviposited, larvae, protonymphs, and deutonymphs, respectively. Additionally, they noticed that, in preferential tests on the leaves, the mites tended to be located in the middle of the untreated leaf, at which point the oviposition was greater.

The same authors also tested compounds from the Neem extract (azadirachtin-A) on females of the two-spotted spider mite [123]. For this case, they introduced bean leaf discs inside Petri dishes with moistened cotton and made applications of the product using a Potter Tower in the middle of the leaf. They concluded that females preferred to be located in the middle of the leaf not treated with the product and, in the same way, they observed that oviposition was higher in females that were located in the untreated areas.

4. Conclusions

In conclusion, 458 records of plant species from 67 plant families (listed in this survey) have repellent or acaricidal effects against the *T. urticae* Koch under laboratory conditions. The efficacy is available at different levels depending on species, extractions (extract or essential oils), plant parts used, and concentrations of test extract/essential oil. Among the most studied botanical families for this purpose are plants from Lamiaceae, Asteraceae, Myrtaceae, and Apiaceae taxons. Extracts from species including *Celosia Trygina* L., *Cassia mimosoides* L., *Clome viscosa* L., *Boscia senagalensis* (Pers.) Lam. Ex. Poir., *Cobretum micranthum* G. Don, *Ipomaea asarifolia* (Desr.) Roem. and Schult., *Cnidoscolus aconitifolius* (Mill) I.M. Johnst., *Azadirachta indica* A. Juss., *Syzygium cumini* (L.) Skeels, *Papaver rhoeas* L., *Plantago major* L., *Ailanthus altissima* (Mill.) Swingle, and *Capsicum annuum* L. exhibited better acaricidal properties with efficacies between 90% and 100% at a concentration range between 0.2% and 1%—comparable to some commercial acaricides. LD_{50} values can be found below 20 µg/mL or 5 µL/L air. Thus, botanical-based preparations can be a good source of effective acaricidal preparations either as extracts or as essential oils. Although the information herein presented only concerns a basic screening of the acaricidal efficacy of botanicals at laboratory (in vitro) levels, several plants could be considered for future research on field evaluations or as sources of acaricide compounds. In this sense, several compounds such as azadirachtin, 10-hydroxygeraniol, terpineols, verbenol, verbenone, carvacrol, plumbagine, linalool, and citral, among others, have been isolated as bioactive acaricidal compounds. In future studies, attention may be focused on acaricidal activity rather than on repellent properties to facilitate two-spotted mite control. However, formulations and application rates similar to those used by farmers must be assessed in order to achieve more predictive results in further field experiments. Sublethal effects must also be relevant in future research, since those effects could produce other subsequent problems or benefits in the control of mites. Finally, more compatibility studies and phytotoxicity as well as extract stability, extraction standardization, and field formulations are required to ensure good results on integrated pest management programs for *T. urticae* Koch control using effective botanicals.

Author Contributions: Conceptualization, R.A.R., D.R., E.C.-B.; methodology, R.A.R.; validation, R.A.R.; formal analysis, R.A.R., D.R., E.C.-B.; investigation, R.A.R.; resources D.R., E.C.-B.; data curation, R.A.R., D.R., E.C.-B.; writing—original draft preparation, R.A.R.; writing—review and editing, D.R., E.C.-B.; supervision, D.R., E.C.-B.; project administration, E.C.-B.; funding acquisition D.R., E.C.-B.

Funding: This research was funded by the Vicerrectoria de Investigaciones at UMNG, grant number INV-CIAS-1788-validity 2016.

Acknowledgments: Authors thank Universidad Militar Nueva Granada (UMNG) for the financial support through the project INV-CIAS-1788.

Conflicts of Interest: The authors declare no conflict of interest. The funders had no role in the design of the study; in the collection, analyses, or interpretation of data; in the writing of the manuscript, or in the decision to publish the results.

Appendix A

Table A1. Compilation of reported studies using plant extracts and essential oils against *T. urticae* Koch under laboratory conditions.

Family	Plant Species	Source	Concentration	Bioassay[a]	*T. urticae* Koch Stage	Effect on *T. urticae* Koch	Identified Compounds	Ref.
Amaranthaceae	*Amaranthus viridis* L.	Whole plant extract	5000 ppm	G	Adults	Mortality between 40 and 60%	-	[108]
Amaranthaceae	*Amaranthus viridis* L.	Whole plant extract	2500 ppm	G	Adults	Mortality between 40 and 60%	-	[108]
Amaranthaceae	*Blepharis linariifolia* Pers.	Whole plant extract	5000 ppm	G	Adults	Mortality between 61 and 80%	-	[108]
Amaranthaceae	*Blepharis linariifolia* Pers.	Whole plant extract	2500 ppm	G	Adults	Mortality between 61 and 80%	-	[108]
Amaranthaceae	*Blepharis* sp.	Whole plant extract	5000 ppm	G	Adults	More than 80% of mortality	-	[108]
Amaranthaceae	*Blepharis* sp.	Whole plant extract	2500 ppm	G	Adults	Mortality between 61 and 80%	-	[108]
Amaranthaceae	*Celosia Trygina* L.	Whole plant extract	5000 ppm	G	Adults	More than 80% of mortality	-	[108]
Amaranthaceae	*Celosia Trygina* L.	Whole plant extract	2500 ppm	G	Adults	More than 80% of mortality	-	[108]
Amaranthaceae	*Chenopodium ambrosioides* Mosyakin et Clemants	Emulsifiable Concentrate	0.50%	A,C	Adults and eggs	94.7% of mortality	-	[41]
Amaranthaceae	*Chenopodium quinoa* Willd.	Seeds extract	6–9% w/v [1.24% w/v (LD$_{50}$)]	E,F	Adult females and nymphs	Mortalities ranged from 30% to 99%	-	[89]
Amaranthaceae	*Kochia scoparia* (L.) Schrad.	-	98.13% (chloroform extraction)	A,E,H	Adult females	92.58% of mortality	-	[52]
Amaryllidaceae	*Allium cepa* L.	Essential oil	-	D	Larvae and adults.	Mortalities of 65% (larvae) and 67% (adults)	-	[88]
Amaryllidaceae	*Allium cepa* L.	Peel fruit extract	1%	G	Adult females	Mortality between 0 and 20%	-	[99]
Amaryllidaceae	*Allium galanthum* Kar. & Kir.	Whole plant extract	1%	G	Adult females	Mortality between 20 and 50%	-	[99]
Amaryllidaceae	*Allium obliquum* L.	Whole plant extract	1%	G	Adult females	Mortality between 50 and 80%	-	[99]
Amaryllidaceae	*Allium sativum* L.	-	7.2 g/L	A,G	Adult females	LD$_{50}$	-	[78]
Amaryllidaceae	*Allium sativum* L.	Bulb extract	7.49 and 13.5 mg/L	E,F	Adult females	LD$_{50}$ and LD$_{90}$ (respectively)	-	[77]
Amaryllidaceae	*Allium sativum* L.	Essential oil	-	D	Larvae and adults	Mortalities of 86% (larvae) and 61% (adults)	-	[88]
Amaryllidaceae	*Pancratium maritimum* L.	Alkaloidal ethanolic extract and bulb essential oil	0.2%, 0.36% and 1.5% respectively			LD$_{50}$	-	[76]
Amaryllidaceae	*Ungernia severtzovii* Regel	Root extract	1%	G	Adult females	Mortality between 20 and 50%	-	[99]
Anacardiaceae	*Cotinus coggygria* Scop.	Essential oil	-	D	Larvae and adults	Mortalities of 58% (larvae) and 58% (adults)	-	[88]
Anacardiaceae	*Cotinus coggygria* Scop.	Aerial part extract	1%	G	Adult females	Mortality between 0 and 20%	-	[99]
Anacardiaceae	*Pistacia lentiscus* L.	Essential oil	-	D	Larvae and adults	Mortalities of 22% (larvae) and 23% (adults)	-	[88]
Annonaceae	*Annona glabra* L.	Seed extract	1000 ppm	D,G	Eggs	No effects	-	[80]

Table A1. Cont.

Family	Plant Species	Source	Concentration	Bioassay[a]	T. urticae Koch Stage	Effect on T. urticae Koch	Identified Compounds	Ref.
Annonaceae	Cananga odorata (Lam.) Hook.F. & Thomson	Essential oil	0.1%	G	Adult females	24.2% of mortality	-	[112]
Annonaceae	Xilopia sericea A.St.-Hill.	Leaves and fruits essential oils	4.08 µL/L	C	Adult females	LD_{50}	α-pinene (0.41% leaves, 17.18% fruits), β-pinene (45.59% fruits), cubenol (57.43% leaves), myrcene (9.13% fruits), between others	[44]
Apiaceae	Ammi visnaga	Seed extract	17 µg/cm^2	D,I	Eggs	LD_{50}	Kheline and visnagine	[65]
Apiaceae	Carum carvi L.	Essential oil	19×10^{-3} µL/mL of air.	J	Adults	100% of mortality	-	[42]
Apiaceae	Carum carvi L.	Essential oil	22.4 µg/cm^3	C,J	Adults	LD_{50}	-	[68]
Apiaceae	Carum carvi L.	Essential oils mixed with Fatty acid potassium salts	570 ppm of essential oil and 2478 ppm of potassium salts	E	Adults	83.4% of mortality	-	[83]
Apiaceae	Conium maculatum L.	Flowers and leaves extract	10–50%	A,B	Adult females	Mortalities of 95.18% and 81.11%, respectively	-	[60]
Apiaceae	Coriandrum sativum L.	Essential oil	19×10^{-3} µL/mL of air	J	Adults	92% of mortality	-	[42]
Apiaceae	Cuminum cyminum L.	Essential oil from seeds	3.74 µL/L air	C,J	Adult females	LD_{50}	α-pinene (29.1%), limonene (22%), 1,8-cineole (17.9%)	[85]
Apiaceae	Daucus carota L.	Essential oil	-	D	Larvae and adults	Mortalities of 5% (larvae) and 3% (adults)	-	[88]
Apiaceae	Deverra scoparia Coss. & Durieu	Essential oil	1.79 and 3.2 mg/L	E	Young females	LD_{50} and LD_{90}, respectively	α-pinene, Δ^3-carene and terpinen-4-ol	[43]
Apiaceae	Deverra scoparia Coss. & Durieu	Essential oil	-	D	larvae and adults	Mortalities of 98% (larvae) and 97% (adults)	-	[88]
Apiaceae	Ferula gumosa Boiss.	Essential oil	6.98 and 6.52 µL/L air 5.75 µL/L	C,J	Eggs and adults, respectively	LD_{50}	β-pinene (50.1%), α-pinene (14.9%), δ-3-carene (6.7%)	[86]
Apiaceae	Foeniculum vulgare Mill.	Seed essential oil vapors	1.17 µL/L (eggs) (females), 1.17%	J	Eggs and adults	LD_{50}	-	[84]
Apiaceae	Foeniculum vulgare Mill.	Fruit essential oil	3.15 µL/L (females)– 1.53 µL/L (eggs)	E	Adults	LD_{50}	-	[83]
Apiaceae	Heracleum persicum Desf. Ex. Fisch.	Fruit essential oils vapors		J	Eggs and adults	LD_{50}	-	[84]

Table A1. Cont.

Family	Plant Species	Source	Concentration	Bioassay[a]	T. urticae Koch Stage	Effect on T. urticae Koch	Identified Compounds	Ref.
Apiaceae	*Heracleum persicum* Desf. Ex. Fisch.	Essential oil	1.53%	E	Adults	LD$_{50}$	-	[83]
Apiaceae	*Smyrnium olusatrum* L.	Inflorescence extract	1.9 and 42.7 µg/mL, respectively for isolated compounds	D	Adult females	LD$_{50}$ (chronic toxicity after 5 days)	Isolation of isofuranodiene and germacrone, separately evaluated	[87]
Apocynaceae	*Vinca erecta* Regel & Schmalh	Aerial part extract	1%	G	Adult females	Mortality between 50 and 80%	-	[99]
Apocynaceae	*Vinca minor* L.	Aerial part extract	1%	G	Adult females	Mortality between 0 and 20%	-	[99]
Araceae	*Arum korolkovii* Regel	Aerial part extract	1%	G	Adult females	Mortality between 0 and 20%	-	[99]
Asclepiadaceae	*Calotropis gigantea* W.T. Aiton	Leaf extract	5000 ppm	G	Adults	Mortality between 61 and 80%	-	[71]
Asclepiadaceae	*Calotropis gigantea* W.T. Aiton	Leaf extract	2500 ppm	G	Adults	Mortality between 40 and 60%	-	[71]
Asteraceae	*Achillea mellifolium* L.	Essential oil from aerial part	1.208% v/v (leaf dipping) and 1.801 µL/L air (fumigation)	G,J	Adult females	LD$_{50}$	Piperitone (12.8%), *p*-cymene (10.6%)	[64]
Asteraceae	*Achillea millefolium* L.	Aerial part extract	1%	G	Adult females	Mortality between 0 and 20%	-	[99]
Asteraceae	*Acroptilon repens* (L.) DC.	Aerial part extract	1%	G	Adult females	Mortality between 50 and 80%	-	[99]
Asteraceae	*Ajania fastigiata* (C. Winkler) Poljakov	Aerial part extract	1%	G	Adult females	Mortality between 0 and 20%	-	[99]
Asteraceae	*Anaphalis rosea-alba* Krasch.	Aerial part extract	1%	G	Adult females	Mortality between 20 and 50%	-	[99]
Asteraceae	*Anthemis nobilis* L.	Essential oil	19 × 10^{-3} µL/mL of air	J	Adults	69% of mortality	-	[42]
Asteraceae	*Anthemis vulgaris* L.	Flower extract	7–50%	A,B	Adult females	Mortalities of 92.37% and 92.34%, respectively	-	[60]
Asteraceae	*Anthemis vulgaris* L.	Leaf extract	13–50%	A,B	Adult females	Mortalities of 82.33% and 76.63%, respectively	-	[60]
Asteraceae	*Artemisia absinthium* L.	Essential oil	19 × 10^{-3} µL/mL of air	J	Adults	97% of mortality	-	[42]
Asteraceae	*Artemisia absinthium* L.	Essential oil	0.043 mg/cm^2	A	-	LD$_{50}$	-	[45]
Asteraceae	*Artemisia absinthium* L.	Aerial part extract	1%	G	Adult females	Mortality between 20 and 50%	-	[99]
Asteraceae	*Artemisia aschurbajewii* C. Winkl.	Aerial part extract	1%	G	Adult females	Mortality between 0 and 20%	-	[99]
Asteraceae	*Artemisia cinae* O. Berg & C.F. Schmidt ex. Pljakov	Leaf extract	1326.53 ppm	A,B	Adult females	LD$_{50}$	-	[60]

Table A1. Cont.

Family	Plant Species	Source	Concentration	Bioassay[a]	T. urticae Koch Stage	Effect on T. urticae Koch	Identified Compounds	Ref.
Asteraceae	Artemisia compacta Fisch. Ex. Besser	Aerial part extract	1%	G	Adult females	Mortality between 20 and 50%	-	[99]
Asteraceae	Artemisia dracunculus L.	Aerial part extract	1%	G	Adult females	Mortality between 20 and 50%	-	[99]
Asteraceae	Artemisia judaica L.	Leaves (acetone extract)	0.56 µg/mL	C,G	Adult females	LD_{50}	-	[39]
Asteraceae	Artemisia panciflora	Aerial part extract	1%	G	Adult females	Mortality between 50 and 80%	-	[99]
Asteraceae	Artemisia vulgaris L.	Leaf extract	15–50%	A,B	Adult females	Mortalities of 54.13% and 75.12%, respectively	-	[60]
Asteraceae	Artemisia vulgaris L.	Aerial part extract	1%	G	Adult females	Mortality between 50 and 80%	-	[99]
Asteraceae	Calendula officinalis L.	Aerial part extract	1%	G	Adult females	Mortality between 50 and 80%	α-basabolol oxide (35.25%), Trans β-farersene (7.76%)	[99]
Asteraceae	Chamomilla recutita L.	Essential oil	0.65–1.17%	C	Adults ggs	LD_{50}	-	[89]
Asteraceae	Chrisanthemum coronarium L.	Essential oil	-	D	Larvae and adults	Mortalities of 88% (larvae) and 93% (adults)	-	[88]
Asteraceae	Handelia trichopylla Heimerl	Aerial part extract	1%	G	Adult females	Mortality between 0 and 20%	-	[99]
Asteraceae	Hertia cheirifolia (L.) Kuntze	Essential oil	3.43 mg/L	E	Adult females	LD_{50} and side-effect over fecundity	-	[46]
Asteraceae	Hertia cheirifolia (L.) Kuntze	Essential oil	-	D	Larvae and adults	Mortalities of 81% (larvae) and 89% (adults)	-	[88]
Asteraceae	Hieracium dschirgalanicum E. Nikit.	Aerial part extract	1%	G	Adult females	Mortality between 0 and 20%	-	[99]
Asteraceae	Inula helenium L.	Aerial part extract	1%	G	Adult females	Mortality between 20 and 50%	-	[99]
Asteraceae	Jurinea capussi Franch.	Aerial part extract	1%	G	Adult females	Mortality between 0 and 20%	-	[99]
Asteraceae	Lamyropappus schakaptaricus Knorr & Tamamsch.	Aerial part extract	1%	G	Adult females	Mortality between 0 and 20%	-	[99]
Asteraceae	Matricaria chamomilla L.	Aerial part extract	1%	G	Adult females	Mortality between 0 and 20%	-	[99]
Asteraceae	Matricaria matricarioides (Less.) Porter	Aerial part extract	1%	G	Adult females	Mortality between 20 and 50%	-	[99]
Asteraceae	Matricaria recutita L.	Aerial part extract	1%	G	Adult females	Mortality between 0 and 20%	-	[99]
Asteraceae	Pseudoglossanthis litwinowii (Tzvel.) R. Kam.	Aerial part extract	1%	G	Adult females	Mortality between 0 and 20%	-	[99]
Asteraceae	Pyrethrum alatavicum O. & B. Fedtsch.	Aerial part extract	1%	G	Adult females	Mortality between 20 and 50%	-	[99]
Asteraceae	Pyrethrum branchanthemoides R. Kam. & Lazkov	Aerial part extract	1%	G	Adult females	Mortality between 20 and 50%	-	[99]
Asteraceae	Pyrethrum cinerariifolium Trev.	Aerial part extract	1%	G	Adult females	Mortality between 20 and 50%	-	[99]

Table A1. Cont.

Family	Plant Species	Source	Concentration	Bioassay[a]	T. urticae Koch Stage	Effect on T. urticae Koch	Identified Compounds	Ref.
Asteraceae	Pyrethrum socotkinae Kovalevsk	Aerial part extract	1%	G	Adult females	Mortality between 50 and 80%	-	[99]
Asteraceae	Pyrethrum sussamyrense Lazkov	Root extract	1%	G	Adult females	Mortality between 0 and 20%	-	[99]
Asteraceae	Santolina africana Jord. & Fourr.	Essential oil	2.35 mg/L	E	Adult females	LD_{50} and side-effect over fecundity	Terpinen-4-ol (54.96%)	[46]
Asteraceae	Santolina africana Jord. & Fourr.	Essential oil	-	D	Larvae and adults	Mortalities of 77% (larvae) and 68% (adults)	-	[88]
Asteraceae	Senecio saposhnikovii Krasch et. Schipcz.	Aerial part extract	1%	G	Adult females	Mortality between 50 and 80%	-	[99]
Asteraceae	Seriphidium herba-album (Asso) Sojak	Essential oil	-	D	Larvae and adults	Mortalities of 54% (larvae) and 37% (adults)	-	[88]
Asteraceae	Tagetes minuta L.	Aerial part extract	1%	G	Adult females	Mortality between 50 and 80%	-	[99]
Asteraceae	Tanacetopsis ferganensis Kovalevsk	Aerial part extract	1%	G	Adult females	Mortality between 20 and 50%	-	[99]
Asteraceae	Tanacetopsis setacea Kovalevsk	Aerial part extract	1%	G	Adult females	Mortality between 0 and 20%	-	[99]
Asteraceae	Tanacetopsis submarginata Kovalevsk	Aerial part extract	1%	G	Adult females	Mortality between 0 and 20%	-	[99]
Asteraceae	Tanacetum boreale Fisch. Ex. DC.	Aerial part extract	1%	G	Adult females	Mortality between 20 and 50%	-	[99]
Asteraceae	Tanacetum pseudoachillea C. Winkl.	Aerial part extract	1%	G	Adult females	Mortality between 50 and 80%	-	[99]
Asteraceae	Tanacetum vulgare L.	Essential oil	4%	A	-	75.6% of mortality	-	[45]
Asteraceae	Tithonia diversifolia Hemsl.	Methanolic extract	150 µg/cm³	D	Adult females	Mortality less than 50%	-	[92]
Asteraceae	Tripleurospermum inodorum Sch. Bip.	Aerial part extract	1%	G	Adult females	Mortality between 0 and 20%	-	[99]
Asteraceae	Xanthium strumarium L.	Fruit extract	9-50%	A,B	Adult females	Mortalities of 68.24% and 85.88%, respectively	-	[60]
Asteraceae	Xanthium strumarium L.	Leaf extract	11-50%	A,B	Adult females	Mortalities of 52.48% and 79.85%, respectively	-	[60]
Berveridaceae	Berberis iliensis Popov	Aerial part extract	1%	G	Adult females	Mortality between 20 and 50%	-	[99]
Bignoniaceae	Jacaranda obtusifolia Bonpl.	Leaf extract	0.06%	C,G	Adult females	Mortality of 64.4%	-	[124]
Boraginaceae	Echium vulgare L.	Aerial part extract	1%	G	Adult females	Mortality between 50 and 80%	-	[99]
Boraginaceae	Onosma visianii Clem.	Root extract	2.6 µg/mL	D	Adult females	LD_{50} (chronic toxicity after 5 days)	Shikonin derivatives (naphthoquinones), i.e., isobutylshikonin and isovalerylshikonin	[93]

Table A1. Cont.

Family	Plant Species	Source	Concentration	Bioassay[a]	T. urticae Koch Stage	Effect on T. urticae Koch	Identified Compounds	Ref.
Brassicaceae	Armoracia rusticana G. Gaertn., B. Mey. & Scherb.	Aerial part extract	1%	G	Adult females	Mortality between 50 and 80%	-	[99]
Brassicaceae	Barbarea vulgaris W.T. Aiton	Aerial part extract	1%	G	Adult females	Mortality between 20 and 50%	-	[99]
Brassicaceae	Capsella bursa-pastoris L.	Aerial part extract	1%	G	Adult females	Mortality between 0 and 20%	-	[99]
Brassicaceae	Cardaria repens Schrenk	Aerial part extract	1%	G	Adult females	Mortality between 0 and 20%	-	[99]
Brassicaceae	Lepidium latifolium L.	Aerial part extract	1%	G	Adult females	Mortality between 50 and 80%	-	[99]
Burseraceae	Boswellia carterii Birdw.	Essential oil	0.1%	G	Adult females	24.8% of mortality	-	[112]
Burseraceae	Commiphora myrrha (Nees) Engl.	Essential oil	0.1%	G	Adult females	22.8% of mortality	-	[112]
Burseraceae	Protium bahianum Daly	Fresh and old resin essential oils	-	J	Adult females	Mortalities of 79.6% (fresh resin) and 59% (old resin)	-	[97]
Caesalpiniaceae	Cassia mimosoides L.	Leaf extract	5000 ppm	G	Adults	More than 80% of mortality	-	[71]
Caesalpiniaceae	Cassia mimosoides L.	Leaf extract	2500 ppm	G	Adults	Mortality between 61 and 80%	-	[71]
Caesalpiniaceae	Cassia occidentalis L.	Whole plant extract	5000 ppm	G	Adults	Mortality between 61 and 80%	-	[71]
Caesalpiniaceae	Cassia occidentalis L.	Whole plant extract	2500 ppm	G	Adults	Mortality between 61 and 80%	-	[71]
Caesalpiniaceae	Cassia tora L.	Whole plant extract	5000 ppm	G	Adults	More than 80% of mortality	-	[71]
Caesalpiniaceae	Cassia tora L.	Whole plant extract	2500 ppm	G	Adults	Mortality between 61 and 80%	-	[71]
Campanulaceae	Codonopsis clematidea Schrenk	Aerial part extract	1%	G	Adult females	Mortality between 50 and 80%	-	[99]
Cannabaceae	Cannabis sativa L.	Essential oil from panicles	0.10%	G	Adult females	83.28% of mortality	β-myrcene (18.5%), trans-caryophyllene (35.6%)	[98]
Cannabaceae	Cannabis sativa L.	Aerial part and root extracts	1%	G	Adult females	Mortality between 50 and 80%	-	[99]
Cannabaceae	Humulus lupulus L.	Flower extract	5-50%	A,B	Adult females	Mortalities of 56.37% and 67.84%, respectively	-	[60]
Cappandaceae	Clome viscosa L.	Whole plant extract	5000 ppm	G	Adults	More than 80% of mortality	-	[71]
Cappandaceae	Clome viscosa L.	Whole plant extract	2500 ppm	G	Adults	Mortality between 61 and 80%	-	[71]
Capparidaceae	Boscia senagalensis (Pers.) Lam. Ex. Poir.	Leaf extract	5000 ppm	G	Adults	More than 80% of mortality	-	[71]
Capparidaceae	Boscia senagalensis (Pers.) Lam. Ex. Poir.	Leaf extract	2500 ppm	G	Adults	Mortality between 61 and 80%	-	[71]
Caprifoliaceae	Sambucus nigra L.	Aerial part extract	1%	G	Adult females	Mortality between 0 and 20%	-	[99]
Caryophyllaceae	Saponaria officinalis L.	Root extract	0.31% (eggs), 1.18% (adulst), 0.91% (oviposition) w/v	I	Eggs, adults and oviposition	LD_{50}	-	[66]

Table A1. Cont.

Family	Plant Species	Source	Concentration	Bioassay[a]	T. urticae Koch Stage	Effect on T. urticae Koch	Identified Compounds	Ref.
Caryophyllaceae	Silene sussamyrica Lazkov	Aerial part extract	1%	G	Adult females	Mortality between 0 and 20%	-	[99]
Chenopodiaceae	Anabasis aphylla L.	Seed and bark extracts	1%	G	Adult females	Mortality between 50 and 80%	-	[99]
Chenopodiaceae	Anthochlamis tianschanica Iljin	Aerial part extract	1%	G	Adult females	Mortality between 20 and 50%	-	[99]
Chenopodiaceae	Chenopodium album (L.) Mosc. Ex. Moq.	Flower and leaf extracts	8–50%.	A,B	Adult females	Mortalities of 96.99% and 91.15%, respectively	-	[60]
Combretaceae	Cobretum micranthum G. Don	Whole plant extract	5000 ppm	G	Adults	More than 80% of mortality	-	[71]
Combretaceae	Cobretum micranthum G. Don	Whole plant extract	2500 ppm	G	Adults	More than 80% of mortality	-	[71]
Combretaceae	Combretum glutinosum Perr. Ex. DC.	Leaf extract	5000 ppm	G	Adults	More than 80% of mortality	-	[71]
Combretaceae	Combretum glutinosum Perr. Ex. DC.	Leaf extract	2500 ppm	G	Adults	More than 80% of mortality	-	[71]
Combretaceae	Combretum glutinosum Perr. Ex. DC.	Stem extract	5000 ppm	G	Adults	Mortality between 61 and 80%	-	[71]
Combretaceae	Combretum glutinosum Perr. Ex. DC.	Stem extract	2500 ppm	G	Adults	Mortality between 40 and 60%	-	[71]
Combretaceae	Guiera senegalensis J.F. Gmel.	Leaf extract	5000 ppm	G	Adults	More than 80% of mortality	-	[71]
Combretaceae	Guiera senegalensis J.F. Gmel.	Leaf extract	2500 ppm	G	Adults	Mortality between 61 and 80%	-	[71]
Combretaceae	Guiera senegalensis J.F. Gmel.	Stem extract	5000 ppm	G	Adults	Mortality between 61 and 80%	-	[71]
Combretaceae	Guiera senegalensis J.F. Gmel.	Stem extract	2500 ppm	G	Adults	Mortality between 40 and 60%	-	[71]
Combretaceae	Piloitigma retilicolin	Whole plant extract	5000 ppm	G	Adults	More than 80% of mortality	-	[71]
Combretaceae	Piloitigma retilicolin	Whole plant extract	2500 ppm	G	Adults	Mortality between 61 and 80%	-	[71]
Convolvulaceae	Convolvulus arvensis L.	Aerial part extract	1%	G	Adult females	Mortality between 20 and 50%	-	[99]
Convolvulaceae	Convolvulus krauseanus Regel. & Schmalh	Root extract	1%	G	Adult females	Mortality between 80 and 100%	-	[99]
Convolvulaceae	Ipomaea asarifolia (Desr.) Roem. & Schult.	Whole plant extract	5000 ppm	G	Adults	More than 80% of mortality	-	[71]
Convolvulaceae	Ipomaea asarifolia (Desr.) Roem. & Schult.	Whole plant extract	2500 ppm	G	Adults	Mortality between 40 and 60%	-	[71]
Convolvulaceae	Ipomaea sp. L.	Whole plant extract	5000 ppm	G	Adults	More than 80% of mortality	-	[71]
Convolvulaceae	Ipomaea sp. L.	Whole plant extract	2500 ppm	G	Adults	Mortality between 61 and 80%	-	[71]

Table A1. Cont.

Family	Plant Species	Source	Concentration	Bioassay[a]	T. urticae Koch Stage	Effect on T. urticae Koch	Identified Compounds	Ref.
Cupressaceae	Cupressus macrocarpa Hartw. ex Gordon	Leaf extract	5.69 µL/L air	A	Adult females	LD_{50}	β-citronellol (35.92%)	[55]
Cupressaceae	Cupressus sempervirens L.	Essential oil	0.1‰	G	Adult females	28.9% of mortality	-	[112]
Cupressaceae	Juniperus communis L.	Essential oil	0.1‰	G	Adult females	42.6% of mortality	-	[112]
Cupressaceae	Juniperus phoenicea L.	Essential oil	-	D	Larvae and adults	Mortalities of 60% (larvae) and 56% (adults)	-	[88]
Cupressaceae	Thuja orientalis L.	Leaf extract	7.51 µL/L air	A	Adult females	LD_{50}	α-pinene (35.49%)	[55]
Elaeagnaceae	Elaeagnus angustifolia L.	Aerial part extract	1%	G	Adult females	Mortality between 20 and 50%	-	[99]
Equisetaceae	Equisetum arvense L.	Aerial part extract	1%	G	Adult females	Mortality between 0 and 20%	-	[99]
Euforbiaceae	Jatropha curcas L.	Leaf extract	0.06%	C,G	Adult females	Mortality of 63.3%	-	[124]
Euphorbiaceae	Chrozophora oblongifolia (Delile) Spreng.	Whole plant extract	312.72 and 206.91 ppm	G	Adult females and larvae	LD_{50}	7-O-β-D-[2″,6″-bis(4-hydroxy-E-cinnamoyl)] glucopyranoside, apigenin 7-O-β-D-glucopyranoside isolated from butanol fration	[125]
Euphorbiaceae	Cnidoscolus aconitifolius (Mill) I.M. Johnst.	Leaf extract	2000 µg/mL	C,G	Adult females	92% of mortality	-	[61]
Euphorbiaceae	Euphorbia ferganensis B. Fedtsch.	Root extract	1%	G	Adult females	Mortality between 0 and 20%	-	[99]
Euphorbiaceae	Euphorbia kansui S.L. Liou S.B. Ho	Root extract	3–5 g/L	C	Adult females	Mortalities of 27% and 55%, respectively	3-O-(2,3-dimethylbutanoyl)-13-dodecanoylingenol y 3-O-(2′E,4′Z-decadienoyl)-ingenol	[100]
Fabaceae	Acacia cyanophylla Lindl.	Essential oil	-	D	Larvae and adults	Mortalities of 58% (larvae) and 26% (adults)	-	[88]
Fabaceae	Ammopiptanthus nanus (M. pop) Cheng	Pod extract	1%	G	Adult females	Mortality between 50 and 80%	-	[99]
Fabaceae	Bowdichia virgilioides Kunth	Leaf extract	0.06% w/v	C,G	Adult females	Mortality of 64.4%	-	[126]
Fabaceae	Gleditschia spp.	Aerial part extract	1%	G	Adult females	Mortality between 0 and 20%	-	[99]
Fabaceae	Glycirrhisa uralensis L.	Aerial part extract	1%	G	Adult females	Mortality between 0 and 20%	-	[99]
Fabaceae	Hedysarum cephalotes Franchet	Whole plant extract	1%	G	Adult females	Mortality between 20 and 50%	-	[99]
Fabaceae	Hedysarum daraut-kurganicum Sultanova	Aerial part extract	1%	G	Adult females	Mortality between 50 and 80%	-	[99]
Fabaceae	Hymenaea courbaril L.	Leaf extract	0.06% w/v	C,G	Adult females	Mortality of 59.4%	-	[126]

Table A1. Cont.

Family	Plant Species	Source	Concentration	Bioassay[a]	T. urticae Koch Stage	Effect on T. urticae Koch	Identified Compounds	Ref.
Fabaceae	Medicago minima L.	Aerial part extract	1%	G	Adult females	Mortality between 50 and 80%	-	[99]
Fabaceae	Melilotus officinalis L.	Aerial part extract	1%	G	Adult females	Mortality between 0 and 20%	-	[99]
Fabaceae	Millettia pinnata L.	Leaf oil	0.004%	C	Adult females	LD$_{50}$ (after 4 days)	-	[101]
Fabaceae	Oxytropis rosea Bunge	Aerial part extract	1%	G	Adult females	Mortality between 20 and 50%	-	[99]
Fabaceae	Sophora korolkovii Koehne.	Aerial part extract	1%	G	Adult females	Mortality between 0 and 20%	-	[99]
Fabaceae	Sophora secundiflora (Ortega) Lag. Ex. DC.	Essential oil	-	D	Larvae and adults	Mortalities of 68% (larvae) and 61% (adults)	-	[88]
Fabaceae	Vicia cracca L.	Aerial part extract	1%	G	Adult females	Mortality between 0 and 20%	-	[99]
Geraniaceae	Pelargonium graveolens L'Her	Leaf extract	12.27 µL/L air	A	Adult females	LD$_{50}$	terpinen-4-ol (20.29%)	[55]
Geraniaceae	Pelargonium graveolens L'Her.	Essential oil	19 × 10^{-3} µL/mL of air.	K	Adults	100% of mortality	-	[42]
Geraniaceae	Pelargonium graveolens L'Her.	Essential oil	-	D	Larvae and adults	Mortalities of 78% (larvae) and 70% (adults)	-	[88]
Geraniaceae	Pelargonium roseum Willd	Essential oil	0.1%.	G	Adult females	30% of mortality	-	[112]
Gramineae	Chrysopogon zizanioides (L.)	Essential oil	18.82 µg/mL	J	Adult females	LD$_{50}$	-	[127]
Gramineae	Cymbopogon citratus (DC.) Stapf	Essential oil	0.1%.	J	Adults	100% of mortality	-	[42]
Gramineae	Cymbopogon citratus (DC.) Stapf	Essential oil	-	G	Adult females	17.8% of mortality	-	[112]
Gramineae	Cymbopogon flexuosus (Nees ex Steud.) W. Watson	Essential oil	17.23 µg/mL	J	Adult females	LD$_{50}$	-	[127]
Gramineae	Cymbopogon Martini (Roxb.) W. Watson	Essential oil	19 × 10^{-3} µL/mL of air	J	Adults	67% of mortality	-	[42]
Gramineae	Cymbopogon nardus (L) Rendle	Essential oil	19 × 10^{-3} µL/mL of air	J	Adults	99% of mortality	-	[42]
Gramineae	Cymbopogon nardus (L) Rendle	Essential oil	22.5 µg/cm^3	C,J	Adults	LD$_{50}$	-	[68]
Gramineae	Cymbopogon winterianus Jowitt ex. Bor	Essential oil	0.1%	G	Adult females	27.6% of mortality	-	[112]
Gramineae	Lolium perenne L.	Leaf and flower methanolic extracts	6-50%	A,B	Adult females	Mortalities of 91.43% and 93.5%, respectively	-	[60]
Guttiferae	Hypericum perforatum L.	Aerial part extract	1%	G	Adult females	Mortality between 50 and 80%	-	[99]
Iridaceae	Iris sogdiana Regel.	Leaf extract	1%	G	Adult females	Mortality between 0 and 20%	-	[99]
Juglandeceae	Juglans regia L.	Leaf extract	12% v/w	C,G	Adult females and nymphs	Mortality between 83 and 90%	-	[128]
Lamiaceae	Acinos thymoides (L.) Moench	Aerial part extract	1%	G	Adult females	Mortality between 50 and 80%	-	[99]

Table A1. Cont.

Family	Plant Species	Source	Concentration	Bioassay[a]	T. urticae Koch Stage	Effect on T. urticae Koch	Identified Compounds	Ref.
Lamiaceae	Ajuga australis R.Br.	Leaf extract	1%	C	-	Mortality between 20 and 49%	-	[103]
Lamiaceae	Callicarpa pedunculata R.Br.	Leaf extract	1%	C	-	Mortality between 20 and 49%	-	[103]
Lamiaceae	Ceratanthus longicornis (F.Muell.) G. Taylor	Leaf extract	1%	C	-	100% of mortality	-	[105]
Lamiaceae	Clerodendrum floribundum R.Br.	Leaf extract	1%	C	-	Mortality between 50 and 89%	-	[103]
Lamiaceae	Clerodendrum inerme (L.) Gaertn.	Leaf extract	1%	C	-	Mortality between 50 and 89%	-	[103]
Lamiaceae	Clerodendrum tomentosum (Vent.) R.Br.	Leaf extract	1%	C	-	Mortality between 50 and 89%	-	[103]
Lamiaceae	Clerodendrum traceyi F. Muell.	Leaf extract	1%	C	-	100% of mortality	-	[103]
Lamiaceae	Faradaya albertissii F. Muell.	Leaf extract	1%	C	-	Less than 20% of mortality	-	[103]
Lamiaceae	Faradaya splendida F. Muell.	Leaf extract	1%	C	-	Mortality between 50 and 89%	-	[103]
Lamiaceae	Glossocarya calcicola Domin.	Leaf extract	1%	C	-	Less than 20% of mortality	-	[103]
Lamiaceae	Glossocarya hemiderma Benth.	Leaf extract	1%	C	-	Mortality between 20 and 49%	-	[103]
Lamiaceae	Gmelina leichardtii (F.Muell.) Benth	Leaf extract	1%	C	-	Mortality between 90 and 99%	-	[103]
Lamiaceae	Hemiandra australis B.J. Conn.	Leaf extract	1%	C	-	Mortality between 20 and 49%	-	[103]
Lamiaceae	Hemiandra leiantha Benth.	Leaf extract	1%	C	-	Less than 20% of mortality	-	[103]
Lamiaceae	Hemiandra pungens R.Br.	Leaf extract	1%	C	-	Mortality between 20 and 49%	-	[103]
Lamiaceae	Hemigenia humilis Benth.	Leaf extract	1%	C	-	Mortality between 20 and 49%	-	[103]
Lamiaceae	Hemigenia sericea Benth.	Leaf extract	1%	C	-	Less than 20% of mortality	-	[103]
Lamiaceae	Hemigenia westringioides Benth.	Leaf extract	1%	C	-	Less than 20% of mortality	-	[103]
Lamiaceae	Hyssopus officinalis L.	Aerial part extract	1%	G	Adult females	Less than 20% of mortality	-	[99]
Lamiaceae	Hyssopus officinalis L.	Essential oil	0.1%.	G	Adult females	28.1% of mortality	-	[112]
Lamiaceae	Lachnostachys eriobotrya (F. Muell.) Druce	Leaf extract	1%	C	-	Less than 20% of mortality	-	[103]
Lamiaceae	Lavandula angustifolia Mill.	Leaf extract	4.93 µL/L	J	Adult females	LD_{50}	1,8-cineole, camphor, β-pinene linalool (37.8%),	[129]
Lamiaceae	Lavandula latifolia Medik.	Essential oil from twigs with leaves andflowers	0.20–0.25% v/v	A,C	Adult females	Mortality between 95 and 100%	1,8-cineole (24.9%), camphor (18.7%)	[56]
Lamiaceae	Lavandula officinalis Chaix	Essential oil	19×10^{-3} µL/mL of air	J	Adults	97% of mortality	-	[42]

Table A1. Cont.

Family	Plant Species	Source	Concentration	Bioassay[a]	T. urticae Koch Stage	Effect on T. urticae Koch	Identified Compounds	Ref.
Lamiaceae	Lavandula officinalis Chaix	Essential oil	-	D	Larvae and adults	Mortalities of 38% (larvae) and 41% (adults)	-	[88]
Lamiaceae	Lavandula vera DC.	Essential oil	0.1%	G	Adult females	26.1% of mortality	-	[112]
Lamiaceae	Leonorus turkestanicus V. Krecz. & Kupr.	Aerial part extract	1%	G	Adult females	Mortality between 50 and 80%	-	[99]
Lamiaceae	Lycopus australis R. Br.	Leaf extract	1%	C	-	Mortality between 20 and 49%	-	[103]
Lamiaceae	Majorana hortensis Moench	Essential oil	1.84–6.26%	C	Adults and eggs (respectively)	LD$_{50}$	terpinen-4-ol (23.86%), p-cymene (23.40%) and sabinene (10.90%)	[89]
Lamiaceae	Melissa officinalis L.	Aerial part extract	1%	G	Adult females	Mortality between 0 and 20%	-	[99]
Lamiaceae	Menta spicata L.	Essential oil	15 µL/L	C,J	Eggs	LD$_{50}$	-	[130]
Lamiaceae	Mentha arvensis L.	Aerial part extract	1%	G	Adult females	Mortality between 80 and 100%	carvone (68.5%)	[99]
Lamiaceae	Mentha longifolia L.	Essential oil	11.08 µg/mL	J	Adult females	LD$_{50}$	-	[127]
Lamiaceae	Mentha piperita L.	Essential oil	19×10^{-3} µL/mL of air	J	Adults	100% of mortality	-	[42]
Lamiaceae	Mentha piperita L.	Essential oil	22.8 µg/cm^3	C,J	Adults	LD$_{50}$	-	[68]
Lamiaceae	Mentha piperita L.	Essential oil	0.1%.	G	Adult females	23.7% of mortality	-	[112]
Lamiaceae	Mentha piperita L.	Essential oil	15.86 µg/mL	J	Adult females	LD$_{50}$	-	[127]
Lamiaceae	Mentha pulegium L.	Essential oil	19×10^{-3} µL/mL of air	J	Adults	100% of mortality	-	[42]
Lamiaceae	Mentha pulegium L.	Essential oil	23.7 µg/cm^3	J	Adults	LD$_{50}$	-	[68]
Lamiaceae	Mentha pulegium L.	Essential oil	-	D	Larvae and adults	Mortalities of 90% (larvae) and 91% (adults)	-	[88]
Lamiaceae	Mentha spicata L.	Essential oil	19×10^{-3} µL/mL of air	J	Adults	100% of mortality	-	[42]
Lamiaceae	Mentha spicata L.	Essential oil	38.8 µg/cm^3	C,J	Adults	LD$_{50}$	-	[68]
Lamiaceae	Mentha spicata L.	essential oil from leaves	7.53 µL/L air	C,J	Adult females	LD$_{50}$	carvone (59.4%), limonene (9.8%), 1,8-cineole (7.4%)	[85]
Lamiaceae	Mentha sylvestris L.	Aerial part extract	1%	G	Adult females	Mortality between 20 and 50%	-	[99]
Lamiaceae	Microcorys capitata (Bartl.) Benth.	Leaf extract	1%	C	-	Less than 20% of mortality	-	[103]
Lamiaceae	Microcorys sp.	Leaf extract	1%	C	-	Less than 20% of mortality	-	[103]
Lamiaceae	Micromeria fruticosa L.	Essential oil vapors	2 µL/L of air	J	Adults and nimphs	96.7% of mortality	-	[105]
Lamiaceae	Micromeria fruticosa L.	Essential oil vapors	2 µL/L	J	Adults and nimphs	96.7% of mortality	-	[105]

Table A1. *Cont.*

Family	Plant Species	Source	Concentration	Bioassay[a]	*T. urticae* Koch Stage	Effect on *T. urticae* Koch	Identified Compounds	Ref.
Lamiaceae	*Nepeta racemosa* L.	Essential oil vapors	2 uL/L of air	J	Adults and nimphs	95% of mortality	-	[105]
Lamiaceae	*Nepeta racemosa* L.	Essential oil vapors	2 µL/L	J	Adults and nimphs	95% of mortality	-	[105]
Lamiaceae	*Ocimum basilicum* L.	Essential oil	19×10^{-3} µL/mL of air	J	Adults	88% of mortality	-	[42]
Lamiaceae	*Ocimum basilicum* L.	Essential oil	39.5 µg/cm^3	C,J	Adults	LD$_{50}$	-	[68]
Lamiaceae	*Ocimum basilicum* L.	Essential oil	0.1%	G	Adult females	21% of mortality	-	[112]
Lamiaceae	*Ocimum basilicum* L.	Essential oil	0.6 µL/L	C,J	Adult females	LD$_{50}$	linalool (65.7%)	[130]
Lamiaceae	*Origanum majorana* L.	Essential oil	19×10^{-3} µL/mL of air	J	Adults	92% of mortality	-	[42]
Lamiaceae	*Origanum majorana* L.	Essential oil	0.1%	G	Adult females	7.1% of mortality	-	[112]
Lamiaceae	*Origanum vulgare* L.	Essential oil	8.52 µL/L air	A	Adult females	LD$_{50}$	pulegone (77.45%)	[55]
Lamiaceae	*Origanum vulgare* L.	Aerial part extract	1%	G	Adult females	Mortality between 20 and 50%	-	[99]
Lamiaceae	*Origanum vulgare* L.	Essential oil vapors	2 uL/L of air	J	Adults and nimphs	95% of mortality	-	[105]
Lamiaceae	*Origanum vulgare* L.	Essential oil vapors	2 µL/L	J	Adults and nimphs	95% of mortality	-	[105]
Lamiaceae	*Otostelgia olgae* (Regel.) Korsch.	Aerial part extract	1%	G	Adult females	Mortality between 0 and 20%	-	[99]
Lamiaceae	*Pityrodia bartlingii* (Lehm.) Benth.	Leaf extract	1%	C	-	Mortality between 50 and 89%	-	[103]
Lamiaceae	*Pityrodia verbascina* (F. Muell.) Benth.	Leaf extract	1%	C	-	Less than 20% of mortality	-	[103]
Lamiaceae	*Plectranthus actites* P.I. Forst.	Leaf extract	1%	C	-	Mortality between 50 and 89%	-	[103]
Lamiaceae	*Plectranthus alloplectus* S.T. Blake	Leaf extract	1%	C	-	Mortality between 20 and 49%	-	[103]
Lamiaceae	*Plectranthus amoenus* P.I. Forst	Leaf extract	1%	C	-	Less than 20% of mortality	-	[103]
Lamiaceae	*Plectranthus apreptus* S.T. Blake	Leaf extract	1%	C	-	Less than 20% of mortality	-	[103]
Lamiaceae	*Plectranthus argentatus* S.T. Blake	Leaf extract	1%	C	-	Less than 20% of mortality	-	[103]
Lamiaceae	*Plectranthus cremnus* B. J. Conn.	Leaf extract	1%	C	-	Mortality between 50 and 89%	-	[103]
Lamiaceae	*Plectranthus diversus* S.T. Blake	Leaf extract	1%	C	-	Mortality between 90 and 99%	-	[103]
Lamiaceae	*Plectranthus fasciculatus* P.I. Forst.	Leaf extract	1%	C	-	Less than 20% of mortality	-	[103]

Table A1. Cont.

Family	Plant Species	Source	Concentration	Bioassay[a]	T. urticae Koch Stage	Effect on T. urticae Koch	Identified Compounds	Ref.
Lamiaceae	Plectranthus foetidus Benth.	Leaf extract	1%	C	-	Less than 20% of mortality	-	[103]
Lamiaceae	Plectranthus glabriflorus P.I. Forst	Leaf extract	1%	C	-	Mortality between 90 and 99%	-	[103]
Lamiaceae	Plectranthus gratus S.T. Blake	Leaf extract	1%	C	-	Less than 20% of mortality	-	[103]
Lamiaceae	Plectranthus graveolens R.Br.	Leaf extract	1%	C	-	Mortality between 50 and 89%	-	[103]
Lamiaceae	Plectranthus habrophyllus P.I. Forst	Leaf extract	1%	C	-	100% of mortality	-	[103]
Lamiaceae	Plectranthus Koonyum Range	Leaf extract	1%	C	-	Mortality between 50 and 89%	-	[103]
Lamiaceae	Plectranthus leiperi P.I. Forst.	Leaf extract	1%	C	-	Mortality between 50 and 89%	-	[103]
Lamiaceae	Plectranthus mirus S.T. Blake	Leaf extract	1%	C	-	Less than 20% of mortality	-	[103]
Lamiaceae	Plectranthus nitidus P.I. Forst.	Leaf extract	1%	C	-	Mortality between 50 and 89%	-	[103]
Lamiaceae	Plectranthus omissus P. I. Forst.	Leaf extract	1%	C	-	Mortality between 50 and 89%	-	[103]
Lamiaceae	Plectranthus parviflorus Willd	Leaf extract	1%	C	-	Mortality between 50 and 89%	-	[103]
Lamiaceae	Plectranthus scutellarioides (L.) R.Br.	Leaf extract	1%	C	-	Mortality between 50 and 89%	-	[103]
Lamiaceae	Plectranthus sp. buchanans Fort	Leaf extract	1%	C	-	Less than 20% of mortality	-	[103]
Lamiaceae	Plectranthus sp. Hann Tableland	Leaf extract	1%	C	-	100% of mortality	-	[103]
Lamiaceae	Plectranthus sp. Pinnacle	Leaf extract	1%	C	-	Less than 20% of mortality	-	[103]
Lamiaceae	Plectranthus spectabilis S.T. Blake	Leaf extract	1%	C	-	Less than 20% of mortality	-	[103]
Lamiaceae	Plectranthus suaveolens S.T. Blake	Leaf extract	1%	C	-	Mortality between 90 and 99%	-	[103]
Lamiaceae	Pogostemon cablin Benth.	Essential oil	0.1%	G	Adult females	20.3% of mortality	-	[112]
Lamiaceae	Premna acuminata R.Br.	Leaf extract	1%	C	-	Mortality between 90 and 99%	-	[103]
Lamiaceae	Premna serratifolia L.	Leaf extract	1%	C	-	100% of mortality	-	[103]
Lamiaceae	Prostanthera incisa Benth.	Leaf extract	1%	C	-	Less than 20% of mortality	-	[103]
Lamiaceae	Prostanthera lasianthos Labill.	Leaf extract	1%	C	-	Mortality between 20 and 49%	-	[103]
Lamiaceae	Prostanthera nivea A. Cunn. Ex. Benth.	Leaf extract	1%	C	-	Less than 20% of mortality	-	[103]
Lamiaceae	Prostanthera rotundifolia R.Br.	Leaf extract	1%	C	-	Less than 20% of mortality	-	[103]

Table A1. Cont.

Family	Plant Species	Source	Concentration	Bioassay[a]	T. urticae Koch Stage	Effect on T. urticae Koch	Identified Compounds	Ref.
Lamiaceae	Prostanthera spinosa F. Muell.	Leaf extract	1%	C	-	Less than 20% of mortality	-	[103]
Lamiaceae	Prostanthera stricta R.T. Baker	Leaf extract	1%	C	-	Less than 20% of mortality	-	[103]
Lamiaceae	Rosmarinus officinalis L.	Essential oil	0.10, 0.15, 0.20, and 0.25%.	A,C	Adult females and eggs	Mortalities of 15%, 79%, 100% and 100% for females, respectively	1,8-cineole (26.7%), camphor (17.5%), α-pinene (18.6%), camphene (11.8%), myrcene (9%), bornyl acetate (4%), β-pinene (2.8%), humulene (0.5%), borneol (1.8%), β-caryophyllene (1.5%), linalool (1%), Verbennone (0.9%), α-terpineol (0.8%)	[53]
Lamiaceae	Rosmarinus officinalis L.	Essential oil	-	D	Larvae and adults	Mortalities of 61% (larvae) and 53% (adults)	-	[88]
Lamiaceae	Rosmarinus officinalis L.	Essential oil	10 mL/L	D	-	LD_{50}	1,8-cineole and α-pinene (mortalities of $88 \pm 4.8\%$ and $32 \pm 4.8\%$, respectively with each compound)	[104]
Lamiaceae	Rosmarinus officinalis L.	Essential oil	0.1%	G	Adult females	11.7% of mortality	-	[112]
Lamiaceae	Salvia desertorum	Aerial part extract	1%	G	Adult females	Mortality between 0 and 20%	-	[99]
Lamiaceae	Salvia fruticosa Mill.	Leaf extract	3.77 μL/L	J	Adult females	LD_{50}	-	[129]
Lamiaceae	Salvia officinalis L.	Essential oil	19×10^{-3} μL/mL of air	J	Adults	100% of mortality	-	[42]

Table A1. Cont.

Family	Plant Species	Source	Concentration	Bioassay[a]	T. urticae Koch Stage	Effect on T. urticae Koch	Identified Compounds	Ref.
Lamiaceae	Salvia officinalis L.	Essential oil	0.10%, 0.15%, 0.20% and 0.25%	A,C	Adult females and eggs	100% of female mortality in all concentrations	α-tujone (42.5), 1,8-cineole (10.3%), camphor (11%), α-pinene (6.7%), camphene (6.5%), β-tujone (6.6%), myrcene (1.4%), bornyl acetate (0.7%), β-pinene (3.4%), humulene (2.4%), viridiflorol (2.2%), borneol (2%), β-caryophyllene (1.5%), cymene (1%)	[53]
Lamiaceae	Salvia officinalis L.	Essential oil	63.7 µg/cm^3	C,J	Adults	LD$_{50}$	-	[68]
Lamiaceae	Salvia officinalis L.	Essential oil	-	D	Larvae and adults	Mortalities of 61% (larvae) and 57% (adults)	-	[88]
Lamiaceae	Salvia sclarea L.	Essential oil	19×10^{-3} µL/mL of air	J	Adults	61% of mortality	-	[42]
Lamiaceae	Salvia sclarea L.	Aerial part extract	1%	G	Adult females	Mortality between 0 and 20%	-	[99]
Lamiaceae	Salvia sclarea L.	Essential oil	0.1%	G	Adult females	71% of mortality	-	[112]
Lamiaceae	Salvia zvedenskyi E. Nikit.	Root extract	1%	G	Adult females	Mortality between 50 and 80%	-	[99]
Lamiaceae	Satureja sahendica Bornm.	Essential oil vapors	0.98 µL/L (females), 0.54 µL/L (eggs)	J	Eggs and adults	LD$_{50}$	-	[84]
Lamiaceae	Satureja sahendica Bornm.	Essential oil	0.54%	E	Adults	LD$_{50}$	-	[83]
Lamiaceae	Scutellaria mollis R. Br.	Leaf extract	1%	C	-	Mortality between 50 and 89%	-	[103]
Lamiaceae	Stachys Tschatkalensis Knorr.	Aerial part extract	1%	G	Adult females	Mortality between 50 and 80%	-	[99]
Lamiaceae	Teucrium racemosum R. Br.	Leaf extract	1%	C	-	Mortality between 20 and 49%	-	[103]
Lamiaceae	Thymbra capitata (L.) Cav.	Essential oil	-	D	Larvae and adults	Mortalities of 61% (larvae) and 52% (adults)	-	[88]
Lamiaceae	Thymus vulgaris L.	Essential oil	19×10^{-3} µL/mL of air	J	Adults	93% of mortality	-	[42]
Lamiaceae	Thymus vulgaris L.	Essential oil	22.7 µg/cm^3	C,J	Adults	LD$_{50}$	-	[68]
Lamiaceae	Thymus vulgaris L.	Essential oil	0.1%	G	Adult females	62.2% of mortality	-	[112]
Lamiaceae	Vitex lignum-vitae Schauer	Leaf extract	1%	C	-	Mortality between 50 and 89%	-	[103]
Lamiaceae	Viticipremna queenslandica Munir	Leaf extract	1%	C	-	Mortality between 90 and 99%	-	[103]

Table A1. *Cont.*

Family	Plant Species	Source	Concentration	Bioassay[a]	*T. urticae* Koch Stage	Effect on *T. urticae* Koch	Identified Compounds	Ref.
Lamiaceae	*Westringia eremicola* A. Cunn. Ex. Benth.	Leaf extract	1%	C	-	Less than 20% of mortality	-	[103]
Lamiaceae	*Westringia glabra* R.Br.	Leaf extract	1%	C	-	Less than 20% of mortality	-	[103]
Lamiaceae	*Westringia saxatilis* B.J. Conn	Leaf extract	1%	C	-	Less than 20% of mortality	-	[103]
Lamiaceae	*Westringia viminalis* B.J. Conn & Tozer	Leaf extract	1%	C	-	Less than 20% of mortality	-	[103]
Lauraceae	*Ziziphora clinopodioides* Lam.	Aerial part extract	1%	G	Adult females	Mortality between 0 and 20%	-	[99]
Lauraceae	*Cinnamomum zeylandicum* Blume	Essential oil	0.1%	G	Adult females	23.6% of mortality	-	[112]
Lauraceae	*Laurus nobilis* L.	Leaf extract	17–50%	A,B	Adult females	Mortalities of 66.11% and 69.72%, respectively	-	[60]
Lauraceae	*Laurus nobilis* L.	Essential oil	-	D	Larvae and adults	Mortalities of 63% (larvae) and 46% (adults)	-	[88]
Lauraceae	*Licaria puchury-major* (Mart.) Kostern.	Essential oil	30.8 µg/mL	J	Adult females	LD$_{50}$	safrole (38.8%), 1,8-cineole (21.7%)	[13]
Lilliaceae	*Convallaria majalis* L.	Root extract	1%	G	Adult females	Mortality between 0 and 20%	-	[99]
Malvaceae	*Abutilon theophasti* Medic.	Aerial part extract	1%	G	Adult females	Mortality between 50 and 80%	-	[99]
Malvaceae	*Corchorus sp.*	Whole plant extract	5000 ppm	G	Adults	Mortality between 61 and 80%	-	[71]
Malvaceae	*Corchorus sp.*	Whole plant extract	2500 ppm	G	Adults	Mortality between 61 and 80%	-	[71]
Malvaceae	*Hybiscus sp.*	Whole plant extract	5000 ppm	G	Adults	More than 80% of mortality	-	[71]
Malvaceae	*Hybiscus sp.*	Whole plant extract	2500 ppm	G	Adults	-	-	[71]
Malvaceae	*Malva pusilla* Sm.	Whole plant extract	1%	G	Adult females	Mortality between 50 and 80%	-	[99]
Meliaceae	*Azadirachta indica* A. Juss.	Commercial formulation	1%.	C,D	Adult females	97.5% of mortality	-	[37]
Meliaceae	*Azadirachta indica* A. Juss.	Leaf extract	5000 ppm	G	Adults	Mortality between 61 and 80%	-	[71]
Meliaceae	*Azadirachta indica* A. Juss.	Leaf extract	2500 ppm	G	Adults	Mortality between 61 and 80%	-	[71]
Meliaceae	*Melia azedarach* L.	Fruit extract	14–50%	A,B	Adult females	Mortalities of 74.57% and 76.45%, respectively	-	[60]
Meliaceae	*Melia azedarach* L.	Essential oil	-	D	Larvae and adults	Mortalities of 77% (larvae) and 75% (adults)	-	[88]
Meliaceae	*Melia azedarach* L.	Acetone and petroleum ether methanolic extracts	-	C	Larvae	Lethal and fecundity effects	-	[107]
Mimosaceae	*Prosopis chinensis* (Molina) Stuntz	Leaf extract	5000 ppm	G	Adults	More than 80% of mortality	-	[71]
Mimosaceae	*Prosopis chinensis* (Molina) Stuntz	Leaf extract	2500 ppm	G	Adults	Mortality between 61 and 80%	-	[71]

Table A1. Cont.

Family	Plant Species	Source	Concentration	Bioassay[a]	T. urticae Koch Stage	Effect on T. urticae Koch	Identified Compounds	Ref.
Mimosaceae	*Prosopis chinensis* (Molina) Stuntz	Stem extract	5000 ppm	G	Adults	Mortality between 61 and 80%	-	[71]
Mimosaceae	*Prosopis chinensis* (Molina) Stuntz	Stem extract	2500 ppm	G	Adults	Mortality between 40 and 60%	-	[71]
Mimosaceae	*Prosopis chinensis* (Molina) Stuntz	Fruit extract	5000 ppm	G	Adults	More than 80% of mortality	-	[71]
Mimosaceae	*Prosopis chinensis* (Molina) Stuntz	Fruit extract	2500 ppm	G	Adults	Mortality between 61 and 80%	-	[71]
Myrtaceae	*Callistemon viminalis* (Sol. ex Gaertn.) G. Don	Leaf extract	40.66 µL/L air	A	Adult females	LD_{50}	1,8-cineole (71.77%)	[55]
Myrtaceae	*Eucalyptus camaldulensis* Dehnh.	Leaf extract	18–50%	A,B	Adult females	Mortalities of 62.61% and 55.57%, respectively	-	[60]
Myrtaceae	*Eucalyptus camaldulensis* Dehnh.	Flower extract	20–50%	A,B	Adult females	Mortalities of 51.91% and 47.15%, respectively	-	[60]
Myrtaceae	*Eucalyptus citriodora* Hook	Essential oil	19×10^{-3} µL/mL of air	J	Adults	100% of mortality	-	[42]
Myrtaceae	*Eucalyptus citriodora* Hook	Essential oil	19.3 µg/cm^3	J	Adults	LD_{50}	-	[68]
Myrtaceae	*Eucalyptus ghomphocephala* A. Cunn. Ex. DC.	Essential oil	-	D	Larvae and adults	Mortalities of 60% (larvae) and 34% (adults)	-	[88]
Myrtaceae	*Eucalyptus globulus* Labill.	Essential oil	19×10^{-3} µL/mL of air	J	Adults	89% of mortality	-	[42]
Myrtaceae	*Eucalyptus globulus* Labill.	Essential oil	0.1%	G	Adult females	19.7% of mortality	-	[112]
Myrtaceae	*Eucalyptus microtheca* F. Muell.	Essential oil vapors from fruits and leaves	1.52 µL/L (females), 5.7 µL/L (eggs)	J	Eggs and adults	LD_{50}	-	[84]
Myrtaceae	*Eucalyptus microtheca* F. Muell.	Fruits and leaves essential oils	0.56% (leaves), 2.36% (fruits)	E	Adults	LD_{50}	-	[83]
Myrtaceae	*Eucalyptus oleosa* L.	Essential oil	2.42 µL/L air	J	Adult females	LD_{50}	1,8-Cineole (31.96%), α-pinene (15.25%), trans-anethole (7.32%)	[132]
Myrtaceae	*Eucalyptus radiata* Sieber ex. DC.	Essential oil	0.1%	G	Adult females	27.9% of mortality	-	[112]
Myrtaceae	*Eucalyptus torquata* L.	Essential oil	3.59 µL/L air	J	Adult females	LD_{50}	1,8-cineole (28.57%), α-pinene (15.74%), globulol (13.11%)	[132]
Myrtaceae	*Eucalyptus* sp.	Essential oil	2.18–7.33%	C	Adults and eggs	LD_{50}	-	[89]
Myrtaceae	*Eugenia caryophyllata* Thunb.	Bud and leaf essential oils	19×10^{-3} µL/mL of air	J	Adults	Mortalities of 80% (buds) and 66% (leaves)	-	[42]

Table A1. Cont.

Family	Plant Species	Source	Concentration	Bioassay[a]	T. urticae Koch Stage	Effect on T. urticae Koch	Identified Compounds	Ref.
Myrtaceae	Eugenia caryophyllata Thunb.	Essential oil	23.6 µg/cm^3	CJ	Adults	LD$_{50}$	-	[68]
Myrtaceae	Melaleuca alternifolia Maiden & Betche ex. Cheel	Essential oil	0.1%	G	Adult females	28.6% of mortality	-	[112]
Myrtaceae	Melaleuca leucadendron L.	Essential oil	0.1%	G	Adult females	23.5% of mortality	-	[112]
Myrtaceae	Melaleuca viridiflora Sol. Ex. Gaertn.	Essential oil	0.1%	G	Adult females	26.8% of mortality	-	[112]
Myrtaceae	Myrtus communis L.	Essential oil	-	D	Larvae and adults	Mortalities of 82% (larvae) and 47% (adults)	-	[88]
Myrtaceae	Pimenta racemosa (Mill.) J.W. Moore	Essential oil	19×10^{-3} µL/mL of air	J	Adults	60% of mortality	-	[42]
Myrtaceae	Syzygium aromaticum (L.) Merr. & L.M. Perry	Essential oil from flower buds	6.13 µL/L air	CJ	Adult females	LD$_{50}$	eugenol (78.5%), β-caryophyllene (13.8%)	[85]
Myrtaceae	Syzygium aromaticum (L.) Merr. & L.M. Perry	Essential oil	0.1%	G	Adult females	41.3% of mortality	-	[112]
Myrtaceae	Syzygium cumini (L.) Skeels	Ethanolic, hexane and ether ethyl acetate extracts	75, 150 and 300 µg/mL	C	Adult females	Mortalities of 98.5% (ethanolic extract), 94% (hexane extract) and 90% (ether-ethyl acetate extract)	-	[108]
Nitrariaceae	Peganum harmala L.	Essential oil	-	D	Larvae and adults	Mortalities of 34% (larvae) and 12% (adults)	-	[88]
Nitrariaceae	Peganum harmala L.	Aerial part extract	1%	G	Adult females	Mortality between 20 and 50%	-	[99]
Nyctaginaceae	Bougainvilleae spectabilis Willd.	Leaf extract	5000 ppm	G	Adults	Mortality between 61 and 80%	-	[71]
Nyctaginaceae	Bougainvilleae spectabilis Willd.	Leaf extract	2500 ppm	G	Adults	Mortality between 61 and 80%	-	[71]
Papaveraceae	Chelidonium majus L.	Aerial part extract	1%	G	Adult females	Mortality between 0 and 20%	-	[99]
Papaveraceae	Papaver paeoninum Schrenk.	Aerial part extract	1%	G	Adult females	Mortality between 0 and 20%	-	[99]
Papaveraceae	Papaver rhoeas L.	Essential oil	-	D	Larvae and adults	Mortalities of 43% (larvae) and 34% (adults)	-	[88]
Papaveraceae	Papaver rhoeas L.	Aerial part extract	1%	G	Adult females	Mortality between 80 and 100%	-	[99]
Papaveraceae	Roemeria refracta DC.	Aerial part extract	1%	G	Adult females	Mortality between 50 and 80%	-	[99]
Pinaceae	Cedrus atlantica (Endl.) Manetti ex. Carrière	Essential oil	0.1%	G	Adult females	12.4% of mortality	-	[112]
Pinaceae	Picea schrenkiana Fisch & Mey.	Leaf extract	1%	G	Adult females	Mortality between 50 and 80%	-	[99]
Pinaceae	Pinus sylvestris L.	Essential oil	0.1%	G	Adult females	50.4% of mortality	-	[112]

Table A1. Cont.

Family	Plant Species	Source	Concentration	Bioassay[a]	T. urticae Koch Stage	Effect on T. urticae Koch	Identified Compounds	Ref.
Piperaceae	Piper aduncum L.	Leaf essential oil compounds	-	J	-	Mortality effect [(E)-nerolidol, α–Humulene and β-caryophyllene] and repellence (β-caryophyllene)	(E)-nerolidol, α–Humulene and β-caryophyllene	[109]
Piperaceae	Piper nigrum L.	Essential oil	0.1%	G	Adult females	22.8% of mortality	-	[112]
Plantaginaceae	Globularia alypum L.	Essential oil	-	D	Larvae and adults	Mortalities of 8% (larvae) and 2% (adults)	-	[88]
Plantaginaceae	Plantago major L.	Aerial part extract	1%	G	Adult females	Mortality between 80 nd 100%	-	[99]
Plumbaginaceae	Limonium tianschanicum Lincz.	Aerial part extract	1%	G	Adult females	Mortality between 20 and 50%	-	[99]
Polygonaceae	Polygonum aviculare L.	Aerial part extract	1%	G	Adult females	Mortality between 20 and 50%	-	[99]
Polygonaceae	Polygonum toktoguilicum Lazkov	Root extract	1%	G	Adult females	Mortality between 0 and 20%	-	[99]
Polygonaceae	Rumex acetosa L.	Aerial part extract	1%	G	Adult females	Mortality between 20 and 50%	-	[99]
Primulaceae	Anagallis arvensis L.	Aerial part extract	1%	G	Adult females	Mortality between 50 and 80%	-	[99]
Ranunculaceae	Aconitum soongaricum Stapf	Aerial part extract	1%	G	Adult females	Mortality between 50 and 80%	-	[99]
Ranunculaceae	Adonis parviflora Fisch.	Aerial part extract	1%	G	Adult females	Mortality between 0 and 20%	-	[99]
Ranunculaceae	Ceratocephallus testiculata (Crantz.) Bess.	Aerial part extract	1%	G	Adult females	Mortality between 0 and 20%	-	[99]
Ranunculaceae	Clematis orientalis L.	Seed extract	1%	G	Adult females	Mortality between 50 and 80%	-	[99]
Ranunculaceae	Clematis songarica Bge.	Seed extract	1%	G	Adult females	Mortality between 20 and 50%	-	[99]
Ranunculaceae	Nigella sativum L.	Seed extract	708.57 ppm	G	Adult females	LD$_{50}$	-	[91]
Ranunculaceae	Ranunculus polyanthemus L.	Aerial part extract	1%	G	Adult females	Mortality between 0 and 20%	-	[99]
Rosaceae	Geum urbanum L.	Aerial part extract	1%	G	Adult females	Mortality between 0 and 20%	-	[99]
Rosaceae	Padus avium Mill.	Leaf extract	1%	G	Adult females	Mortality between 0 and 20%	-	[99]
Rosaceae	Prunus laurocerasus L.	Leaves, flower and seed extract	12% v/w	A,C	Eggs and adult females	Mortality between 37 and 100%	-	[54]
Rubiaceae	Boirerrio radiata	Whole plant extract	5000 ppm	G	Adults	Mortality between 40 and 60%	-	[71]
Rubiaceae	Boirerrio radiata	Whole plant extract	2500 ppm	G	Adults	Mortality between 40 and 60%	-	[71]
Rubiaceae	Galium verum L.	Aerial part extract	1%	G	Adult females	Mortality between 50 and 80%	-	[99]
Rubiaceae	Gardenia jasminoides J. Ellis	Fruits extract	10000 ppm	I,J	Adult females and nymphs	49 and 66% of mortality, respectively	-	[133]
Rutaceae	Citrus aurantium L.	Essential oil	19×10^{-3} μL/mL of air	J	Adults	68% of mortality	-	[42]
Rutaceae	Citrus aurantium L.	Essential oil	-	D	Larvae and adults	Mortalities of 63% (larvae) and 55% (adults)	-	[88]
Rutaceae	Citrus aurantium L.	Fruit epicarp essential oil	1%	I,J	Adult females	Repellent effect due to all 27 identified compounds	d-limonene	[67]

Table A1. Cont.

Family	Plant Species	Source	Concentration	Bioassay[a]	T. urticae Koch Stage	Effect on T. urticae Koch	Identified Compounds	Ref.
Rutaceae	Citrus aurantium L. var. Armara	Essential oil	0.1%	G	Adult females	21.4% of mortality	-	[112]
Rutaceae	Citrus bergamia Risso & Poit.	Essential oil	19×10^{-3} µL/mL of air	J	Adults	87% of mortality	-	[42]
Rutaceae	Citrus bergamia Risso & Poit.	Essential oil	0.1%	G	Adult females	11% of mortality	-	[112]
Rutaceae	Citrus limon (L.) Burm. F.	Essential oil	0.1%	G	Adult females	34.9% of mortality	-	[112]
Rutaceae	Citrus paradisi Macfad	Essential oil	6.96 µL/L air	A	Adult females	LD_{50}	limonene (74.29%)	[55]
Rutaceae	Citrus paradisi Macfad.	Essential oil	0.1%	G	Adult females	30.6% of mortality	-	[112]
Rutaceae	Citrus sinensis Osbeck	Essential oil	19×10^{-3} µL/mL of air	J	Adults	61% of mortality	-	[42]
Rutaceae	Citrus sinensis Osbeck	Fruit epicarp essential oil	1%	I,J	Adult females	Repellent effect due to all 27 identified compounds	d-limonene	[67]
Rutaceae	Citrus sinensis Osbeck	Essential oil	0.1%	G	Adult females	45.6% of mortality	-	[112]
Rutaceae	Haplophyllum tuberculatum (Forssk.) A. Juss.	Essential oil	-	D	Larvae and adults	Mortalities of 94% (larvae) and 93% (adults)	-	[88]
Rutaceae	Ruta chalepensis L.	Essential oil	-	D	Larvae and adults	Mortalities of 66% (larvae) and 61% (adults)	-	[88]
Rutaceae	Zanthoxylum armatum DC.	Leaf extract	5000 and 10000 ppm	G	Adults	Mortalities of 36% and 39%, respectively	-	[111]
Santalaceae	Santalum sp.	Essential oil	0.1%	G	Adult females	87.2% of mortality and fecundity decrease	-	[112]
Scrophulariaceae	Calceolaria andina Benth	Two extract compounds	80 and 30 ppm	G	-	LD_{50}	2-(1,1-dimethylprop-2-enyl)-3-hydroxi-1,4-naphthoquinone and 2-acetoxy-3-(1,1-dimethylprop-2-enyl)-1,4-naphthoquinone	[113]
Scrophulariaceae	Verbascum thapsus L.	Aerial part extract	1%	G	Adult females	Mortality between 20 and 50%	-	[99]
Simarubaceae	Ailanthus altissima (Mill.) Swingle	Leaf extract	1%	G	Adult females	Mortality between 80 and 100%	-	[99]
Simarubaceae	Quassia sp.	Aerial part extract	10000 ppm and 47 ppm (chaparinone)	C	-	LD_{50} (chaparinone)	Chaparinone quasinoid	[114]
Solanaceae	Capsicum annuum L.	Aerial part extract	1%	G	Adult females	Mortality between 80 and 100%	-	[99]
Solanaceae	Capsicum annuum L.	Fruit extract	-	B,J	Adult females	45% of mortality	-	[116]
Solanaceae	Capsicum baccatum L.	Fruit extract	-	B,J	Adult females	Repellent effect	-	[116]
Solanaceae	Capsicum chinense Jacq.	Fruit extract	-	B,J	Adult females	Repellent effect	-	[116]
Solanaceae	Capsicum frutescens L.	Fruit extract	-	B,J	Adult females	Repellent effect	-	[116]

Table A1. Cont.

Family	Plant Species	Source	Concentration	Bioassay[a]	T. urticae Koch Stage	Effect on T. urticae Koch	Identified Compounds	Ref.
Solanaceae	Datura stramonium L.	Seed and leaf extracts	167.25 (leaves) and 145.75 (seeds) mg/L	C,D	Adults	Mortalities of 98% (leaves) and 25% (seeds)	-	[5]
Solanaceae	Hyoscyamus niger L.	Aerial part extract	1%	G	Adult females	Mortality between 50 and 80%.	-	[99]
Solanaceae	Lycopersicon hirsutum Dunal	-	-	J	Adult females	Repellent activity	Dihidrofarnesoic acid	[115]
Solanaceae	Solanum nigrum L.	Flower and leaf extracts	12–50%	A,B	Adult females	Mortalities of 69.88% and 79.36%, respectively	-	[60]
Solanaceae	Solanum nigrum L.	Fruit extract	19–50%	A,B	Adult females	Mortalities of 68.78% and 53.29%, respectively	-	[60]
Solanaceae	Solanum nigrum L.	Leaf extract	1%	G	Adult females	Mortality between 20 and 50%	-	[99]
Solanaceae	Solanum nigrum L.	Leaf extract	279.69 µg/mL	C,G	Adult females	LD_{50} after 72 h	-	[134]
Sterculiaceae	Waltheria indica L.	Whole plant extract	5000 ppm	G	Adults	Mortality between 61 and 80%	-	[71]
Sterculiaceae	Waltheria indica L.	Whole plant extract	2500 ppm	G	Adults	Mortalities of 64.11% and 73.25%, respectively	-	[71]
Styracaceae	Styrax officinalis L.	Seed cover extract	16–50%	A,B	Adult females	Mortalities of 68.17% and 31.28%, respectively	-	[60]
Styracaceae	Styrax officinalis L.	Seed extract	21–50%	A,B	Adult females		-	[60]
Umbelliferae	Angelica tschimganica (Korov.) B. Tikhom.	Aerial part extract	1%	G	Adult females	Mortality between 20 and 50%	-	[99]
Umbelliferae	Conium maculatum L.	Seed extract	1%	G	Adult females	Mortality between 20 and 50%	-	[99]
Umbelliferae	Dorema microcarpum Korov.	Aerial part extract	1%	G	Adult females	Mortality between 20 and 50%	-	[99]
Umbelliferae	Ferula foetida (Bunge) Regel.	Root extract	1%	G	Adult females	Mortality between 0 and 20%	-	[99]
Umbelliferae	Ferula foetidissima Regel. & Schmahl.	Root extract	1%	G	Adult females	Mortality between 20 and 50%	-	[99]
Umbelliferae	Ferula inciso-serrata M. Pimen. & Baranova	Aerial part and root extracts	1%	G	Adult females	Mortality between 0 and 20%	-	[99]
Umbelliferae	Heracleum dissectum Ledeb.	Aerial part extract	1%	G	Adult females	Mortality between 20 and 50%	-	[99]
Umbelliferae	Mediasia macrophylla (Regel. & Schmahl.) M. Pimen.	Aerial part extract	1%	G	Adult females	Mortality between 50 and 80%	-	[99]
Umbelliferae	Prangos lipskyi Korov	Root extract	1%	G	Adult females	Mortality between 80 and 100%	-	[99]
Urticaceae	Urtica pilulifera L.	Essential oil	-	D	Larvae and adults	Mortalities of 49% (larvae) and 46% (adults)	-	[88]
Valerianaceae	Valeriana officinalis L.	Aerial part extract	1%	G	Adult females	Mortality between 20 and 50%.	-	[99]
Verbenaceae	Lantana camara L.	Essential oil	-	D	Larvae and adults	Mortalities of 2% (larvae) and 1% (adults)	-	[88]
Verbenaceae	Lippia origanoides H.B.K.	Essential oil	25.1 µg/mL	J	Adult females	LD_{50}	carvacrol (48.31%), p-cymene (9.11%), thymol (8.78%)	[135]

Table A1. Cont.

Family	Plant Species	Source	Concentration	Bioassay[a]	T. urticae Koch Stage	Effect on T. urticae Koch	Identified Compounds	Ref.
Verbenaceae	Lippia sidoides Cham.	Essential oil	0.01 µL/L (extract). 0.001 µL/L (thymol). 3.02 µL/L (p-cymene). 0.08 µL/L (β-caryophyllene) and 0.036 µL/L (carvacrol)	J	Adult females	LD_{50}	Thymol, p-cymene, β-caryophyllene and carvacrol	[117]
Zingiberaceae	Elettaria cardamomum (L.) Maton	Essential oil	19×10^{-3} µL/mL of air	J	Adults	87% of mortality	-	[42]
Zingiberaceae	Zingiber officinale Rosc.	Essential oil	0.1%	G	Adult females	11.9% of mortality	-	[112]
Zygophyllaceae	Balanites aegyptiaca (L.) Delile	Leaf extract	5000 ppm	G	Adults	More than 80% of mortality	-	[71]
Zygophyllaceae	Balanites aegyptiaca (L.) Delile	Leaf extract	2500 ppm	G	Adults	Mortality between 61 and 80%	-	[71]

[a] This column includes the bioassays used in each study, which are coded as follows: A = slide dip, B = petri dish, C = leaf disc direct (LDD); D = leaf disc residue (LDR); E = leaf disc direct potter tower (LDD-PT); F = leaf disc residue potter tower (DR-PTM); G = leaf disc residue dipping (LDR-D); H = leaf absorption test; I = whole plant direct; J = filter paper diffussion (fumigant).

References

1. Kumari, S.; Chauhan, U.; Kumari, A.; Nadda, G. Comparative toxicities of novel and conventional acaricides against different stages of *Tetranychus urticae* Koch (Acarina: Tetranychidae). *J. Saudi Soc. Agric. Sci.* **2017**, *16*, 191–196. [CrossRef]
2. Van Leeuwen, T.; Tirry, L.; Yamamoto, A.; Nauen, R.; Dermauw, W. The economic importance of acaricides in the control of phytophagous mites and an update on recent acaricide mode of action research. *Pestic. Biochem. Physiol.* **2015**, *121*, 12–21. [CrossRef] [PubMed]
3. Landeros, J.; Ail, C.E.; Cerna, E.; Ochoa, Y.; Guevara, L.; Aguirre, L.A. Susceptibility and resistance mechanisms of *Tetranychus urticae* (Acariformes: Tetranychidae) in greenhouse roses. *Rev. Colomb. Entomol.* **2010**, *36*, 5–9.
4. Hoy, M.A. *Agricultural Acarology: Introduction to Integrated Mite Management*; CRC Press: Boca Ratón, FL, USA, 2011.
5. Kumral, N.A.; Çobanoğlu, S.; Yalcin, C. Acaricidal, repellent and oviposition deterrent activities of *Datura stramonium* L. against adult *Tetranychus urticae* (Koch). *J. Pest Sci.* **2009**, *83*, 173–180. [CrossRef]
6. Breeuwer, J.A.J.; Jacobs, G. Wolbachia: Intracellular manipulators of mite reproduction. *Exp. Appl. Acarol.* **1996**, *20*, 421–434. [CrossRef] [PubMed]
7. Meena, N.K.; Rampal; Barman, D.; Medhi, R.P. Biology and seasonal abundance of the two-spotted spider mite, *Tetranychus urticae*, on orchids and rose. *Phytoparasitica* **2013**, *41*, 597–609. [CrossRef]
8. Tehri, K. A review on reproductive strategies in two spotted spider mite, *Tetranychus urticae* Koch (Acari: Tetranychidae). *J. Entomol. Zool. Stud.* **2014**, *2*, 35–39.
9. Grbić, M.; Van Leeuwen, T.; Clark, R.M.; Rombauts, S.; Rouzé, P.; Grbić, V.; Osborne, E.J.; Dermauw, W.; Ngoc, P.C.T.; Ortego, F.; et al. The genome of *Tetranychus urticae* reveals herbivorous pest adaptations. *Nature* **2011**, *479*, 487–492. [CrossRef]
10. Reis, P.R.; Silva, E.A.; Zacarias, M.S. Controle biológico de ácaros em cultivos protegidos. *Inf. Agropecuário* **2005**, *26*, 58–68.
11. Mohankumar, S.; Balakrishnan, N.; Samiyappan, R. Biotechnological and molecular approaches in the management of non-insect pests of crop plants. In *Integrated Pest Management*; Elsevier: Amsterdam, The Netherlands, 2014; pp. 337–369.
12. İnak, E.; Alpkent, Y.N.; Çobanoğlu, S.; Dermauw, W.; Van Leeuwen, T. Resistance incidence and presence of resistance mutations in populations of *Tetranychus urticae* from vegetable crops in Turkey. *Exp. Appl. Acarol.* **2019**, *78*, 343–360. [CrossRef]
13. Demaeght, P.; Osborne, E.J.; Odman-Naresh, J.; Grbić, M.; Nauen, R.; Merzendorfer, H.; Clark, R.M.; Van Leeuwen, T. High resolution genetic mapping uncovers chitin synthase-1 as the target-site of the structurally diverse mite growth inhibitors clofentezine, hexythiazox and etoxazole in *Tetranychus urticae*. *Insect Biochem. Mol. Biol.* **2014**, *51*, 52–61. [CrossRef] [PubMed]
14. Ferreira, C.B.S.; Andrade, F.H.N.; Rodrigues, A.R.S.; Siqueira, H.A.A.; Gondim, M.G.C. Resistance in field populations of *Tetranychus urticae* to acaricides and characterization of the inheritance of abamectin resistance. *Crop Prot.* **2015**, *67*, 77–83. [CrossRef]
15. Snyder, M.J.; Glendinning, J.I. Causal connection between detoxification enzyme activity and consumption of a toxic plant compound. *J. Comp. Physiol. A* **1996**, *179*, 255–261. [CrossRef] [PubMed]
16. Stumpf, N.; Nauen, R. Biochemical markers linked to abamectin resistance in *Tetranychus urticae* (Acari: Tetranychidae). *Pestic. Biochem. Physiol.* **2002**, *72*, 111–121. [CrossRef]
17. Pavlidi, N.; Tseliou, V.; Riga, M.; Nauen, R.; Van Leeuwen, T.; Labrou, N.E.; Vontas, J. Functional characterization of glutathione S-transferases associated with insecticide resistance in *Tetranychus urticae*. *Pestic. Biochem. Physiol.* **2015**, *121*, 53–60. [CrossRef] [PubMed]
18. Pavlidi, N.; Khalighi, M.; Myridakis, A.; Dermauw, W.; Wybouw, N.; Tsakireli, D.; Stephanou, E.G.; Labrou, N.E.; Vontas, J.; Van Leeuwen, T. A glutathione-S-transferase (TuGSTd05) associated with acaricide resistance in *Tetranychus urticae* directly metabolizes the complex II inhibitor cyflumetofen. *Insect Biochem. Mol. Biol.* **2017**, *80*, 101–115. [CrossRef] [PubMed]
19. Merzendorfer, H. ABC transporters and their role in protecting insects from pesticides and their metabolites. In *Advances in Insect Physiology*; Elsevier: Amsterdam, The Netherlands, 2014; pp. 1–72.

20. Dermauw, W.; Van Leeuwen, T. The ABC gene family in arthropods: Comparative genomics and role in insecticide transport and resistance. *Insect Biochem. Mol. Biol.* **2014**, *45*, 89–110. [CrossRef] [PubMed]
21. Dermauw, W.; Wybouw, N.; Rombauts, S.; Menten, B.; Vontas, J.; Grbic, M.; Clark, R.M.; Feyereisen, R.; Van Leeuwen, T. A link between host plant adaptation and pesticide resistance in the polyphagous spider mite *Tetranychus urticae*. *Proc. Natl. Acad. Sci.* **2012**, *110*, 113–122. [CrossRef] [PubMed]
22. Van Leeuwen, T.; Dermauw, W.; Grbic, M.; Tirry, L.; Feyereisen, R. Spider mite control and resistance management: Does a genome help? *Pest Manag. Sci.* **2012**, *69*, 156–159. [CrossRef] [PubMed]
23. Van Leeuwen, T.; Vontas, J.; Tsagkarakou, A.; Dermauw, W.; Tirry, L. Acaricide resistance mechanisms in the two-spotted spider mite *Tetranychus urticae* and other important Acari: A review. *Insect Biochem. Mol. Biol.* **2010**, *40*, 563–572. [CrossRef] [PubMed]
24. Demaeght, P.; Dermauw, W.; Tsakireli, D.; Khajehali, J.; Nauen, R.; Tirry, L.; Vontas, J.; Lümmen, P.; Van Leeuwen, T. Molecular analysis of resistance to acaricidal spirocyclic tetronic acids in *Tetranychus urticae*: CYP392E10 metabolizes spirodiclofen, but not its corresponding enol. *Insect Biochem. Mol. Biol.* **2013**, *43*, 544–554. [CrossRef] [PubMed]
25. Kwon, D.H.; Choi, J.Y.; Je, Y.H.; Lee, S.H. The overexpression of acetylcholinesterase compensates for the reduced catalytic activity caused by resistance-conferring mutations in *Tetranychus urticae*. *Insect Biochem. Mol. Biol.* **2012**, *42*, 212–219. [CrossRef] [PubMed]
26. Flood, J.; Day, R. Managing risks from pests in global commodity networks–policy perspectives. *Food Secur.* **2016**, *8*, 89–101. [CrossRef]
27. Pimentel, D. Environmental and economic costs of the application of pesticides primarily in the united states. *Environ. Dev. Sustain.* **2005**, *7*, 229–252. [CrossRef]
28. Price, J.F.; Legard, D.E.; Chandler, C.K. Two-spotted spider mite resistance to abamectin miticide on strawberry and strategies for resistance management. *Acta Hortic.* **2002**, 683–685. [CrossRef]
29. García-Marí, F.; Gonzalez-Zamora, J.E. Biological control of *Tetranychus urticae* (Acari: Tetranychidae) with naturally occurring predators in strawberry plantings in Valencia, Spain. *Exp. Appl. Acarol.* **1999**, *23*, 487–495. [CrossRef]
30. Leite, L.G.; Smith, L.; Moraes, G.J.; Roberts, D.W. In vitro production of hyphal bodies of the mite pathogenic fungus *Neozygites floridana*. *Mycologia* **2000**, *92*, 201–207. [CrossRef]
31. Isman, M.B. Botanical insecticides, deterrents, and repellents in modern agriculture and an increasingly regulated world. *Annu. Rev. Entomol.* **2006**, *51*, 45–66. [CrossRef]
32. Attia, S.; Grissa, K.L.; Lognay, G.; Bitume, E.; Hance, T.; Mailleux, A.C. A review of the major biological approaches to control the worldwide pest *Tetranychus urticae* (Acari: Tetranychidae) with special reference to natural pesticides. *J. Pest Sci.* **2013**, *86*, 361–386. [CrossRef]
33. Monsreal-Ceballos, R.J.; Ruiz-Sánchez, E.; Sánchez-Borja, M.; Ballina-Gómez, H.S.; González-Moreno, A.; Reyes-Ramírez, A. Effects of commercial botanical insecticides in *Tamarixia radiata*, an ectoparasitoid of *Diaphorina citri*. *Ecosistemas y Recur. Agropecu.* **2017**, *4*, 589. [CrossRef]
34. Duso, C.; Malagnini, V.; Pozzebon, A.; Castagnoli, M.; Liguori, M.; Simoni, S. Comparative toxicity of botanical and reduced-risk insecticides to Mediterranean populations of *Tetranychus urticae* and *Phytoseiulus persimilis* (Acari Tetranychidae, Phytoseiidae). *Biol. Control* **2008**, *47*, 16–21. [CrossRef]
35. Spollen, K.M.; Isman, M.B. Acute and sublethal effects of a Neem insecticide on the commercial biological control agents *Phytoseiulus persimilis* and *Amblyseius cucumeris* (Acari: Phytoseiidae) and *Aphidoletes aphidimyza* (Diptera: Cecidomyiidae). *J. Econ. Entomol.* **1996**, *89*, 1379–1386. [CrossRef]
36. Bernardi, D.; Botton, M.; da Cunha, U.S.; Bernardi, O.; Malausa, T.; Garcia, M.S.; Nava, D.E. Effects of azadirachtin on *Tetranychus urticae* (Acari: Tetranychidae) and its compatibility with predatory mites (Acari: Phytoseiidae) on strawberry. *Pest Manag. Sci.* **2013**, *69*, 75–80. [CrossRef] [PubMed]
37. Brito, H.M.; Gondim, M.G.C.G., Jr.; de Oliveira, J.V.; da Câmara, C.A.G. Toxicidade de formulações de nim (*Azadirachta indica* A. Juss.) ao ácaro-rajado e a *Euseius alatus* De Leon e *Phytoseiulus macropilis* (Banks) (Acari: Phytoseiidae). *Neotrop. Entomol.* **2006**, *35*, 500–505. [CrossRef] [PubMed]
38. Yanar, D. Side effects of different doses of azadirachtin on predatory mite *Metaseiulus occidentalis* (Nesbitt) (acari: Phytoseiidae) under laboratory conditions. *Appl. Ecol. Environ. Res.* **2019**, *17*, 3433–3440. [CrossRef]
39. El-Sharabasy, H.M. Acaricidal activities of *Artemisia judaica* L. extracts against *Tetranychus urticae* Koch and its predator *Phytoseiulus persimilis* Athias-Henriot (Tetranychidae: Phytoseiidae). *J. Biopestic.* **2010**, *3*, 514–519.

40. Lima, D.B.; Melo, J.W.S.; Guedes, N.M.P.; Gontijo, L.M.; Guedes, R.N.C.; Gondim, M.G.C., Jr. Bioinsecticide-predator interactions: Azadirachtin behavioral and reproductive impairment of the coconut mite predator *Neoseiulus baraki*. *PLoS ONE* **2015**, *10*, e0118343. [CrossRef]
41. Chiasson, H.; Bostanian, N.J.; Vincent, C. Acaricidal properties of a *Chenopodium*-Based Botanical. *J. Econ. Entomol.* **2004**, *97*, 1373–1377. [CrossRef]
42. Choi, W.-I.; Lee, S.-G.; Park, H.-M.; Ahn, Y.-J. Toxicity of plant essential oils to *Tetranychus urticae* (Acari: Tetranychidae) and *Phytoseiulus persimilis* (Acari: Phytoseiidae). *J. Econ. Entomol.* **2004**, *97*, 553–558. [CrossRef]
43. Attia, S.; Grissa, K.L.; Lognay, G.; Heuskin, S.; Mailleux, A.C.; Hance, T. Chemical composition and acaricidal properties of *Deverra scoparia* essential oil (Araliales: Apiaceae) and blends of its major constituents against *Tetranychus urticae* (Acari: Tetranychidae). *J. Econ. Entomol.* **2011**, *104*, 1220–1228. [CrossRef]
44. Pontes, W.J.T.; de Oliveira, J.C.S.; da Câmara, C.A.G.; Júnior, M.G.C.G.; de Oliveira, J.V.; Schwartz, M.O.E. Atividade acaricida dos óleos essencias de folhas e frutos de *Xylopia sericea* sobre o ácaro rajado (*Tetranychus urticae* Koch). *Quim. Nova* **2007**, *30*, 838. [CrossRef]
45. Chiasson, H.; Bélanger, A.; Bostanian, N.; Vincent, C.; Poliquin, A. Acaricidal properties of *Artemisia absinthium* and *Tanacetum vulgare* (Asteraceae) essential oils obtained by three methods of extraction. *J. Econ. Entomol.* **2001**, *94*, 167–171. [CrossRef] [PubMed]
46. Attia, S.; Grissa, K.L.; Mailleux, A.C.; Heuskin, S.; Lognay, G.; Hance, T. Acaricidal activities of *Santolina africana* and *Hertia cheirifolia* essential oils against the two-spotted spider mite (*Tetranychus urticae*). *Pest Manag. Sci.* **2012**, *68*, 1069–1076. [CrossRef] [PubMed]
47. Rassem, H.H.A.; Nour, A.H.; Yunus, R.M. Techniques for extraction of essential oils from plants: A review. *Aust. J. Basic Appl. Sci.* **2016**, *10*, 117–127.
48. Kabir, K.H.; Chapman, R.B.; Penman, D.R. Miticide bioassays with spider mites (Acari: Tetranychidae): Effect of test method, exposure period and mortality criterion on the precision of response estimates. *Exp. Appl. Acarol.* **1993**, *17*, 695–708. [CrossRef]
49. Walker, W.F.; Boswell, A.L.; Smith, F.F. Resistance of spider mites to acaricides: Comparison of slide dip and leaf dip methods. *J. Econ. Entomol.* **1973**, *66*, 549–550. [CrossRef]
50. Voss, G. Ein neues akarizid-austestungsverfahren für spinnmilben. *Anz. Schädlingskd* **1961**, *34*, 76–77. [CrossRef]
51. Dittrich, V. A comparative study of toxicological test methods on a population of the two-spotted spider mite (*Tetranychus telarius*). *J. Econ. Entomol.* **1962**, *55*, 644–648. [CrossRef]
52. Shi, G.L.; Zhao, L.L.; Liu, S.Q.; Cao, H.; Clarke, S.R.; Sun, J.H. Acaricidal activities of extracts of *Kochia scoparia* against *Tetranychus urticae*, *Tetranychus cinnabarinus*, and *Tetranychus viennensis* (Acari: Tetranychidae). *J. Econ. Entomol.* **2006**, *99*, 858–863. [CrossRef]
53. Laborda, R.; Manzano, I.; Gamón, M.; Gavidia, I.; Pérez-Bermúdez, P.; Boluda, R. Effects of *Rosmarinus officinalis* and *Salvia officinalis* essential oils on *Tetranychus urticae* Koch (Acari: Tetranychidae). *Ind. Crops Prod.* **2013**, *48*, 106–110. [CrossRef]
54. Akyazi, R.; Soysal, M.; Hassan, E. Toxic and repellent effects of *Prunus laurocerasus* L. (Rosaceae) extracts against *Tetranychus urticae* Koch (Acari: Tetranychidae). *Türk. Entomol. Derg.* **2015**, *39*, 367–380. [CrossRef]
55. Mahmoud, N.F.; Badawy, M.E.I.; Marei, A.E.-S.M.; Abdelgaleil, S.A.M. Acaricidal and antiacetylcholinesterase activities of essential oils from six plants growing in Egypt. *Int. J. Acarol.* **2019**, *45*, 245–251. [CrossRef]
56. Laborda, R.; Manzano, I.; Gamon, M.; Gavidia, I.; Boluda, R.; Perez-Bermudez, P. Spike lavender essential oil reduces the survival rate and fecundity of two-spotted spider mite, *Tetranychus urticae* (Acari: Tetranychidae). *J. Agric. Sci. Technol.* **2018**, *20*, 1013–1023.
57. Eldoksch, H.A.; Ayad, F.A.; El-Sebae, A.-K.H. Acaricidal activity of plant extracts and their main terpenoids on the two-spotted spider mite *Tetranychus Urticae* (Acari: Tetranychidae). *Alexandria Sci. Exch. J.* **2009**, *30*, 344–349.
58. Helle, W.; Overmeer, W. Toxicological test methods. In *Spider Mites. Their Biology, Natural Enemies and Control*.; Helle, W., Sabelis, M., Eds.; Elsevier: Amsterdam, The Netherlands, 1985; Volume 1, pp. 391–395.
59. Kabir, K.H.; Chapman, R.B. Operational and biological factors influencing responses of spider mites (acari: Tetranychidae) to propargite by using the petri dish-potter tower method. *J. Econ. Entomol.* **1997**, *90*, 272–277. [CrossRef]

60. Yanar, D.; Kadıoğlu, I.; Gökçe, A. Acaricidal effects of different plant parts extracts on two-spotted spider mite (*Tetranychus urticae* Koch). *Afr. J. Biotechnol.* **2011**, *10*, 11745–11750.
61. Numa, S.; Rodríguez, L.; Rodríguez, D.; Coy-Barrera, E. Susceptibility of *Tetranychus urticae* Koch to an ethanol extract of *Cnidoscolus aconitifolius* leaves under laboratory conditions. *Springerplus* **2015**, *4*, 338. [CrossRef]
62. Potter, C. An improved laboratory apparatus for applying direct sprays and surface films, with data on the electrostatic charge on atomized spray fluids. *Ann. Appl. Biol.* **1952**, *39*, 1–28. [CrossRef]
63. Bostanian, N.J.; Beudjekian, S.; McGregor, E.; Racette, G. A modified excised leaf disc method to estimate the toxicity of slow- and fast-acting reduced-risk acaricides to mites. *J. Econ. Entomol.* **2009**, *102*, 2084–2089. [CrossRef]
64. Ebadollahi, A.; Jalali-Sendi, J.; Razmjou, J. Toxicity and phytochemical profile of essential oil from Iranian *Achillea mellifolium* L. against *Tetranychus urticae* Koch (Acari: Tetranychidae). *Toxin Rev.* **2016**, *35*, 24–28. [CrossRef]
65. Pavela, R. Acaricidal properties of extracts and major furanochromenes from the seeds of *Ammi visnaga* Linn. against *Tetranychus urticae* Koch. *Ind. Crops Prod.* **2015**, *67*, 108–113. [CrossRef]
66. Pavela, R. Extract from the roots of *Saponaria officinalis* as a potential acaricide against *Tetranychus urticae*. *J. Pest Sci.* **2017**, *90*, 683–692. [CrossRef]
67. da Camara, C.A.G.; Akhtar, Y.; Isman, M.B.; Seffrin, R.C.; Born, F.S. Repellent activity of essential oils from two species of *Citrus* against *Tetranychus urticae* in the laboratory and greenhouse. *Crop Prot.* **2015**, *74*, 110–115. [CrossRef]
68. Han, J.; Choi, B.R.; Lee, S.G.; Il Kim, S.; Ahn, Y.J. Toxicity of plant essential oils to acaricide-susceptible and -resistant *Tetranychus urticae* (Acari: Tetranychidae) and *Neoseiulus californicus* (Acari: Phytoseiidae). *J. Econ. Entomol.* **2010**, *103*, 1293–1298. [CrossRef] [PubMed]
69. Barua, C.C.; Talukdar, A.; Begum, S.A.; Lahon, L.C.; Sarma, D.K.; Pathak, D.C.; Borah, P. Antinociceptive activity of methanolic extract of leaves of *Achyranthes aspera* Linn. (Amaranthaceae) in animal models of nociception. *Indian J. Exp. Biol.* **2010**, *48*, 817–821. [PubMed]
70. Monzote, L.; Stamberg, W.; Staniek, K.; Gille, L. Toxic effects of carvacrol, caryophyllene oxide, and ascaridole from essential oil of *Chenopodium ambrosioides* on mitochondria. *Toxicol. Appl. Pharmacol.* **2009**, *240*, 337–347. [CrossRef]
71. Hiremath, I.G.; Ahn, Y.J.; Kim, S.I.; Choi, B.R.; Cho, J.R. Insecticidal and acaricidal activities of african plant extracts against the brown planthopper and two-spotted spider mite. *Korean J. Appl. Entomol.* **1995**, *34*, 200–205.
72. Harder, M.J.; Tello, V.E.; Giliomee, J.H. The Acaricidal effect of ethanolic extracts of *Chenopodium quinoa* Willd. on *Tetranychus urticae* Koch (Acari: Tetranychidae). *Afr. Entomol.* **2016**, *24*, 50–60. [CrossRef]
73. Renard-Nozaki, J.; Kim, T.; Imakura, Y.; Kihara, M.; Kobayashi, S. Effect of alkaloids isolated from Amaryllidaceae on *Herpes simplex* virus. *Res. Virol.* **1989**, *140*, 115–128. [CrossRef]
74. Weniger, B.; Italiano, L.; Beck, J.-P.; Bastida, J.; Bergoñon, S.; Codina, C.; Lobstein, A.; Anton, R. Cytotoxic activity of Amaryllidaceae alkaloids. *Planta Med.* **1995**, *61*, 77–79. [CrossRef]
75. Ho, S.H.; Koh, L.; Ma, Y.; Huang, Y.; Sim, K.Y. The oil of garlic, *Allium sativum* L. (Amaryllidaceae), as a potential grain protectant against *Tribolium castaneum* (Herbst) and *Sitophilus zeamais* Motsch. *Postharvest Biol. Technol.* **1996**, *9*, 41–48. [CrossRef]
76. Abbassy, M.A.; El-Gougary, O.A.; El-Hamady, S.; Sholo, M.A. Insecticidal, acaricidal and synergistic effects of soosan, *Pancratium maritimum* extracts and constituents. *J. Egypt. Soc. Parasitol.* **1998**, *28*, 197–205. [PubMed]
77. Attia, S.; Grissa, K.L.; Mailleux, A.C.; Lognay, G.; Heuskin, S.; Mayoufi, S.; Hance, T. Effective concentrations of garlic distillate (*Allium sativum*) for the control of *Tetranychus urticae* (Tetranychidae). *J. Appl. Entomol.* **2012**, *136*, 302–312. [CrossRef]
78. Geng, S.; Chen, H.; Zhang, J.; Tu, H. Bioactivity of garlic-straw extracts against the spider mites, *Tetranychus urticae* and *T. viennensis*. *J. Agric. Urban Entomol.* **2014**, *30*, 38–48. [CrossRef]
79. Rabelo, S.V.; de Sousa Siqueira Quintans, J.; Costa, E.V.; da Silva Almeida, J.R.G.; Júnior, L.J.Q. Annona species (Annonaceae) oils. In *Essential Oils in Food Preservation, Flavor and Safety*; Elsevier: Amsterdam, The Netherlands, 2016; pp. 221–229.
80. Ohsawa, K.; Atsuzawa, S.; Mitsui, T.; Yamamoto, I. Isolation and insecticidal activity of three acetogenins from seeds of pond apple, *Annona glabra* L. *J. Pestic. Sci.* **1991**, *16*, 93–96. [CrossRef]

81. Tunçtürk, M.; Özgökçe, F. Chemical composition of some Apiaceae plants commonly used inherby cheese in Eastern Anatolia. *Turkish J. Agric. For.* **2015**, *39*, 55–62. [CrossRef]
82. Benelli, G.; Flamini, G.; Fiore, G.; Cioni, P.L.; Conti, B. Larvicidal and repellent activity of the essential oil of *Coriandrum sativum* L. (Apiaceae) fruits against the filariasis vector *Aedes albopictus* Skuse (Diptera: Culicidae). *Parasitol. Res.* **2013**, *112*, 1155–1161. [CrossRef] [PubMed]
83. Tsolakis, H.; Ragusa, S. Effects of a mixture of vegetable and essential oils and fatty acid potassium salts on *Tetranychus urticae* and *Phytoseiulus persimilis*. *Ecotoxicol. Environ. Saf.* **2008**, *70*, 276–282. [CrossRef]
84. Amizadeh, M.; Hejazi, M.J.; Saryazdi, G.A. Fumigant toxicity of some essential oils on *Tetranychus urticae* (Acari: Tetranychidae). *Int. J. Acarol.* **2013**, *39*, 285–289. [CrossRef]
85. Kheradmand, K.; Beynaghi, S.; Asgari, S.; Garjan, A.S. Toxicity and repellency effects of three plant essential oils against two-spotted spider mite, *Tetranychus urticae* (Acari: Tetranychidae). *J. Agr. Sci. Tech.* **2015**, *17*, 1223–1232.
86. Fatemikia, S.; Abbasipour, H.; Saeedizadeh, A. Phytochemical and acaricidal study of the Galbanum, *Ferula gumosa* Boiss. (Apiaceae) essential oil against *Tetranychus urticae* Koch (Tetranychidae). *J. Essent. Oil Bear. Plants* **2017**, *20*, 185–195. [CrossRef]
87. Benelli, G.; Pavela, R.; Canale, A.; Nicoletti, M.; Petrelli, R.; Cappellacci, L.; Galassi, R.; Maggi, F. Isofuranodiene and germacrone from *Smyrnium olusatrum* essential oil as acaricides and oviposition inhibitors against *Tetranychus urticae*: Impact of chemical stabilization of isofuranodiene by interaction with silver triflate. *J. Pest Sci.* **2017**, *90*, 693–699. [CrossRef]
88. Attia, S.; Grissa, K.L.; Ghrabi, Z.G.; Mailleux, A.C.; Lognay, G.; Hance, T. Acaricidal activity of 31 essential oils extracted from plants collected in Tunisia. *J. Essent. Oil Res.* **2012**, *24*, 279–288. [CrossRef]
89. Afify, A.E.-M.M.R.; Ali, F.S.; Turky, A.F. Control of *Tetranychus urticae* Koch by extracts of three essential oils of chamomile, marjoram and Eucalyptus. *Asian Pac. J. Trop. Biomed.* **2012**, *2*, 24–30. [CrossRef]
90. Aslan, I.; Kordali, S.; Çalmaşur, Ö. Toxicity of the vapours of *Artemisia absinthium* essential oils to *Tetranychus urticae* Koch and *Bemisia tabasi* (Genn.). *Fresenius Environ. Bull.* **2005**, *14*, 415–417.
91. Derbalah, A.S.; Keratrum, A.Y.; El-Dewy, M.E.; El-Shamy, E.H. Efficacy of some insecticides and plant extracts against *Tetranychus urticae* under laboratory conditions. *Egypt. J. Plant Prot. Res.* **2013**, *1*, 47–70.
92. Pavela, R.; Dall'acqua, S.; Sut, S.; Baldan, V.; Kamte, S.L.N.; Nya, P.C.B.; Cappellacci, L.; Petrelli, R.; Nicoletti, M.; Canale, A.; et al. Oviposition inhibitory activity of the Mexican sunflower *Tithonia diversifolia* (Asteraceae) polar extracts against the two-spotted spider mite *Tetranychus urticae* (Tetranychidae). *Physiol. Mol. Plant Pathol.* **2018**, *101*, 85–92. [CrossRef]
93. Sut, S.; Pavela, R.; Kolarčik, V.; Cappellacci, L.; Petrelli, R.; Maggi, F.; Dall'Acqua, S.; Benelli, G. Identification of *Onosma visianii* roots extract and purified shikonin derivatives as potential acaricidal agents against *Tetranychus urticae*. *Molecules* **2017**, *22*, E1002. [CrossRef]
94. Carretero, M.E.; López-Pérez, J.L.; Abad, M.J.; Bermejo, P.; Tillet, S.; Israel, A.; Noguera-P, B. Preliminary study of the anti-inflammatory activity of hexane extract and fractions from *Bursera simaruba* (Linneo) Sarg. (Burseraceae) leaves. *J. Ethnopharmacol.* **2008**, *116*, 11–15. [CrossRef]
95. Rama, K.S.; Chandrasekar, R.M.; Rani, S.; Pullaiah, T. Bioactive principles and biological properties of essential oils of burseraceae: A review. *J. Pharmacogn. Phytochem.* **2016**, *5*, 247–258.
96. Baratta, M.T.; Dorman, H.J.D.; Deans, S.G.; Figueiredo, A.C.; Barroso, J.G.; Ruberto, G. Antimicrobial and antioxidant properties of some commercial essential oils. *Flavour Fragr. J.* **1998**, *13*, 235–244. [CrossRef]
97. Pontes, W.J.T.; de Oliveira, J.C.S.; da Camara, C.A.G.; Lopes, A.C.H.R.; Gondim, M.G.C.; de Oliveira, J.V.; Schwartz, M.O.E. Composition and acaricidal activity of the resin's essential oil of *Protium bahianum* Daly against two spotted spider mite (*Tetranychus Urticae*). *J. Essent. Oil Res.* **2007**, *19*, 379–383. [CrossRef]
98. Górski, R.; Sobieralski, K.; Siwulski, M. The effect of hemp essential oil on mortality *Aulacorthum solani* Kalt. and *Tetranychus urticae* Koch. *Ecol. Chem. Eng. S.* **2016**, *23*, 505–511. [CrossRef]
99. Chermenskaya, T.D.; Stepanycheva, E.A.; Shchenikova, A.V.; Chakaeva, A.S. Insectoacaricidal and deterrent activities of extracts of Kyrgyzstan plants against three agricultural pests. *Ind. Crops Prod.* **2010**, *32*, 157–163. [CrossRef]
100. Le Dang, Q.; Choi, Y.H.; Choi, G.J.; Jang, K.S.; Park, M.S.; Park, N.-J.; Lim, C.H.; Kim, H.; Ngoc, L.H.; Kim, J.-C. Pesticidal activity of ingenane diterpenes isolated from *Euphorbia kansui* against *Nilaparvata lugens* and *Tetranychus urticae*. *J. Asia Pac. Entomol.* **2010**, *13*, 51–54. [CrossRef]

101. Islam, T.; Biswas, M.J.H.; Howlader, M.T.H.; Ullah, M.S. Laboratory evaluation of *Beauveria bassiana*, some plant oils and insect growth regulators against two-spotted spider mite, *Tetranychus urticae* Koch (Acari: Tetranychidae). *Persian, J. Acarol.* **2017**, *6*, 203–211.
102. Shah, G.; Shri, R.; Panchal, V.; Sharma, N.; Singh, B.; Mann, A.S. Scientific basis for the therapeutic use of *Cymbopogon citratus*, stapf (Lemon grass). *J. Adv. Pharm. Technol. Res.* **2011**, *2*, 3. [CrossRef]
103. Rasikari, H.L.; Leach, D.N.; Waterman, P.G.; Spooner-hart, R.N.; Basta, A.H.; Banbury, L.K.; Forster, P.I. Acaricidal and cytotoxic activities of extracts from selected genera of Australian Lamiaceae. *J. Econ. Entomol.* **2005**, *98*, 1259–1266. [CrossRef]
104. Miresmailli, S.; Bradbury, R.; Isman, M.B. Comparative toxicity of *Rosmarinus officinalis* L. essential oil and blends of its major constituents against *Tetranychus urticae* Koch (Acari: Tetranychidae) on two different host plants. *Pest Manag. Sci.* **2006**, *62*, 366–371. [CrossRef]
105. Çalmaşur, Ö.; Aslan, İ.; Şahin, F. Insecticidal and acaricidal effect of three Lamiaceae plant essential oils against *Tetranychus urticae* Koch and *Bemisia tabaci* Genn. *Ind. Crops Prod.* **2006**, *23*, 140–146. [CrossRef]
106. Carpinella, M.C.; Defago, M.T.; Valladares, G.; Palacios, S.M. Antifeedant and insecticide properties of a limonoid *Melia azedarach* (Meliaceae) with potential use for pest management. *J. Agric. Food Chem.* **2003**, *51*, 369–374. [CrossRef]
107. Ismail, S. Selectivity and joint action of *Melia azedarach* L. fruit extracts with certain acaricides to *Tetranychus urticae* Koch and *Stethorus gilvifrons* Mulsant. *Ann. Agric. Sci.* **1997**, *35*, 605–618.
108. Afify, A.E.-M.M.R.; El-Beltagi, H.S.; Fayed, S.A.; Shalaby, E.A. Acaricidal activity of different extracts from *Syzygium cumini* L. Skeels (Pomposia) against *Tetranychus urticae* Koch. *Asian Pac. J. Trop. Biomed.* **2011**, *1*, 359–364. [CrossRef]
109. Chaveerach, A.; Mokkamul, P.; Sudmoon, R.; Tanee, T. Ethnobotany of the genus Piper (Piperaceae) in Thailand. *Ethnobot. Res. Appl.* **2006**, *4*, 223–231. [CrossRef]
110. Araújo, M.J.C.; Câmara, C.A.G.; Born, F.S.; Moraes, M.M.; Badji, C.A. Acaricidal activity and repellency of essential oil from *Piper aduncum* and its components against *Tetranychus urticae*. *Exp. Appl. Acarol.* **2012**, *57*, 139–155. [CrossRef]
111. Tewary, D.K.; Bhardwaj, A.; Shanker, A. Pesticidal activities in five medicinal plants collected from mid hills of western Himalayas. *Ind. Crops Prod.* **2005**, *22*, 241–247. [CrossRef]
112. Roh, H.S.; Lim, E.G.; Kim, J.; Park, C.G. Acaricidal and oviposition deterring effects of santalol identified in sandalwood oil against two-spotted spider mite, *Tetranychus urticae* Koch (Acari: Tetranychidae). *J. Pest Sci.* **2011**, *84*, 495–501. [CrossRef]
113. Khambay, B.P.S.; Batty, D.; Cahill, M.; Denholm, I.; Mead-Briggs, M.; Vinall, S.; Niemeyer, H.M.; Simmonds, M.S.J. Isolation, Characterization, and biological activity of naphthoquinones from *Calceolaria andina* L. *J. Agric. Food Chem.* **1999**, *47*, 770–775. [CrossRef]
114. Latif, Z.; Craven, L.; Hartley, T.G.; Kemp, B.R.; Potter, J.; Rice, M.J.; Waigh, R.D.; Waterman, P.G. An insecticidal quassinoid from the new Australian species *Quassia* sp. aff. *bidwillii*. *Biochem. Syst. Ecol.* **2000**, *28*, 183–184. [CrossRef]
115. Snyder, J.C.; Guo, Z.; Thacker, R.; Goodman, J.P.; Pyrek, J.S. 2,3-dihydrofarnesoic acid, a unique terpene from trichomes of *Lycopersicon hirsutum*, repels spider mites. *J. Chem. Ecol.* **1993**, *19*, 2981–2997. [CrossRef]
116. Antonious, G.F.; Meyer, J.E.; Snyder, J.C. Toxicity and repellency of hot pepper extracts to spider mite, *Tetranychus urticae* Koch. *J. Environ. Sci. Heal. Part B* **2006**, *41*, 1383–1391. [CrossRef]
117. Cavalcanti, S.C.H.; Dos, S.; Niculau, E.; Blank, A.F.; Câmara, C.A.G.; Araújo, I.N.; Alves, P.B. Composition and acaricidal activity of *Lippia sidoides* essential oil against two-spotted spider mite (*Tetranychus urticae* Koch). *Bioresour. Technol.* **2010**, *101*, 829–832. [CrossRef]
118. Lee, S.; Tsao, R.; Peterson, C.; Coats, J.R. Insecticidal activity of monoterpenoids to western corn rootworm (Coleoptera: Chrysomelidae), two-spotted spider mite (Acari: Tetranychidae), and house fly (Diptera: Muscidae). *J. Econ. Entomol.* **1997**, *90*, 883–892. [CrossRef]
119. Martínez-Villar, E.; Sáenz-De-Cabezón, F.J.; Moreno-Grijalba, F.; Marco, V.; Pérez-Moreno, I. Effects of azadirachtin on the two-spotted spider mite, *Tetranychus urticae* (Acari: Tetranychidae). *Exp. Appl. Acarol.* **2005**, *35*, 215–222. [CrossRef]
120. Han, J.; Kim, S.-I.; Choi, B.-R.; Lee, S.-G.; Ahn, Y.-J. Fumigant toxicity of lemon eucalyptus oil constituents to acaricide-susceptible and acaricide-resistant *Tetranychus urticae*. *Pest Manag. Sci.* **2011**, *67*, 1583–1588. [CrossRef]

121. Akhtar, Y.; Isman, M.B.; Lee, C.-H.; Lee, S.-G.; Lee, H.-S. Toxicity of quinones against two-spotted spider mite and three species of aphids in laboratory and greenhouse conditions. *Ind. Crops Prod.* **2012**, *37*, 536–541. [CrossRef]
122. Marčić, D.; Međo, I. Acaricidal activity and sublethal effects of an oxymatrine-based biopesticide on two-spotted spider mite (Acari: Tetranychidae). *Exp. Appl. Acarol.* **2014**, *64*, 375–391. [CrossRef]
123. Marčić, D.; Međo, I. Sublethal effects of azadirachtin-A (NeemAzal-T/S) on *Tetranychus urticae* (Acari: Tetranychidae). *Syst. Appl. Acarol.* **2015**, *30*, 25.
124. Numa, S.; Rodríguez-Coy, L.; Rodríguez, D.; Coy-Barrera, E. Laboratory screening of six botanicals for acaricidal activity against two-spotted spider mite, *Tetranychus urticae* (Koch). *Biopestic. Int.* **2017**, *13*, 13–20.
125. Mostafa, M.E.; Alshamy, M.M.; Abdelmonem, A.; Abdel-Mogib, M. Acaricidal activity of *Chrozophora oblongifolia* on the two spotted spider mite, *Tetranychus urticae* Koch. *J. Entomol. Nematol.* **2017**, *9*, 23–28.
126. Numa, S.; Rodríguez-Coy, L.; Rodríguez, D.; Coy-Barrera, E. Examination of the acaricidal effect of a set of colombian native plants-derived extracts against *Tetranychus urticae* Koch under laboratory conditions. *J. Biopestic.* **2018**, *11*, 30–37.
127. Reddy, S.G.E.; Dolma, S.K. Acaricidal activities of essential oils against two-spotted spider mite, *Tetranychus urticae* Koch. *Toxin Rev.* **2017**, *37*, 62–66. [CrossRef]
128. Erdogan, P.; Yilmaz, B.S. Acaricidal activity of extracts of *Juglans regia* L. on *Tetranychus urticae* Koch (Acari: Tetranychidae). *J. Food Sci. Eng.* **2017**, *7*, 202–208.
129. Chrysargyris, A.; Laoutari, S.; Litskas, V.D.; Stavrinides, M.C.; Tzortzakis, N. Effects of water stress on lavender and sage biomass production, essential oil composition and biocidal properties against *Tetranychus urticae* (Koch). *Sci. Hortic. Amst.* **2016**, *213*, 96–103. [CrossRef]
130. Pavela, R.; Stepanycheva, E.; Shchenikova, A.; Chermenskaya, T.; Petrova, M. Essential oils as prospective fumigants against *Tetranychus urticae* Koch. *Ind. Crops Prod.* **2016**, *94*, 755–761. [CrossRef]
131. Zevedo, S.G.; Mar, J.M.; da Silva, L.S.; França, L.P.; Machado, M.B.; Tadei, W.P.; Bezerra, J.D.A.; dos Santos, A.L.; Sanches, E.A.; Sanches, E.A. Bioactivity of *Licaria puchury-major* essential oil against *Aedes aegypti*, *Tetranychus urticae* and *Cerataphis lataniae*. *Rec. Nat. Prod.* **2018**, *12*, 229–238. [CrossRef]
132. Ebadollahi, A.; Sendi, J.J.; Maroufpoor, M.; Rahimi-Nasrabadi, M. Acaricidal potentials of the terpene-rich essential oils of two Iranian *Eucalyptus* species against *Tetranychus urticae* Koch. *J. Oleo Sci.* **2017**, *66*, 307–314. [CrossRef]
133. Wagan, T.A.; Cai, W.; Hua, H. Repellency, toxicity, and anti-oviposition of essential oil of *Gardenia jasminoides* and its four major chemical components against whiteflies and mites. *Sci. Rep.* **2018**, *8*, 9375. [CrossRef]
134. Numa, S.; Rodríguez-Coy, L.; Rodríguez, D.; Coy-Barrera, E. Effect of acaricidal activity of *Solanum nigrum* on *Tetranychus urticae* Koch under laboratory conditions. *Afr. J. Biotechnol.* **2015**, *15*, 363–369.
135. Mar, J.M.; Silva, L.S.; Azevedo, S.G.; França, L.P.; Goes, A.F.; dos Santos, A.L.; Bezerra, J.D.A.; Rita de Cássia, S.N.; Machado, M.B.; Sanches, E.A. *Lippia origanoides* essential oil: An efficient alternative to control *Aedes aegypti*, *Tetranychus urticae* and *Cerataphis lataniae*. *Ind. Crops Prod.* **2018**, *111*, 292–297. [CrossRef]

© 2019 by the authors. Licensee MDPI, Basel, Switzerland. This article is an open access article distributed under the terms and conditions of the Creative Commons Attribution (CC BY) license (http://creativecommons.org/licenses/by/4.0/).

MDPI
St. Alban-Anlage 66
4052 Basel
Switzerland
Tel. +41 61 683 77 34
Fax +41 61 302 89 18
www.mdpi.com

Plants Editorial Office
E-mail: plants@mdpi.com
www.mdpi.com/journal/plants

Lightning Source UK Ltd.
Milton Keynes UK
UKHW020830170922
408975UK00002B/71